Fuzzy Systems - Theory and Applications

Edited by Constantin Volosencu

Published in London, United Kingdom

IntechOpen

Supporting open minds since 2005

Fuzzy Systems - Theory and Applications
http://dx.doi.org/10.5772/intechopen.87799
Edited by Constantin Volosencu

Contributors
Akanksha Singh, Rabah Mellah, Hocine Khati, Hand Talem, Said Guermah, Sugiyarto Surono, Mustika Sari, Feng Liu, Atimad Harir, Said Melliani, Lalla Saadia Chadli, Vakeel A. Khan, Mobeen Ahmad, Masood Alam, Jeydson Lopes Da Silva, Daoud Raid, Al-khashab M.Basheer Ismael Yaareb, Ahmed Omer, Vijaylakshmi S. S Jigajinni, JaeHyuk Cho, Janani Bharatraj, Welitom Ttatom Pereira da Silva, Marco Antonio Almeida de Souza, Hedda Schmidtke, Chilaka Ranga, Teruvai Manoj, Mahuya Deb, Chuan-qiang Fan, Wei-he Xie

First published in London, United Kingdom, 2022 by IntechOpen
IntechOpen is the global imprint of INTECHOPEN LIMITED, registered in England and Wales, registration number: 11086078, 5 Princes Gate Court, London, SW7 2QJ, United Kingdom
Printed in Croatia

British Library Cataloguing-in-Publication Data
A catalogue record for this book is available from the British Library

Additional hard and PDF copies can be obtained from orders@intechopen.com

Fuzzy Systems - Theory and Applications
Edited by Constantin Volosencu
p. cm.
Print ISBN 978-1-83962-291-5
Online ISBN 978-1-83962-292-2
eBook (PDF) ISBN 978-1-83962-293-9

For EU product safety concerns:
IN TECH d.o.o., Prolaz Marije Krucifikse Kozulić 3, 51000 Rijeka, Croatia, info@intechopen.com or visit our website at intechopen.com.

Meet the editor

Prof. Dr. Constantin Volosencu graduated as an engineer from the "Traian Vuia" Politechnic Institute Timisoara, Romania, and obtained a doctorate from the Politehnica University of Timisoara where he is currently a full professor in the Department of Automation and Applied Informatics. He is the author of ten books, seven book chapters, twenty-seven patents, and more than 160 journal and conference papers. He has also edited eleven books. Dr. Volosencu is a manager of research grants, an editor in chief and member of international journal editorial boards, a former plenary speaker, a member of scientific committees, and a chair at international conferences. His research is in the fields of control systems, control of electric drives, fuzzy control systems, neural network applications, fault detection and diagnosis, sensor network applications, monitoring of distributed parameter systems, and power ultrasound applications. He has developed automation equipment for machine tools, spooling machines, high-power ultrasound processes, and more.

Contents

Preface

The treatment of systems using fuzzy logic has developed in recent decades, with this mathematical theory being an interesting tool for researchers to solve complex scientific and technical problems. Fuzzy systems have found applications in various sectors of human activity, such as industry, business, finance, medicine, and more. A fuzzy system architecture is built on four concepts: fuzzification, knowledge base, inference, and defuzzification. Fuzzy logic allows for a systemic mathematical treatment using acceptable human reasoning. The fuzzy systems thus developed can in turn deal with the uncertainties in real applications.

The book addresses specialists who are interested in applications of fuzzy systems. It covers some fuzzy theoretical concepts and includes new methodologies and applications of fuzzy systems. Published studies demonstrate the ability of fuzzy systems in various practical fields. Researchers in different domains have developed new fuzzy concepts, applications, and tools that enhance human understanding and improve the specialist's ability to analyze, design, and implement high-performance algorithms. The book presents applications that focus on the methodologies used, with case studies as well as implementation and testing issues.

The book includes eight sections: "Fuzzy Mathematics"; "Decision Making"; "Clustering"; "Adaptive Neural Fuzzy Inference Systems"; "Process Monitoring"; "Control Systems"; "Green Infrastructure"; and "Applications in Medicine." These sections include fifteen chapters that illustrate the research in the abovementioned domains. Section 1 includes four chapters that present theoretical studies on the solutions of fuzzy fractional differential equations, Pythagorean fuzzy descriptor systems, fuzzy sequence spaces, and context logic as an extension of fuzzy logic. Section 2 contains two theoretical studies about dual hesitant fuzzy sets and their application to multi-attribute decision making and the application of intuitionistic fuzzy sets to evaluate employee performance. Section 3 contains a theoretical study on data clustering for fuzzifier value derivation. Section 4 contains a study to improve characteristics of ANFIS with application in robotics. Section 5 contains two chapters on health monitoring of power transformers and aircraft fuel systems. Section 6 contains two theoretical studies on fuzzy logic controller with particle swarm optimization tuning and the use of fuzzy logic in the control of solar energy installations. Section 7 contains two studies on the application of Zadeh's Z-numbers to the location of smoke towers and on the fuzzy modelling of urban water supply crisis. Section 8 contains a chapter addressing the application of fuzzy logic in the diagnosis of human diseases.

The editor thanks the contributing authors as well as the staff at IntechOpen, especially Author Service Manager Kristina Kardum Cvitan.

Constantin Voloşencu
Department of Automation and Applied Informatics,
"Politehnica" University Timişoara,
Timişoara, Romania

Section 1

Fuzzy Mathematics

Chapter 1

Existence, Uniqueness and Approximate Solutions of Fuzzy Fractional Differential Equations

Atimad Harir, Said Melliani and Lalla Saadia Chadli

Abstract

In this paper, the Cauchy problem of fuzzy fractional differential equations $T_\gamma u(t) = F(t, u(t))$, $u(t_0) = u_0$, with fuzzy conformable fractional derivative (γ-differentiability, where $\gamma \in (0, 1]$) are introduced. We study the existence and uniqueness of solutions and approximate solutions for the fuzzy-valued mappings of a real variable, we prove some results by applying the embedding theorem, and the properties of the fuzzy solution are investigated and developed. Also, we show the relation between a solution and its approximate solutions to the fuzzy fractional differential equations of order γ.

Keywords: fuzzy conformable fractional derivative, fuzzy fractional differential equations, existence and uniqueness of solution, approximate solutions, Cauchy problem of fuzzy fractional differential equations

1. Introduction

In this paper, we will study Fuzzy solutions to

$$T_\gamma u(t) = F(t, u(t)), \quad u(t_0) = u_0, \ \ \gamma \in (0, 1], \tag{1}$$

where subject to initial condition u_0 for fuzzy numbers, by the use of the concept of conformable fractional H-differentiability, we study the Cauchy problem of fuzzy fractional differential equations for the fuzzy valued mappings of a real variable. Several import-extant results are obtained by applying the embedding theorem in [1] which is a generalization of the classical embedding results [2, 3].

In Section 2 we recall some basic results on fuzzy number. In Section 3 we introduce some basic results on the conformable fractional differentiability [4, 5] and conformable integrability [5, 6] for the fuzzy set-valued mapping in [7]. In Section 4 we show the relation between a solution and its approximate solution to the Cauchy problem of the fuzzy fractional differential equation, and furthermore, and we prove the existence and uniqueness theorem for a solution to the Cauchy problem of the fuzzy fractional differential equation.

2. Preliminaries

We now recall some definitions needed in throughout the paper. Let us denote by $R_{\mathcal{F}}$ the class of fuzzy subsets of the real axis $\{u : R \to [0,\ 1]\}$ satisfying the following properties:

i. u is normal: there exists $x_0 \in R$ with $u(x_0) = 1$,

ii. u is convex fuzzy set: for all $x, t \in R$ and $0 < \lambda \leq 1$, it holds that

$$u(\lambda x + (1 - \lambda)t) \geq \min\{u(x), u(t)\}, \tag{2}$$

iii. u is upper semicontinuous: for any $x_0 \in R$, it holds that

$$u(x_0) \geq \lim_{x \to x_0} u(x), \tag{3}$$

iv. $[u]^0 = cl\{x \in R | u(x) > 0\}$ is compact.

Then $R_{\mathcal{F}}$ is called the space of fuzzy numbers see [8]. Obviously, $R \subset R_{\mathcal{F}}$. If u is a fuzzy set, we define $[u]^\alpha = \{x \in R | u(x) \geq \alpha\}$ the α-level (cut) sets of u, with $0 < \alpha \leq 1$. Also, if $u \in R_{\mathcal{F}}$ then α-cut of u denoted by $[u]^\alpha = [u_1^\alpha, u_2^\alpha]$.

Lemma 1 see [9] *Let $u, v : R_{\mathcal{F}} \to [0, 1]$ be the fuzzy sets. Then $u = v$ if and only if $[u]^\alpha = [v]^\alpha$ for all $\alpha \in [0, 1]$.*

For $u, v \in R_{\mathcal{F}}$ and $\lambda \in R$ the sum $u + v$ and the product λu are defined by

$$[u + v]^\alpha = [u_1^\alpha + v_1^\alpha, u_2^\alpha + v_2^\alpha], \tag{4}$$

$$[\lambda u]^\alpha = \lambda[u]^\alpha = \{[\lambda u_1^\alpha, \lambda u_2^\alpha],\ \lambda \geq 0;\ [\lambda u_2^\alpha, \lambda u_1^\alpha],\ \lambda < 0, \tag{5}$$

$\forall \alpha \in [0, 1]$. Additionally if we denote $\hat{0} = \chi_{\{0\}}$, then $\hat{0} \in R_{\mathcal{F}}$ is a neutral element with respert to $+$.

Let $d : R_{\mathcal{F}} \times R_{\mathcal{F}} \to R_+ \cup \{0\}$ by the following equation:

$$d(u, v) = \sup_{\alpha \in [0,1]} d_H([u]^\alpha,\ [v]^\alpha),\ forall\ u, v \in R_{\mathcal{F}}, \tag{6}$$

where d_H is the Hausdorff metric defined as:

$$d_H([u]^\alpha,\ [v]^\alpha) = \max\{|u_1^\alpha - v_1^\alpha|,\ u_2^\alpha - v_2^\alpha||\} \tag{7}$$

The following properties are well-known see [10]:

$$d(u + w, v + w) = d(u, v) \quad and \quad d(u, v) = d(v, u), \quad \forall\ u, v, w \in R_{\mathcal{F}}, \tag{8}$$

$$d(ku, kv) = |k| d(u, v), \quad \forall k \in R,\ \ u, v \in R_{\mathcal{F}} \tag{9}$$

$$d(u + v, w + e) \leq d(u, w) + d(v, e), \quad \forall\ u, v, w, e \in R_{\mathcal{F}}, \tag{10}$$

and $(R_{\mathcal{F}}, d)$ is a complete metric space.

Definition 1 The mapping $u : [0, a] \to R_{\mathcal{F}}$ for some interval $[0, a]$ is called a fuzzy process. Therefore, its α-level set can be written as follows:

$$[u(t)]^\alpha = [u_1^\alpha(t), u_2^\alpha(t)],\ \ t \in [0, a],\ \ \alpha \in [0, 1]. \tag{11}$$

Theorem 1.1 [11] Let $u : [0, a] \to R_{\mathcal{F}}$ *be Seikkala differentiable and denote* $[u(t)]^{\alpha} = [u_1^{\alpha}(t), u_2^{\alpha}(t)]$. *Then, the boundary function* $u_1^{\alpha}(t)$ *and* $u_2^{\alpha}(t)$ *are differentiable and*

$$[u'(t)]^{\alpha} = \left[(u_1^{\alpha})'(t),\ (u_2^{\alpha})'(t)\right],\ \alpha \in [0, 1]. \tag{12}$$

Definition 2 [12] Let $u : [0, a] \to R_{\mathcal{F}}$. The fuzzy integral, denoted by $\int_b^c u(t)dt, b, c \in [0, a]$, is defined levelwise by the following equation:

$$\left[\int_b^c u(t)dt\right]^{\alpha} = \left[\int_b^c u_1^{\alpha}(t)dt,\ \int_b^c u_2^{\alpha}(t)dt\right], \tag{13}$$

for all $0 \le \alpha \le 1$. In [12], if $u : [0, a] \to R_{\mathcal{F}}$ is continuous, it is fuzzy integrable.
Theorem 1.2 [13] If $u \in R_{\mathcal{F}}$, *then the following properties hold*:

i. $$[u]^{\alpha_2} \subset [u]^{\alpha_1},\ \ if\ \ 0 \le \alpha_1 \le \alpha_2 \le 1; \tag{14}$$

ii. $\{\alpha_k\} \subset [0, 1]$ *is a nondecreasing sequence which converges to* α *then*

$$[u]^{\alpha} = \bigcap_{k \ge 1} [u]^{\alpha_k}. \tag{15}$$

Conversely if $A_{\alpha} = \{[u_1^{\alpha}, u_2^{\alpha}];\ \ \alpha \in (0, 1]\}$ *is a family of closed real intervals verifying* (*i*) *and* (*ii*), *then* $\{A_{\alpha}\}$ *defined a fuzzy number* $u \in R_{\mathcal{F}}$ *such that* $[u]^{\alpha} = A_{\alpha}$.
From [1], we have the following theorems:
Theorem 1.3 There exists a real Banach space X *such that* $\mathbb{R}_{\mathcal{F}}$ *can be the embedding as a convex cone C with vertex* 0 *into X. Furthermore, the following conditions hold*:

i. the embedding j is isometric,

ii. addition in X induces addition in $\mathbb{R}_{\mathcal{F}}$, i.e, for any $u, v \in \mathbb{R}_{\mathcal{F}}$,

iii. multiplication by a nonnegative real number in X induces the corresponding operation in $\mathbb{R}_{\mathcal{F}}$, i.e., for any $u \in \mathbb{R}_{\mathcal{F}}$,

iv. C-C is dense in X,

v. C is closed.

3. Fuzzy conformable fractional differentiability and integral

Definition 3 [4] *Let* $F : (0, a) \to \mathbb{R}_{\mathcal{F}}$ *be a fuzzy function.* γ^{th} *order "fuzzy conformable fractional derivative" of F is defined by*

$$T_{\gamma}(F)(t) = \lim_{\varepsilon \to 0^+} \frac{F(t + \varepsilon t^{1-\gamma}) \ominus F(t)}{\varepsilon} = \lim_{\varepsilon \to 0^+} \frac{F(t) \ominus F(t - \varepsilon t^{1-\gamma})}{\varepsilon}. \tag{16}$$

for all $t > 0, \gamma \in (0, 1)$. *Let* $F^{(\gamma)}(t)$ *stands for* $T_{\gamma}(F)(t)$. *Hence*

$$F^{(\gamma)}(t) = \lim_{\varepsilon \to 0^+} \frac{F(t + \varepsilon t^{1-\gamma}) \ominus F(t)}{\varepsilon} = \lim_{\varepsilon \to 0^+} \frac{F(t) \ominus F(t - \varepsilon t^{1-\gamma})}{\varepsilon}. \tag{17}$$

If F is γ- differentiable in some $(0,\ a)$, and $\lim_{t\to 0^+} F^{(\gamma)}(t)$ exists, then

$$F^{(\gamma)}(0) = \lim_{t \to 0^+} F^{(\gamma)}(t) \tag{18}$$

and the limits (in the metric d).

Remark 1 *From the definition, it directly follows that if F is γ-differentiable then the multivalued mapping F_α is γ-differentiable for all $\alpha \in [0,1]$ and*

$$T_\gamma F_\alpha = \left[F^{(\gamma)}(t)\right]^\alpha, \tag{19}$$

where $T_\gamma F_\alpha$ is denoted from the conformable fractional derivative of F_α of order γ. The converse result does not hold, since the existence of Hukuhara difference $[u]^\alpha \ominus [v]^\alpha,\ \alpha \in [0,1]$ does not imply the existence of H-difference $u \ominus v$.

Theorem 1.4 [4] Let $\gamma \in (0,1]$.

If F is differentiable and F is γ-differentiable then

$$T_\gamma F(t) = t^{1-\gamma} F'(t) \tag{20}$$

Theorem 1.5 [5, 14] If $F : (0,a) \to \mathbb{R}_\mathcal{F}$ is γ-differentiable then it is continuous.

Remark 2 If $F : (0,a) \to \mathbb{R}_\mathcal{F}$ is γ-differentiable and $F^{(\gamma)}$ for all $\gamma \in (0,1]$ is continuous, then we denote $F \in C^1((0,a), \mathbb{R}_\mathcal{F})$.

Theorem 1.6 [5, 14] Let $\gamma \in (0,1]$ and if $F, G : (0,a) \to \mathbb{R}_\mathcal{F}$ are γ-differentiable and $\lambda \in \mathbb{R}$ then

$$T_\gamma(F+G)(t) = T_\gamma(F)(t) + T_\gamma(G)(t) \quad and \quad T_\gamma(\lambda F)(t) = \lambda T_\gamma(F)(t). \tag{21}$$

Definition 4 [5] *Let $F \in C((0,\ a),\ \mathbb{R}_\mathcal{F}) \cap L^1((0,\ a),\ \mathbb{R}_\mathcal{F})$, Define the fuzzy fractional.*

integral for $a \geq 0$ and $\gamma \in (0,1]$.

$$I_\gamma^a(F)(t) = I_1^a(t^{\gamma-1}F)(t) = \int_a^t \frac{F}{s^{1-\gamma}}(s)ds, \tag{22}$$

where the integral is the usual Riemann improper integral.

Theorem 1.7 [5] $T_\gamma I_\gamma^a(F)(t)$, for $t \geq a$, where F is any continuous function in the domain of I_γ^a.

Theorem 1.8 [5] Let $\gamma \in (0,1]$ and F be γ-differentiable in $(0,a)$ and assume that the conformable derivative $F^{(\gamma)}$ is integrable over $(0,a)$. Then for each $s \in (0,a)$ we have

$$F(s) = F(a) + I_\gamma^a F^{(\gamma)}(t) \tag{23}$$

4. Existence and uniqueness solution to fuzzy fractional differential equations

In this section we state the main results of the paper, i.e. we will concern ourselves with the question of the existence theorem of approximate solutions by using the embedding results on fuzzy number space $(\mathbb{R}_{\mathcal{F}}, d)$ and we prove the uniqueness theorem of solution for the Cauchy problem of fuzzy fractional differential equations of order $\gamma \in (0,1]$.

4.1 Solution and its approximate solutions

Assume that $F : (0,a) \times \mathbb{R}_{\mathcal{F}} \to \mathbb{R}_{\mathcal{F}}$ is continuous $C(((0,a) \times \mathbb{R}_{\mathcal{F}}), \mathbb{R}_{\mathcal{F}})$. Consider the fractional initial value problem

$$T_\gamma(u)(t) = F(t, u(t)), \quad u(t_0) = u_0, \tag{24}$$

where $u_0 \in \mathbb{R}_{\mathcal{F}}$ and $\gamma \in (0,1]$.

From Theorems (1.5), (1.7) and (1.8), it immediately follows:

Theorem 1.9 A mapping $u : (0,a) \to \mathbb{R}_{\mathcal{F}}$ is a solution to the problem (24) if and only if it is continuous and satisfies the integral equation

$$u(t) = u_0 + \int_{t_0}^{t} s^{\gamma-1} F(s, u(s)) ds \tag{25}$$

for all $t \in (0,a)$ and $\gamma \in (0,1]$.

In the following we give the relation between a solution and its approximate solutions.

We denote $\Delta_0 = [t_0, t_0 + \theta] \times B(u_0, \mu)$ where θ, μ be two positive real numbers $u_0 \in \mathbb{R}_{\mathcal{F}}$, $B(u_0, \mu) = \{x \in \mathbb{R}_{\mathcal{F}} | d(u, u_0) \leq \mu\}$.

Theorem 1.10 Let $\gamma \in (0,1]$ and $F \in C(\Delta_0, \mathbb{R}_{\mathcal{F}})$, $\eta \in (0, \theta)$, $u_n \in C^1([t_0, t_0 + \eta], \ B(u_0, \mu))$ such that

$$ju_n^{(\gamma)}(t) = jF(t, u_n(t)) + B_n(t), \quad u_n(t_0) = u_0, \ \ \|B_n(t)\| \leq \varepsilon_n \tag{26}$$
$$\forall t \in [t_0, t_0 + \eta], \ \ n = 1, 2, \ldots.$$

where $\varepsilon_n > 0$, $\varepsilon_n \to 0$, $B_n(t) \in C([t_0, t_0 + \eta], X)$, and j s the isometric embedding from $(\mathbb{R}_{\mathcal{F}}, d)$ onto its range in the Banach space X. For each $t \in [t_0, t_0 + \eta]$ there exists an $\beta > 0$ such that the H-differences $u_n(t + \varepsilon t^{1-\gamma}) \ominus u_n(t)$ and $u_n(t) \ominus u_n(t - \varepsilon t^{1-\gamma})$ exist for all $0 \leq \varepsilon < \beta$ and $n = 1, 2, \ldots$. If we have

$$d(u_n(t), u(t)) \to 0 \tag{27}$$

uniform convergence (u.c) for all $t \in [t_0, t_0 + \eta]$, $n \to \infty$, then $u \in C^1([t_0, t_0 + \eta], B(u_0, \mu))$ and

$$T_\gamma(u(t)) = F(t, u(t)), \quad u(t_0) = u_0, \ \ t \in [t_0, t_0 + \eta]. \tag{28}$$

Proof: By (27) we know that $u(t) \in C([t_0, t_0 + \eta], B(u_0, \mu))$. For fixed $t_1 \in [t_0, t_0 + \eta]$ and any $t \in [t_0, t_0 + \eta]$, $t > t_1$, denote $\varepsilon = h t_1^{\gamma-1}$ and $\forall \ \gamma \in (0,1]$

$$G(t,n) = \frac{ju_n\left(t_1 + \varepsilon t_1^{1-\gamma}\right) - ju_n(t_1)}{\varepsilon} - jF(t_1, u_n(t_1)) - B_n(t_1). \tag{29}$$

$$= \frac{ju_n(t_1 + h) - ju_n(t_1)}{h t_1^{\gamma-1}} - jF(t_1, u_n(t_1)) - B_n(t_1). \tag{30}$$

$$= t_1^{1-\gamma} \frac{ju_n(t) - ju_n(t_1)}{t - t_1} - jF(t_1, u_n(t_1)) - B_n(t_1). \tag{31}$$

It is well know that

$$\lim_{t\to t_1} G(t,n) = j(u_n)^{(\gamma)}(t_1) - jF(t_1,u_n(t_1)) - B_n(t_1) \tag{32}$$

$$= j(u_n)^{(\gamma)}(t_1) - jF(t_1,u_n(t_1)) - B_n(t_1) = \Theta \in X \tag{33}$$

$$\lim_{n\to\infty} G(t,n) = t_1^{1-\gamma}\frac{ju(t) - ju(t_1)}{t - t_1} - jF(t_1,u(t_1)) \tag{34}$$

From $F \in C^1(\Delta_0, \mathbb{R}_{\mathcal{F}})$, is know that for any $\varepsilon > 0$, there exists $\beta_1 > 0$ such that

$$d(F(t,v), F(t_1,u(t_1))) < \frac{\varepsilon}{4} \tag{35}$$

whenever $t_1 < t < t_1 + \beta_1$ and $d(v, u(t_1)) < \beta_1$ with $v \in B(u_0, \mu)$ Take natural number $N > 0$ such hat

$$\varepsilon_n < \frac{\varepsilon}{4}, d(u_n(t), u(t)) < \frac{\beta_1}{2} \quad for\ any\ \ n > N, \ \ t \in [t_0, t_0 + \eta] \tag{36}$$

Take $\beta > 0$ such that $\beta < \beta_1$ and

$$d(u(t), u(t_1)) < \frac{\beta_1}{2} \tag{37}$$

whenever $t_1 < t < t_1 + \beta$.
By the definition of $G(t,n)$ and (26), we have $\forall\gamma \in (0,1]$

$$ju_n\left(t_1 + \varepsilon t_1^{1-\gamma}\right) - ju_n(t_1) - (\varepsilon)j(u_n)^{(\gamma)}(t_1) = (\varepsilon)jF(t_1,u_n(t_1)) \tag{38}$$

$$t_1^{1-\gamma}(ju_n(t) - ju_n(t_1)) - (t - t_1)t_1^{1-\gamma}j(u_n)'(t_1) = (t - t_1)jF(t_1,u_n(t_1)) \tag{39}$$

We choose $\psi \in X^*$ such that $\|\psi\| = 1$ and for all $\gamma \in (0,1]$

$$\psi\left(t_1^{1-\gamma}(ju_n(t) - ju_n(t_1)) - (t - t_1)t_1^{1-\gamma}j(u_n)'(t_1)\right) \tag{40}$$

$$= \|t_1^{1-\gamma}(ju_n(t) - ju_n(t_1)) - (t - t_1)t_1^{1-\gamma}j(u_n)'(t_1)\| \tag{41}$$

Let $t_1^{1-\gamma}\varphi(t) = t_1^{1-\gamma}\psi(ju_n(t)) - (t - t_1)t_1^{1-\gamma}j(u_n)'(t_1)$, consequently

$$t_1^{1-\gamma}\varphi'(t) = t_1^{1-\gamma}\psi(\ ju_n'(t)) - t_1^{1-\gamma}j(u_n)'(t_1) \tag{42}$$

hence

$$\|t_1^{1-\gamma}(ju_n(t) - ju_n(t_1)) - (t - t_1)t_1^{1-\gamma}j(u_n)'(t_1)\| \tag{43}$$

$$= t_1^{1-\gamma}(\varphi(t) - \varphi(t_1)) = t_1^{1-\gamma}\varphi'(\hat{t})(t - t_1) \tag{44}$$

$$= \psi\left(t_1^{1-\gamma}(\ ju_n'(\hat{t}) - ju_n'(t_1))\right)(t - t_1) \tag{45}$$

$$\le \|\psi\|\|t_1^{1-\gamma}(ju_n'(\hat{t}) - ju_n'(t_1))\|(t - t_1) \tag{46}$$

$$= \|t_1^{1-\gamma}(\ ju_n'(\hat{t}) - ju_n'(t_1))\|(t - t_1), \tag{47}$$

where $t_1 \le \hat{t} \le t$. In view of (39), we have

$$\|G(t,n)\| \leq \|t_1^{1-\gamma}(ju_n'(\hat{t}) - ju_n'(t_1))\|, \quad t_1 \leq \hat{t} \leq t. \tag{48}$$

From (36) and (37) we know that

$$d(u(\hat{t}), u(t_1)) < \frac{\beta_1}{2} \tag{49}$$

and

$$d(u_n(\hat{t}), u(t_1)) \leq d(u_n(\hat{t}), u(\hat{t})) + d(u(\hat{t}), u(t_1)) \tag{50}$$

$$< \frac{\beta_1}{2} + \frac{\beta_1}{2} = \beta_1 \tag{51}$$

Hence by (35) and (48) we have for all $\gamma \in (0,1]$.

$$\|G(t,n)\| \leq \|t_1^{1-\gamma}(ju_n'(\hat{t}) - ju_n'(t_1))\| \tag{52}$$

$$= \|jF(\hat{t}, u_n(\hat{t})) + B_n(\hat{t}) - jF(t_1, u_n(t_1)) - B_n(t_1)\| \tag{53}$$

$$\leq \|jF(\hat{t}, u_n(\hat{t})) - jF(t_1, u(t_1))\| \tag{54}$$

$$+\|jF(t_1, u(t_1)) - jF(t_1, u_n(t_1))\| + 2\varepsilon_n \tag{55}$$

$$\leq d(jF(\hat{t}, u_n(\hat{t})) - jF(t_1, u(t_1))) \tag{56}$$

$$+d(jF(t_1, u(t_1)) - jF(t_1, u_n(t_1))) + 2\varepsilon_n \tag{57}$$

$$< \frac{\varepsilon}{4} + \frac{\varepsilon}{4} + 2\varepsilon_n < \varepsilon \tag{58}$$

whenever $n > N$ and $t_1 < t < t_1 + \beta$.

Let $n \to \infty$, and applying (34), we have

$$\|t_1^{1-\gamma}\frac{ju(t) - ju(t_1)}{t - t_1} - jF(t_1, u(t_1))\| \leq \varepsilon, \quad t_1 < t < t_1 + \beta. \tag{59}$$

On the other hand, from the assumption of Theorem (1.9), there exists an $\beta(t_1) \in (0, \beta)$ such that the H-differences $u_n(t) \ominus u_n(t_1)$ exist for all $t \in [t_1, t_1 + \beta(t_1)]$ and $n = 1, 2, \ldots$.

Now let $v_n(t) = u_n(t) \ominus u_n(t_1)$ we verify that the fuzzy number-valued sequence $\{v_n(t)\}$ uniformly converges on $[t_1, t_1 + \beta(t_1)]$. In fact, from the assumption $d(u_n(t), u(t)) \to 0$ u.c. for all $t \in [t_0, t_0 + \eta]$, we know

$$d(v_n(t), v_m(t)) = d(v_n(t) + u_n(t_1), v_m(t) + u_n(t_1)) \tag{60}$$

$$\leq d(u_n(t), u_m(t)) + d(u_m(t), v_m(t) + u_n(t_1)) \tag{61}$$

$$= d(u_n(t), u_m(t)) + d(v_m(t) + u_m(t_1), v_m(t) + u_n(t_1)) \tag{62}$$

$$= d(u_n(t), u_m(t)) + d(u_m(t_1), u_n(t_1)) \tag{63}$$

$$\to \quad u.c \ \forall t \in [t_1, t_1 + \beta(t_1)] \ \ n, m \to \infty. \tag{64}$$

Since $(\mathbb{R}_{\mathcal{F}}, d)$ is complete, there exists a fuzzy number-valued mapping $v(t)$ such that $\{v_n(t)\}$ u.c to $v(t)$ on $[t_1, t_1 + \beta(t_1)]$ as $n \to \infty$.

In addition, we have

$$d(u(t_1) + v(t), u(t)) \leq d(u(t_1) + v(t), u_n(t_1 + v_n(t_1)) + d(u_n(t_1 + v_n(t), u(t)) \tag{65}$$

$$\leq d(u(t_1) + v(t), u(t_1) + v_n(t)) \tag{66}$$

$$+d(u(t_1)+v_n(t),u_n(t_1)+v_n(t))+d(u_n(t),u(t)) \tag{67}$$

$$=d(v_n(t),u(t))+d(u_n(t_1),u(t_1))+d(u_n(t),u(t)) \tag{68}$$

$$\forall t\in[t_1,t_1+\beta(t_1)].$$

Let $n\to\infty$. It follows that

$$u(t_1)+v(t)\equiv u(t)\quad for all\ \ t\in[t_1,t_1+\beta(t_1)]. \tag{69}$$

Hence the H-difference $u(t)\ominus u(t_1)$ exist for all $t\in[t_1,t_1+\beta(t_1)]$.
Thus from (59) we have for all $\gamma\in(0,1]$.

$$d\left(\frac{u\left(t_1+t_1^{1-\gamma}\varepsilon\right)\ominus u(t_1)}{\varepsilon},F(t_1,u(t_1))\right)\le\varepsilon,\quad t\in[t_1,t_1+\beta(t_1)]. \tag{70}$$

So, $\lim_{\varepsilon\to0^+}u\left(t_1+t_1^{1-\gamma}\varepsilon\right)\ominus u(t_1)/\varepsilon=F(t_1,u(t_1))$. Similarty, we have

$$\lim_{\varepsilon\to0^-}\frac{u\left(t_1+t_1^{1-\gamma}\varepsilon\right)\ominus u(t_1)}{\varepsilon}=F(t_1,u(t_1)).$$

Hence $u^{(\gamma)}(t_1)$ exists and

$$u^{(\gamma)}(t_1)=F(t_1,u(t_1)). \tag{71}$$

from $t_1\in[t_0,t_0+\eta]$ is arbitrary, we know that Eq. (28) holds true and $u\in C^1([t_0,t_0+\eta],B(u_0,\mu))$. The proof is concluded.

Lemma 2 For all $t\in[t_0,t_0+\eta]$, $n=1,2,\ldots$ and $\gamma\in(0,1]$.
If we replace Eq. (26) by

$$ju_{n+1}(t)=jF(t,u_n(t))+B_n(t),\quad u_n(t_0)=u_0,\ \ \|B_n(t)\|\le\varepsilon_n, \tag{72}$$

retain other assumptions, then the conclusions also hold true.

Proof: This is completely similar to the proof of Theorem (1.10), hence itis omitted here.

4.2 Uniqueness solution

In this section, by using existence theorom of approximate solutions, and the embedding results on fuzzy number space $(\mathbb{R}_{\mathcal{F}},d)$, we give the existence and uniqueness theorem for the Cauchy problem of the fuzzy fractional differential equations of order γ.

Theorem 1.11

i. Let $F\in C(\Delta_0,\mathbb{R}_{\mathcal{F}})$ and $d\big(F(t,u),\hat{0}\big)\le\sigma$ for all $(t,u)\in\Delta_0$.

ii. $G\in C([t_0,t_0+\theta]\times[0,\mu],\mathbb{R})$, $G(t,0)\equiv0$, and $0\le G(t,y)\le\sigma_1$, for all $t\in[t_0,t_0+\theta], 0\le y\le\mu$ such that $G(t,y)$ is noncreasing on y the fractional initial value problem

$$T_\gamma y(t)=G(t,y(t)),\quad y(t_0)=0 \tag{73}$$

has only the solution $y(t)\equiv0$ on $[t_0,t_0+\theta]$.

iii. $d(F(t,u),F(t,v)) \leq G(t,d(u,v))$ for all $(t,u),(t,v) \in \Delta_0$, and $d(u,v) \leq \mu$.

Then the Cauchy problem (28) has unique solution $u \in C^1([t_0,t_0+\eta],B(u_0,\mu))$ on $[t_0,t_0+\eta]$, where $\eta = \min\{\theta,\mu/\sigma,\mu/\sigma_1\}$, and the successive iterations

$$u_{n+1}(t) = u_0 + \int_{t_0}^{t} s^{\gamma-1}F(s,u_n(s))ds \tag{74}$$

uniformly converge to $u(t)$ on $[t_0,t_0+\eta]$.

Proof: In the proof of Theorem 4.1 in [15], taking the conformable derivative $u^{(\gamma)}$ for all $\gamma \in (0,1]$, using theorem (1.4) and properties (9), then we obtain the proof of Theorem (1.11).

Example 1 Let $L > 0$ is a constant, taking $G(t,y) = Ly$ in the proof of Theorem (4.2), then obtain the proof of Corollary 4.1 in [15] where $\sigma_1 = L\mu$, hence $\eta = \min\{\theta,\mu/\sigma,1/L\}$. Then the Cauchy problem (28) has unique solution $u \in C^1([t_0,t_0+\eta],B(\Delta_0,\mu))$, and the successive iterations (74) uniformly converge to $u(t)$ on $[t_0,t_0+\eta]$.

5. Conclusion

In this work, we introduce the concept of conformable differentiability for fuzzy mappings, enlarging the class of γ-differentiable fuzzy mappings where $\gamma \in (0,1]$. Subsequently, by using the γ-differentiable and embedding theorem, we study the Cauchy problem of fuzzy fractional differential equations for the fuzzy valued mappings of a real variable. The advantage of the γ-differentiability being also practically applicable, and we can calculate by this derivative the product of two functions because all fractional derivatives do not satisfy see [4].

On the other hand, we show and prove the relation between a solution and its approximate solutions to the Cauchy problem of the fuzzy fractional differential equation, and the existence and uniqueness theorem for a solution to the problem (1) are proved.

For further research, we propose to extend the results of the present paper and to combine them the results in citeref for fuzzy conformable fractional differential equations.

Conflict of interest

The authors declare no conflict of interest.

Author details

Atimad Harir*, Said Melliani and Lalla Saadia Chadli
Laboratory of Applied Mathematics and Scientific Computing, Sultan Moulay Slimane University, Beni Mellal, Morocco

*Address all correspondence to: atimad.harir@gmail.com

References

[1] Kaleva O. The Cauchy problem for fuzzy DiFFerentiel equations. Fuzzy Sets and Systems. 1990;**35**:366-389

[2] Radstrom H. An embedding theorem for spaces of convex set. Proceedings of American Mathematical Society. 1952;**3**: 165-169

[3] Casting C, Valadier M. Convex Analysis and Measurable Multifunctions. Berlin: Springer-Verlag; 1977

[4] Harir A, Melliani S, Chadli LS. Fuzzy generalized conformable fractional derivative. In: Advances in Fuzzy Systems, Volume 2020, Article ID 1954975, 7 Pages. 2019. DOI: https://doi.org/10.1155/2020/1954975

[5] Harir A, Melliani S, Chadli LS, Fuzzy conformable fractional semigroups of operators. International Journal of Differential Equations. Received 6 August 2020; Revised 16 October 2020; Accepted 21 October 2020

[6] Harir A, Melliani S, Chadli LS. Fuzzy fractional evolution equations and fuzzy solution operators. In: Advances in Fuzzy Systems, Volume 2019, Article ID 5734190, 10 Pages. 2019. DOI: https://doi.org/10.1155/2019/5734190

[7] Kaleva O. Fuzzy differential equations. Fuzzy Sets and Systems. 1987;**24**:301-317

[8] Diamond P, Kloeden PE. Metric Spaces of Fuzzy Sets: Theory and Applica- Tions. Singapore: World Scienific; 1994

[9] Goo HY, Park JS. On the continuity of the Zadeh extensions. J. Chungcheong Math. Soc. 2007;**20**(4): 525-533

[10] Puri ML, Ralescu DA. Differentials of fuzzy functions. Journal of Mathematical Analysis and Applications. 1983;**91**:552-558

[11] Kaleva O. A note on fuzzy differential equations. Nonlinear Analysis. 2006;**64**:895-900

[12] Song S, Guo L, Feng C. Global existence of solutions to fuzzy differential equations. Fuzzy Sets and Systems. 2000;**115**:371-376

[13] Negoita CV, Ralescu DA. Applications of Fuzzy Sets to System Analysis, Birkhauser. Basel; 1975

[14] Harir A, Melliani S, Chadli LS. Solutions of fuzzy fractional differential equations. Iranian Journal of Mathematical Sciences and Informatics.

[15] Congxin W, Shiji S. Approximate solutions, existence, and uniqueness ofthe Cauchy problem of fuzzy differential equations. Journal of Mathematic Analysis and Applications. 1996;**202**:629-644

Chapter 2

(α, β)−Pythagorean Fuzzy Numbers Descriptor Systems

Chuan-qiang Fan, Wei-he Xie and Feng Liu

Abstract

By using pythagorean fuzzy sets and T-S fuzzy descriptor systems, the new (α, β)-pythagorean fuzzy descriptor systems are proposed in this paper. Their definition is given firstly, and the stability of this kind of systems is studied, the relation of (α, β)-pythagorean fuzzy descriptor systems and T-S fuzzy descriptor systems is discussed. The (α, β)-pythagorean fuzzy controller and the stability of (α, β)-pythagorean fuzzy descriptor systems are deeply researched. The (α, β)-pythagorean fuzzy descriptor systems can be better used to solve the problems of actual nonlinear control. The (α, β)-pythagorean fuzzy descriptor systems will be a new research direction, and will become a universal method to solve practical problems. Finally, an example is given to illustrate effectiveness of the proposed method.

Keywords: Pythagorean fuzzy sets, T-S fuzzy descriptor systems, stability

1. Introduction

Pythagorean fuzzy sets [1–4] were proposed by Yager in 2013, are a new tool to deal with vagueness. Pythagorean fuzzy sets maintain the advantages of both membership and non-membership, but the value range of membership function and non-membership function is expanded from triangle to quarter circle. The expansion of the value area makes the amount of information of pythagorean fuzzy sets expand 1.57 times that of the intuitionistic fuzzy sets, and ensures that intuitionistic fuzzy sets are all pythagorean fuzzy sets. They can be used to characterize the uncertain information more sufficiently and accurately than intuitionistic fuzzy sets. Pythagorean fuzzy sets have attracted great attention of a great many scholars that have been extended to new fields and these extensions have been used in many areas such as decision making, aggregation operators, and information measures. Due to theirs wide scope of description cases are very common in diverse real-life issue, pythagorean fuzzy sets have given a boost to the management of vagueness caused by fuzzy scope. Pythagorean fuzzy sets have provided two novel algorithms in decision making problems under Pythagorean fuzzy environment.

Takagi-Sugeno (T-S) fuzzy systems [5–9] has been applied on intelligent computing research and complex nonlinear systems. T-S fuzzy systems have also been extended to new fields and these extensions have been used in many areas by a great many scholars. However, the membership functions of T-S fuzzy systems cannot make full use of the all uncertain message in the premise conditions. So we decide to study the new (α,β)-pythagorean fuzzy descriptor systems in order to solve practical control problems more easily and feasible.

IntechOpen

The advantages of (α, β)-pythagorean fuzzy descriptor systems are the following:

1. Pythagorean fuzzy sets maintain the advantages of both membership and non-membership, but the value range of membership function and non-membership function is expanded from triangle to quarter circle. The expansion of the value area makes the amount of information of pythagorean fuzzy sets expand 1.57 times that of the intuitionistic fuzzy sets. They can be used to characterize the uncertain information more suffificiently and accurately than intuitionistic fuzzy sets.

2. The membership function and non-membership function of pythagorean fuzzy sets can be easy to be defined. The value ranges of membership function and non-membership function are also more consistent with objective reality and many hesitant problems and people's thinking.

3. Pythagorean fuzzy sets can ensure that intuitionistic fuzzy sets are all pythagorean fuzzy sets, i.e. intuitionistic fuzzy sets are the special examples of pythagorean fuzzy sets. So intuitionistic fuzzy control systems can be changed into (0,1)-pythagorean fuzzy control systems.

4. (α,β)-pythagorean fuzzy descriptor systems are a broader generalization of T-S fuzzy descriptor systems i.e. T-S fuzzy descriptor systems are the special examples of (α, β)-pythagorean fuzzy descriptor systems.

5. We can judge the degree of weight in the control process according to the value of membership function and non-membership function of the rules. By setting the values of α and β, we decide whether the rules will participate in the final calculation, thereby reducing the calculation process and improving the control efficiency and effectiveness.

6. In fact, (α, β)-pythagorean fuzzy descriptor systems are consistent with the control methods of human being. This method is to imitate the control process of people and also solves the most difficult problem for humans.

The rest of this paper is organized as follows: In Section 1, the basic concepts of T-S fuzzy descriptor systems are introduced. In Section 2, (α,β)-pythagorean fuzzy descriptor systems are firstly proposed. Then the relationship of T-S fuzzy descriptor systems and (α,β)-pythagorean fuzzy descriptor systems are discussed in Section 3. (α,β)-pythagorean fuzzy controller and the stability of (α,β)-pythagorean fuzzy descriptor systemsare deeply researched in Section 4. In Section 5, a numbers examples is given to show the corollaries are corrected. We discussed in detail the effects of controls in several cases. Through this practical example, we find that the selection of pythagorean fuzzy membership functions in the premise conditions of the rules has a great influence on the control effect. Therefore, the choice of pythagorean fuzzy membership functions must be determined after more tests, and we can not completely believe the original given functions. Finally, the conclusion is given in Section 6.

Notations: Throughout this paper, R^n and $R^{n\times m}$denote respectively the n dimensional Euclidean space and $n \times m$ dimensional Euclidean space. PFS denotes pythagorean fuzzy set.

2. Preliminaries

This section will briefly introduce some baisc definitions and theorems on pythagorean fuzzy sets and T-S fuzzy descriptor systems.

Definition 1.1 [1–4] Let X be a universe of discourse. A PFS P in X is given by.

$$P = \{ <x, \mu_P(x), \nu_P(x)> | x \in X\},$$

where μ_P: $X \rightarrow [0,1]$ denotes the degree of membership and ν_P: $X \rightarrow [0,1]$ denotes the degree of non-membership of the element $x \in X$ to the set P, respectively, with the condition that $0 \leq (\mu_P(x))^2 + (\nu_P(x))^2 \leq 1$. The degree of indeterminacy $\pi_P(x) = 1 - (\mu_P(x))^2 - (\nu_P(x))^2$.

For convenience, a pythagorean fuzzy number $(\mu_P(x), \nu_P(x))$ denoted by $p = (\mu_P, \nu_P)$.

Definition 1.2 [10, 11] T-S fuzzy descriptor systems are as follows:

Rule i: if $x_1(t)$ is F_1^i and...and $x_n(t)$ is F_n^i, then.

$$E\dot{x}(t) = A_i x(t) + B_i \mu(t)$$
$$y(t) = C_i x(t) + D_i \mu(t)$$

Where $x(t) = [x_1(t), x_2(t), \cdots, x_n(t)]^T \in R^n$ and $\mu(t) \in R^m$ are the state and control input, respectively; A_i, B_i, C_i and D_i are known real constant matrices with appropriate dimension;

$E$is a singular matrix; $F_1^i, F_2^i, \cdots, F_n^i \ (i = 1, 2, \ldots, r)$ are the fuzzy sets.

By fuzzy blending, the overall fuzzy model is inferred as follows.

$$E\dot{x}(t) = A(t)x(t) + B(t)\mu(t)$$
$$y(t) = C(t)x(t) + D(t)\mu(t)$$

where

$$A(t) = \sum_{i=1}^{r} h_i(x(t))A_i, B(t) = \sum_{i=1}^{r} h_i(x(t))$$
$$B_i, C(t) = \sum_{i=1}^{r} h_i(x(t))C_i, D(t) = \sum_{i=1}^{r} h_i(x(t))D_i,$$

and $h_i(x(t))$ is the normalized grade of membership, given as.

$$h_i(x(t)) = \frac{\omega_i(x(t))}{\sum_{i=1}^{r} \omega_i(x(t))}, \omega_i(x(t)) = \Pi_{i=1}^{n} \mu_{ij}(x_j(t)),$$

which is satisfying

$$0 \leq h_i(x(t)) \leq 1, \sum_{i=1}^{r} h_i(x(t)) = 1,$$

$\mu_{ij}(x_j(t))$ is the grade of membership function of $x_j(t)$ in F_j^i.

3. $(\alpha, \beta)-$pythagorean fuzzy descriptor systems

As T-S fuzzy descriptor systems are very familiar to us, and pythagorean fuzzy sets are a new tool to deal with vagueness. So we decide to study the new (α, β)-

pythagorean fuzzy descriptor systems in order to solve practical control problems more easily and feasible. Next, the related definitions of $(\alpha, \beta)-$pythagorean fuzzy descriptor systems are gradually given.

Definition 2.1 $(\alpha, \beta)-$pythagorean fuzzy descriptor systems are as follows:

Rule i: if $x_1(t)$ is P_1^i and...and $x_n(t)$ is P_n^i, then.

$$E\dot{x}(t) = A_i x(t) + B_i \mu(t) \tag{1}$$

$$y(t) = C_i x(t) + D_i \mu(t) \tag{2}$$

where $x(t) = [x_1(t)), x_2(t), \cdots, x_n(t)]^T \in R^n$ and $\mu(t) \in R^m$ are the state vector and the control input vector, respectively; $y(t)$ is the measurable output vector; A_i, B_i, C_i and D_i are known real constant matrices with appropriate dimension; E is a singular matrix; $P_1^i, P_2^i, \ldots, P_n^i$ $(i = 1, 2, \ldots, r)$ are all pythagorean fuzzy sets.

By fuzzy blending, the overall fuzzy model is inferred as follows.

$$E\dot{x}(t) = A(t)x(t) + B(t)\mu(t)$$

$$y(t) = C(t)x(t) + D(t)\mu(t)$$

where

$$A(t) = \sum_{i=1}^{r} h_i(x(t))A_i, B(t) = \sum_{i=1}^{r} h_i(x(t))B_i,$$

$$C(t) = \sum_{i=1}^{r} h_i(x(t))C_i, D(t) = \sum_{i=1}^{r} h_i(x(t))D_i,$$

and $h_i(x(t))$ is the normalized grade of membership, given as.

$$h_i(x(t)) = \frac{h_{i(\alpha,\beta)}(x(t))}{\sum_{i=1}^{r} h_{i(\alpha,\beta)}(x(t))}, i = 1, 2, 3, \cdots, r;$$

where

$$h_{i(\alpha,\beta)}(x(t)) = \begin{cases} h_i^1(x(t)) & when\ h_i^1(x(t)) \geq \alpha\ or\ h_i^2(x(t)) \leq \beta \\ 0 & else \end{cases}, \alpha + \beta \leq 1, i = 1, 2, 3, \cdots, r;$$

$$h_i^1(x(t)) = \frac{\mu_{P^i}(x(t))}{\sum_{i=1}^{r} \mu_{P^i}(x(t))}, h_i^2(x(t)) = \frac{\nu_{P^i}(x(t))}{\sum_{i=1}^{r} \nu_{P^i}(x(t))},$$

where $h_i^1(x(t))$ and $h_i^2(x(t))$ are respectively positive and negative membership functions.

$$\sum_{i=1}^{r} h_{i1}(x(t)) = 1, \sum_{i=1}^{r} h_{i2}(x(t)) = 1;$$

$$\mu_{P^i}(x_j(t)) = \prod_{j=1}^{r} \mu_{P_j^i}(x_j(t)), \nu_{P^i}(x_j(t)) = \prod_{j=1}^{r} \nu_{P_j^i}(x_j(t));$$

$\mu_{P_i^j}(x_j(t))$ and $\nu_{P_j^i}(x_j(t))$ is the membership function value of $x_j(t)$ that belongs and does not belong to the intuitionistic fuzzy numbers set P_j^i.

Remark 2.1:

1. We can judge the degree of weight in the control process according to the value of the positive and negative membership functions of the rules. By setting the values of α and β, we decide whether the rules will participate in the final calculation, thereby reducing the calculation process and improving the control efficiency and effectiveness.

2. In fact, (α,β)-pythagorean fuzzy descriptor systems are consistent with the control methods of human being. People generally proceed appropriate control at one point by the past experience, i.e. people's decisions are decided and implemented at roughly one point. This method is to imitate the control process of people

3. The relations between (α,β)-pythagorean fuzzy descriptor systems and T-S fuzzy descriptor systems

Firstly, the relation of T-S fuzzy descriptor systems and (α,β)-pythagorean fuzzy descriptor systems is studied through an example.

When $\alpha = 0, \beta = 1,$ then

$$h_i(x(t)) = h_{i(\alpha,\beta)}(x(t)) = h_{i1}(x(t)) = \frac{\mu_i^M(x(t))}{\sum_{i=1}^r \mu_i^M(x(t))}, h_{i2}(x(t)) = 0, \mu_i^M(x(t)) = \prod_{j=1}^n \mu_{ij}{}^M(x_j(t)).$$

Then the special (0,1)-pythagorean fuzzy descriptor systems are T-S fuzzy descriptor systems. In other words, T-S fuzzy descriptor systems are all the special (0,1)-pythagorean fuzzy descriptor systems. Therefore, it is easy to get the following Theorem 3.1.

Theorem 3.1 T-S fuzzy descriptor systems are all the (α,β)-pythagorean fuzzy descriptor systems.

Proof:It is so easy, so omit.

4. (α,β)−pythagorean fuzzy numbers controller

Now we continue to study the feedback control and stability of pythagorean fuzzy descriptor systems according to the traditional research path of the control systems.

Suppose.

Rule i: if $x_1(t)$ is $P_1^i(x_1(t))$ and ... and $x_n(t)$ is $P_n^i(x_n(t))$, then.

$$u(x(t)) = \sum_{i=1}^r h_i(x(t)))G_i x(t) \tag{3}$$

where G_i (i = 1,2,..., r) are the state feedback-gains matrices.

$$h_i(x(t)) = \frac{h_{i(\alpha,\beta)}(x(t))}{\sum_{i=1}^r h_{i(\alpha,\beta)}(x(t))}, i = 1, 2, 3, \cdots, r;$$

where

$$h_{i(\alpha,\beta)}(x(t)) = \begin{cases} h_i^1(x(t)) & when\ h_i^1(x(t)) \geq \alpha\ or\ h_i^2(x(t)) \leq \beta \\ 0 & else \end{cases}, \alpha+\beta \leq 1, i = 1,2,3,\cdots,r;$$

$$h_i^1(x(t)) = \frac{\mu_{P^i}(x(t))}{\sum_{i=1}^{r}\mu_{P^i}(x(t))}, h_i^2(x(t)) = \frac{\nu_{p^i}(x(t))}{\sum_{i=1}^{r}\nu_{p^i}(x(t))},$$

where $h_i^1(x(t))$ and $h_i^2(x(t))$are respectively positive and negative membership functions.

$$\sum_{i=1}^{r} h_{i_1}(x(t)) = 1, \sum_{i=1}^{r} h_{i_2}(x(t)) = 1;$$

$$\mu_{P^i}(x_j(t)) = \prod_{j=1}^{r} \mu_{P^i_j}(x_j(t)), \nu_{P^i}(x_j(t)) = \prod_{j=1}^{r} \nu_{P^i_j}(x_j(t));$$

$\mu_{P^j_i}(x_j(t))$ and $\nu_{P^i_j}(x_j(t))$ is the membership function value of $x_j(t)$ that belongs and does not belong to the intuitionistic fuzzy numbers set P^i_j.

If we take (3) into (1, 2), we can get.

$$E\dot{x}(t) = \sum_{i=1}^{r}\sum_{j=1}^{r} h_i(x(t))h_j(x(t))(A_i + B_iG_j)x(t) \tag{4}$$

$$y(t) = \sum_{i=1}^{r}\sum_{j=1}^{r} h_i(x(t))h_j(x(t))(C_i + D_iG_j)x(t) \tag{5}$$

The system stability is guaranteed by determining the feedback gains G_j.

Basic LMI-based stability conditions guaranteeing the stability of the above control system in the form of (4, 5) are given in the following theorem.

Theorem 4.1 The system (3) is asymptotically stable, if there exist matrices $N_j \in R^{m\times n}$ (j = 1,2,3,..., r) and $K = K^T \in R^{n\times n}$ such that the following LMIs are satisfied:

$$K > 0 \tag{6}$$

$$E^TK = K^TE \geq 0 \tag{7}$$

$$Q_{ij} = A_iK^{-1} + K^{-1}A_i{}^T + B_iN_j + N_j{}^TB_i{}^T < 0 \forall i,j \tag{8}$$

where the feedback gains are defined as $G_j = N_jK$ for all j.

Proof: Considering the quadratic Lyapunov function.

$$V(x(t)) = x^T(t)E^TKx(t),$$

where $0 < K = K^T \in R^{n\times n}$.

then

$$\dot{V}(x(t)) = \dot{x}^T(t)E^TKx(t) + x^T(t)E^TK\dot{x}(t) = (E\dot{x}(t))^TKx(t) + x^T(t)K^T(E\dot{x}(t))$$
$$= \sum_{i=1}^{r}\sum_{j=1}^{r} h_ih_jx^T(t)KK^{-1}\{A^T{}_iK + K^TN_j^TB_i^TK + K^TA_i + K^TB_iN_jK\}K^{-1}Kx(t),$$

let $Z = Kx(t)$, then

$$\dot{V}(x(t)) = \sum_{i=1}^{r}\sum_{j=1}^{r} h_i h_j x^T(t) K K^{-1}\left\{A^T{}_i K + K N_j^T B_i^T K + K A_i + K B_i N_j K\right)\} K^{-1} K x(t)$$
$$= \sum_{i=1}^{r}\sum_{j=1}^{r} h_i h_j Z\left(K^{-1} A^T{}_i + N_j^T B_i^T + A_i K^{-1} + B_i N_j\right) Z.$$

As $Q_{ij} = A_i K^{-1} + K^{-1} A_i{}^T + B_i N_j + N_j{}^T B_i{}^T < 0$, so the system (3) is asymptotically stable.

5. Simulation example

Example 5.1: Considering an inverted pendulum, subject to parameter uncertainties [12–15] as the nonlinear plant to be controlled. The dynamic equation for the inverted pendulum is given by.

$$\ddot{\theta}(t) = \frac{g \sin(\theta(t)) - a m_p L \dot{\theta}(t)^2 \sin(2\theta(t))/2 - a \cos(\theta(t))\mu(t)}{4L/3 - a m_p L \cos^2(\theta(t))}$$

Where $\theta(t)$is the angular displacement of the pendulum, $g = 9.8\ \mathrm{m/s^2}$ is the acceleration due to gravity, $m_p \in [m_{p_{\min}}, m_{p_{\max}}] = [2,3]kg$ is the mass of the pendulum, $M_c \in [M_{\min}, M_{\max}] = [8, 12]$.

Kg is the mass of the cart, $a = 1/(m_p + M_c)$, $2\,L = 1$ m is the length of the pendulum, and $u(t)$is the force (in newtons) applied to the cart. The inverted pendulum is considered working in the operating domain characterized by $x_1 = \theta(t) \in [-5\pi/12, 5\pi/12]$ and $x_2 = \dot{\theta}(t) \in [-5,5]$.

Rule 1: If $x_1(t)$ is M_1^1, $x_2(t)$ is M_2^1, then

$$\begin{pmatrix} \dot{x}_1(t) \\ \dot{x}_2(t) \end{pmatrix} = \begin{pmatrix} 0 & 1 \\ 10.0078 & 0 \end{pmatrix}\begin{pmatrix} x_1(t) \\ x_2(t) \end{pmatrix} + \begin{pmatrix} 0 \\ -0.1765 \end{pmatrix}\mu(t);$$

Rule 2: If $x_1(t)$ is M_1^2, $x_2(t)$ is M_2^2, then

$$\begin{pmatrix} \dot{x}_1(t) \\ \dot{x}_2(t) \end{pmatrix} = \begin{pmatrix} 0 & 1 \\ 10.0078 & 0 \end{pmatrix}\begin{pmatrix} x_1(t) \\ x_2(t) \end{pmatrix} + \begin{pmatrix} 0 \\ -0.0261 \end{pmatrix}\mu(t);$$

Rule 3: If $x_1(t)$ is M_1^3, $x_2(t)$ is M_2^3, then

$$\begin{pmatrix} \dot{x}_1(t) \\ \dot{x}_2(t) \end{pmatrix} = \begin{pmatrix} 0 & 1 \\ 18.4800 & 0 \end{pmatrix}\begin{pmatrix} x_1(t) \\ x_2(t) \end{pmatrix} + \begin{pmatrix} 0 \\ -0.1765 \end{pmatrix}\mu(t);$$

Rule 4: If $x_1(t)$ is M_1^4, $x_2(t)$ is M_2^4, then

$$\begin{pmatrix} \dot{x}_1(t) \\ \dot{x}_2(t) \end{pmatrix} = \begin{pmatrix} 0 & 1 \\ 18.4800 & 0 \end{pmatrix}\begin{pmatrix} x_1(t) \\ x_2(t) \end{pmatrix} + \begin{pmatrix} 0 \\ -0.0261 \end{pmatrix}\mu(t);$$

Next, according to the ideas based on the principles of interpolation and interval coverage, we firstly change the interval-valued T-S fuzzy model of inverted pendulum into the special (α,β)- pythagorean fuzzy descriptor systems of inverted pendulum as follows.

Rule 1: If $x_1(t)$ is $P_1^1(x_1(t))$, $x_2(t)$ is $P_2^1(x_2(t))$, then

$$\begin{pmatrix} \dot{x}_1(t) \\ \dot{x}_2(t) \end{pmatrix} = \begin{pmatrix} 0 & 1 \\ 10.0078 & 0 \end{pmatrix} \begin{pmatrix} x_1(t) \\ x_2(t) \end{pmatrix} + \begin{pmatrix} 0 \\ -0.1765 \end{pmatrix} \mu(t);$$

Rule 2: If $x_1(k)$ is $P_1^2(x_1(t))$, $x_2(t)$ is $P_2^2(x_2(t))$, then

$$\begin{pmatrix} \dot{x}_1(t) \\ \dot{x}_2(t) \end{pmatrix} = \begin{pmatrix} 0 & 1 \\ 10.0078 & 0 \end{pmatrix} \begin{pmatrix} x_1(t) \\ x_2(t) \end{pmatrix} + \begin{pmatrix} 0 \\ -0.0261 \end{pmatrix} \mu(t);$$

Rule 3: If $x_1(t)$ is $P_1^3(x_1(k))$, $x_2(t)$ is $P_2^3(x_2(t))$, then

$$\begin{pmatrix} \dot{x}_1(t) \\ \dot{x}_2(t) \end{pmatrix} = \begin{pmatrix} 0 & 1 \\ 18.4800 & 0 \end{pmatrix} \begin{pmatrix} x_1(t) \\ x_2(t) \end{pmatrix} + \begin{pmatrix} 0 \\ -0.1765 \end{pmatrix} \mu(t);$$

Rule 4: If $x_1(t)$ is $P_1^4(x_1(k))$, $x_2(t)$ is $P_2^4(x_2(t))$, then

$$\begin{pmatrix} \dot{x}_1(t) \\ \dot{x}_2(t) \end{pmatrix} = \begin{pmatrix} 0 & 1 \\ 18.4800 & 0 \end{pmatrix} \begin{pmatrix} x_1(t) \\ x_2(t) \end{pmatrix} + \begin{pmatrix} 0 \\ -0.0261 \end{pmatrix} \mu(t);$$

According to the theorem 4.1, we can get.

$$K = \begin{pmatrix} 1/7 & -1/7 \\ -1/7 & 8/7 \end{pmatrix}, K^{-1} = \begin{pmatrix} 8 & 1 \\ 1 & 1 \end{pmatrix}, N_1 = N_3 = (\,100 \quad 100\,), N_2 = N_4 = (\,1000 \quad 1000\,),$$

So the above (α,β)-pythagorean fuzzy descriptor systems of inverted pendulum is asymptotically stable, and the state feedback-gains matrices $G_1 = G_3 = (\,0 \quad 100\,)$, $G_2 = G_4 = (\,0 \quad 1000\,)$.

The first case, suppose $x_1(0) = -\frac{11\pi}{29}, x_2(0) = -0.88, \alpha = 0.3, \beta = 0.25$, then take the variable $x_1(t)$ as the main factor of the control, and according to **Table 2** we can control in three steps, i.e. $x_1(0) = -\frac{11\pi}{29} \to x_1(t_1) \to x_1(t_2) \approx 0$ and $0 < t_1 \leq t_2$.

When $x_1(0) = -\frac{11\pi}{29}, x_2(0) = -0.88$, and $\alpha = 0.30, \beta = 0.25$, according to **Table 2** we can get $\mu_{P_1^1}(x_1(0)) = \mu_{P_1^2}(x_1(0)) = 0.69$, $\nu_{P_1^1}(x_1(0)) = \nu_{P_1^2}(x_1(0)) = 0.72$, $\mu_{P_1^3}(x_1(0)) = \mu_{P_1^4}(x_1(0)) = 0$, $\nu_{P_1^3}(x_1(0)) = \nu_{P_1^4}(x_1(0)) = 1$, $\mu_{P_2^1}(x_2(0)) = \mu_{P_2^2}(x_2(0)) = 0.02$, $\nu_{P_2^1}(x_2(0)) = \nu_{P_2^2}(x_2(0)) = 1$, $\mu_{P_2^3}(x_2(0)) = \mu_{P_2^4}(x_2(0)) = 0.40$, $\nu_{P_2^3}(x_2(0)) = \nu_{P_2^4}(x_2(0)) = 0.92$, noteworthy, $\mu_{P_1^1}(x_1(0)) + \nu_{P_1^1}(x_1(0)) = \mu_{P_1^2}(x_1(0)) + \nu_{P_1^2}(x_1(0)) = 1.41 > 1$, $\mu_{P_2^3}(x_2(0)) + \nu_{P_2^3}(x_2(0)) = \mu_{P_2^4}(x_2(0)) + \nu_{P_2^4}(x_2(0)) = 1.32 > 1$. Then according to Definition 2.1, taking it one step further, we can get $h_1^1 = 0.49$, $h_1^2 = 0.22$, $h_2^1 = 0.49$, $h_2^2 = 0.22$, $h_3^1 = 0.01$, $h_3^2 = 0.28$, $h_4^1 = 0.01$, $h_4^2 = 0.28$, so $h_1 = 0.5$, $h_2 = 0.5$, $h_3 = 0$, $h_4 = 0$, according to 4.2, so the overall fuzzy model of the (0.30,0.25)- pythagorean fuzzy descriptor systems is.

$$\begin{pmatrix} \dot{x}_1(t) \\ \dot{x}_2(t) \end{pmatrix} = \begin{pmatrix} 0 & 0.25 \\ 2.5019 & -13.929 \end{pmatrix} \begin{pmatrix} x_1(t) \\ x_2(t) \end{pmatrix},$$

The solution of the systems is $x(t) = \begin{pmatrix} -\frac{11}{29}\pi + 0.25t \\ 0.2172 - 0.045t - 1.0972e^{-13.93t} \end{pmatrix}$

When $x_1(4) \approx -0.19103$,$x_2(4) \approx 0.0372$, and $\alpha = 0.30, \beta = 0.25$, according to **Table 2**, we can get $\mu_{P_1^1}(x_1(4)) = \mu_{P_1^2}(x_1(4)) = 0.03$, $\nu_{P_1^1}(x_1(4)) = \nu_{P_1^2}(x_1(4)) = 1$,$\mu_{P_1^3}(x_1(4)) = \mu_{P_1^4}(x_1(4)) = 0.20$, $\nu_{P_1^3}(x_1(4)) = \nu_{P_1^4}(x_1(4)) = 0.98$, $\mu_{P_2^1}(x_2(4)) = \mu_{P_2^2}(x_2(4)) = 0.50$, $\nu_{P_2^1}(x_2(4)) = \nu_{P_2^2}(x_2(4)) = 0.87$, $\mu_{P_2^3}(x_2(4)) = \mu_{P_2^4}(x_2(4)) = 0$, $\nu_{P_2^3}(x_2(4)) = \nu_{P_2^4}(x_2(4)) = 1$, noteworthy, $\mu_{P_1^1}(x_1(4)) + \nu_{P_1^1}(x_1(4)) = \mu_{P_1^2}(x_1(4)) + \nu_{P_1^2}(x_1(4)) = 1.03 > 1$, $\mu_{P_2^1}(x_2(4)) + \nu_{P_2^1}(x_2(4)) = \mu_{P_2^2}(x_2(4)) + \nu_{P_2^2}(x_2(4)) = 1.37 > 1$.Then according to Definition 2.1, taking it one step further, we can get $h_1^1 = 0.49$, $h_1^2 = 0.23$, $h_2^1 = 0.49$, $h_2^2 = 0.23$, $h_3^1 = 0.01$, $h_3^2 = 0.27$, $h_4^1 = 0.01$, $h_4^2 = 0.27$, so $h_1 = 0.50$, $h_2 = 0.50$, $h_3 = 0$, $h_4 = 0$, then according to 4.2, so the overall fuzzy model of the (0.3,0.3)- pythagorean fuzzy descriptor systems is.

$$\begin{pmatrix} \dot{x}_1(t-4) \\ \dot{x}_2(t-4) \end{pmatrix} = \begin{pmatrix} 0 & 0.25 \\ 2.5019 & -13.929 \end{pmatrix} \begin{pmatrix} x_1(t-4) \\ x_2(t-4) \end{pmatrix},$$

The solution of the systems is $x(t) = \begin{pmatrix} -0.19103 + 0.25t \\ 0.03754 - 0.045t - 0.00034e^{-13.93t} \end{pmatrix}$

When $x_1(4.764) \approx -0.0000344$, $x_2(4.764) \approx 0.00323$, so the overall fuzzy model of the (0.30, 0.25)-pythagorean fuzzy descriptor systems is E-asymptotic stability. But it takes a shorter time (**Figure 1**).

The second case(interval-valued T-S fuzzy model of inverted pendulum), suppose $x_1(0) = -\frac{11\pi}{29}$,$x_2(0) = -0.88$, then take the variable $x_1(t)$ as the main factor of the control, and according to **Table 1** we can control in three steps, i.e. $x_1(0) = -\frac{11\pi}{29} \to x_1(1) \to x_1(t) \approx 0$.

When $x_1(0) = -\frac{11\pi}{29} \in \left[-\frac{11\pi}{29}, 0\right)$, $x_2(0) = -0.88 \in [-0.88, 0)$, and $\lambda_1 = \lambda_2 = \frac{1}{2}$, according to **Table 1**, we can get $h_1 = 0.26$, $h_2 = 0.58$, $h_3 = 0.05$, $h_4 = 0.11$, according to theorem 4.1, so the overall interval-valued fuzzy model of the interval-valued fuzzy descriptor systems is

$$\begin{pmatrix} \dot{x}_1(t) \\ \dot{x}_2(t) \end{pmatrix} = \begin{pmatrix} 0 & 1 \\ -48.561 & -59.925 \end{pmatrix} \begin{pmatrix} x_1(t) \\ x_2(t) \end{pmatrix},$$

The solution of the systems is $x(t) = \begin{pmatrix} -1.2229e^{-0.82t} + 0.0319e^{-59.1t} \\ 1.0048e^{-0.82t} - 1.8848e^{-59.1t} \end{pmatrix}$;

When $x_1(1) = -0.5386 \in [-0.5386, 0)$, $x_2(1) = 0.4425 \in [0, 0.4425)$, and $\lambda_1 = \lambda_2 = \frac{1}{2}$, according to **Table 1**, we can get $h_1 = 0.32$, $h_2 = 0.25$, $h_3 = 0.24$, $h_4 = 0.19$, according to 4.2, the overall interval-valued fuzzy model of the interval-valued fuzzy descriptor systems is

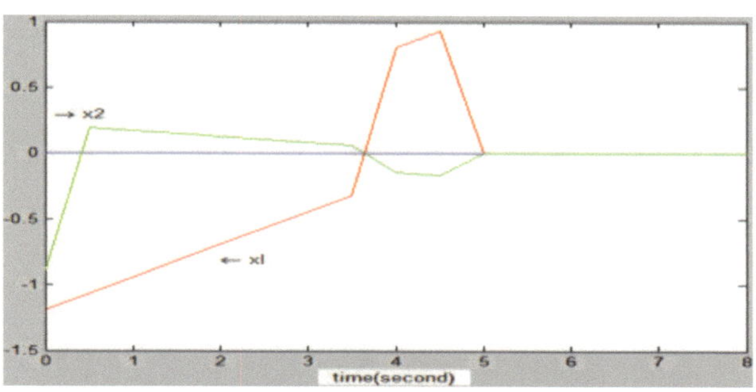

Figure 1.
$x_1(t)$ and $x_2(t)$ under the (0.30,0.25)- pythagorean fuzzy descriptor systems.

Left membership functions	Right membership functions
$M^1_{M_1^1}(x_1) = 1 - e^{-\frac{x_1^2}{1.2}}$	$M^3_{M_1^1}(x_1) = 1 - 0.23e^{-\frac{x_1^2}{0.25}}$
$M^1_{M_1^2}(x_1) = 1 - e^{-\frac{x_1^2}{1.2}}$	$M^3_{M_1^2}(x_1) = 1 - 0.23e^{-\frac{x_1^2}{0.25}}$
$M^1_{M_1^3}(x_1) = 0.23e^{-\frac{x_1^2}{0.25}}$	$M^3_{M_1^3}(x_1) = e^{-\frac{x_1^2}{1.2}}$
$M^1_{M_1^4}(x_1) = 0.23e^{-\frac{x_1^2}{0.25}}$	$M^3_{M_1^4}(x_1) = e^{-\frac{x_1^2}{1.2}}$
$M^1_{M_2^1}(x_2) = 0.5e^{-\frac{x_2^2}{0.25}}$	$M^3_{M_2^1}(x_2) = e^{-\frac{x_2^2}{1.5}}$
$M^1_{M_2^2}(x_2) = 1 - e^{-\frac{x_2^2}{1.5}}$	$M^3_{M_2^2}(x_2) = 1 - 0.5e^{-\frac{x_2^2}{0.25}}$
$M^1_{M_2^3}(x_2) = 0.5e^{-\frac{x_2^2}{0.25}}$	$M^3_{M_2^3}(x_2) = e^{-\frac{x_2^2}{1.5}}$
$M^1_{M_2^4}(x_2) = 1 - e^{-\frac{x_2^2}{1.5}}$	$M^3_{M_2^4}(x_2) = 1 - 0.5e^{-\frac{x_2^2}{0.25}}$

Table 1.
The membership functions of the IT-2 T-S fuzzy model of inverted pendulum.

Membership functions	Non-membership functions
$\mu_{P_1^1}(x_1) = 1 - e^{-\frac{x_1^2}{1.2}}$	$\nu_{P_1^1}(x_1) = \sqrt{1 - \left(1 - e^{-\frac{x_1^2}{1.2}}\right)^2}$
$\mu_{P_1^2}(x_1) = 1 - e^{-\frac{x_1^2}{1.2}}$	$\nu_{P_1^2}(x_1) = \sqrt{1 - \left(1 - e^{-\frac{x_1^2}{1.2}}\right)^2}$
$\mu_{P_1^3}(x_1) = 0.23e^{-\frac{x_1^2}{0.25}}$	$\nu_{P_1^3}(x_1) = \sqrt{1 - \left(0.23e^{-\frac{x_1^2}{0.25}}\right)^2}$
$\mu_{P_1^4}(x_1) = 0.23e^{-\frac{x_1^2}{0.25}}$	$\nu_{P_1^4}(x_1) = \sqrt{1 - \left(0.23e^{-\frac{x_1^2}{0.25}}\right)^2}$
$\mu_{P_2^1}(x_2) = 0.5e^{-\frac{x_2^2}{0.25}}$	$\nu_{P_2^1}(x_2) = \sqrt{1 - \left(0.5e^{-\frac{x_2^2}{0.25}}\right)^2}$
$\mu_{P_2^2}(x_2) = 1 - e^{-\frac{x_2^2}{1.5}}$	$\nu_{P_2^2}(x_2) = \sqrt{1 - \left(1 - e^{-\frac{x_2^2}{1.5}}\right)^2}$
$\mu_{P_2^3}(x_2) = 0.5e^{-\frac{x_2^2}{0.25}}$	$\nu_{P_2^3}(x_2) = \sqrt{1 - \left(0.5e^{-\frac{x_2^2}{0.25}}\right)^2}$
$\mu_{P_2^4}(x_2) = 1 - e^{-\frac{x_2^2}{1.5}}$	$\nu_{P_2^4}(x_2) = \sqrt{1 - \left(1 - e^{-\frac{x_2^2}{1.5}}\right)^2}$

Table 2.
The membership functions and non-membership functions of (α,β)-pythagorean fuzzy descriptor systems of inverted pendulum.

,

$$\begin{pmatrix} \dot{x}_1(t) \\ \dot{x}_2(t) \end{pmatrix} = \begin{pmatrix} 0 & 1 \\ -44.49 & -58.141 \end{pmatrix} \begin{pmatrix} x_1(t) \\ x_2(t) \end{pmatrix}$$

The solution of the systems is $x(t) = \begin{pmatrix} -0.4654e^{-0.78(t)} - 0.0732e^{-57.37(t)} \\ 0.4174e^{-0.78(t)} + 0.0251e^{-57.37(t)} \end{pmatrix}$;

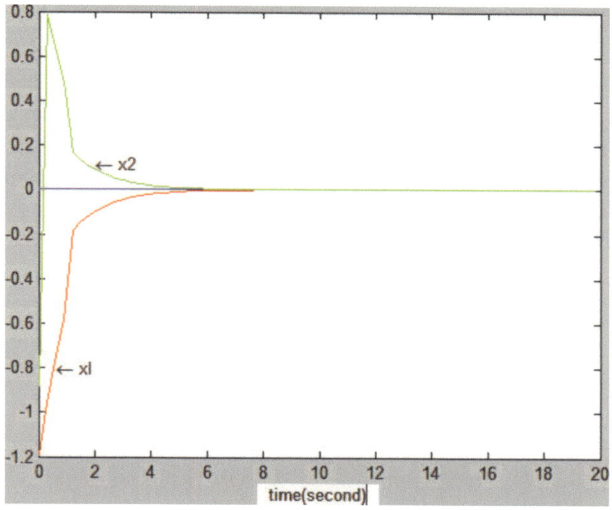

Figure 2.
$x_1(t)$ and $x_2(t)$ under the IT2 T-S fuzzy descriptor systems.

When $x_1(9.6) \approx -0.0006$, $x_2(9.6) \approx -0.0005$, so the IT2 T-S fuzzy descriptor system of the inverted pendulum will to be stable too.

Thus the stable control time of the (0.30,0.25)-pythagorean fuzzy descriptor systems of inverted pendulum is **4.836** second shorter than the interval-valued T-S fuzzy descriptor systems of the inverted pendulum (**Figure 2**).

Remark 5.1: In this way, the (0.30,0.25)-pythagorean fuzzy descriptor systems can get the better effect than the control effect of interval-valued T-S fuzzy model of inverted pendulum. It is easy to see that the (0.30,0.25)-pythagorean fuzzy descriptor systems has the best control, and can reduce the number of rules and thus reduce the amount of calculations.

In this way, it can get the better effect than the control effect of interval-valued T-S fuzzy model of inverted pendulum. Because the feedback more or less needs a little time, when the system carries out feedback instructions, but the time has gone, so the feedback that have been given are also lagging and out of date. $(\alpha, \beta)-$ pythagorean fuzzy descriptor systems can be closer to the actual, and easy to control the error range. The new control method is more convenient and feasible!

6. Conclusions

In this paper, the new $(\alpha, \beta)-$pythagorean fuzzy descriptor systemsare firstly introduced, and more consistent with the human way of thinking and more likely to be set up and more convenient for popularization. The new$(\alpha, \beta)-$pythagorean fuzzy descriptor systems is very simply and quickly. We can do not know the control principle, but we can directly achieve good control effect. The new theory can be studied in parallel to the basic framework of the original theories and easy to promote the old theories and achieve good results. In addition, we can judge the degree of weight in the control process according to the value of the positive and negative membership functions of the rules. By setting the values ofαandβ, we decide whether the rules will participate in the final calculation, thereby reducing the number of the rules and the calculation process, and improving the control efficiency and effectiveness. Otherwise, T-S fuzzy descriptor systems are the special examples of $(\alpha, \beta)-$pythagorean fuzzy descriptor systems. $(\alpha, \beta)-$pythagorean fuzzy controller and the stability of$(\alpha, \beta)-$pythagorean fuzzy descriptor systems are deeply researched. At last, a numbers example is given to show the corollaries are corrected.

But the theoretical part of the new systems need to be in-depth studied, and specific applications are also to be further developed. For example, $(\alpha, \beta)-$ pythagorean fuzzy descriptor systems can also be used as the model of autonomous learning in order to establish intelligent control, and can be used well in unmanned driving in the future. So$(\alpha, \beta)-$pythagorean fuzzy descriptor systems is just to meet the reality requirements.

Author details

Chuan-qiang Fan[1], Wei-he Xie[2] and Feng Liu[3*]

1 College of Sciences, Liaoning Shihua University, Fushun, China

2 Department of Naval Gun, Dalian Naval Academy, Dalian, China

3 Dalian Survey Group of the National Statistics Bureau, Dalian, China

*Address all correspondence to: 77325344@qq.com

References

[1] Yager R R, Abbasov A M. Pythagorean membership grades, complex numbers, and decision making [J]. Int Journal of Intelligence Systems, 2013,28(5):436–452.

[2] Yager R R. Pythagorean membership grades in multicriteria decision making [J]. IEEE Trans on Fuzzy Systems, 2014,22(4):958–965.

[3] Grag H. A new generalized pythagorean fuzzy information aggregation using einstein operations and its applicaion to decision making [J]. Int J of Intelligence Systems, 2016,31 (9):886–920.

[4] Wei G W. Pythagorean fuzzy interaction aggregation operators and their application to multiple attribute decision making [J]. Journal of Intelligent and Fuzzy Systems, 207,33 (4):2119–2132.

[5] K. Tanaka and M. Sugeno. Stability analysis and design of Fuzzy control systems, Fuzzy sets syst., 1992,45(2): 135–156.

[6] M. Sugeno. Fuzzy control, Nikkan kougyou shinbunsha publisher,Tokyo, 1988.

[7] Taniguchi T, Tanaka K, Yanafuji K and Wang H O. Fuzzy Descriptor Systems: Stability Analysis and Design via LMIs [J], Proc of ACC, 1827–1831, 1999.

[8] Taniguchi T, Tanaka K and Wang H O. Fuzzy Descriptor Systems and Nonlinear Model Following Control, IEEE Trans on Fuzzy System, vol.8, 442–452, 2000.

[9] Zadeh L. A. The Concept of a linguistic variable and its application to approximate reasoning-I. Information Sciences, 1975.8(1):199–249.

[10] SHENG L, MA X Y. Stability analysis and controller design of discrete interval type-2 fuzzy systems. Asian Journal of Control, 2014,16(4):1091–1104

[11] Atanassov K, Gargov G, Interval valued intuitionistic fuzzy sets. Fuzzy Sets Syst, 1989,31:343–349.

[12] K.T. Atanassov, Operators over interval valued intuitionistic fuzzy sets, Fuzzy Sets and Systems, 1994,64(2): 159–174.

[13] Liu Feng, Yuan Xue-hai. Fuzzy number intuitionistic fuzzy set, Fuzzy systems and mathematics, 2007,21(1): 88–91.

[14] Fan Chuan-qiang, Distance between fuzzy number intuitionitic fuzzy sets, Journal of liaoning shihua university, 2010,30(2):85–88.

[15] H.K. Lam, et al., Control design for interval type-2 fuzzy systems under imperfect premise matching, IEEE Trans. on Industrial Electronics, 2014,61:956–968.

Chapter 3

A Study of Fuzzy Sequence Spaces

Vakeel A. Khan, Mobeen Ahmad and Masood Alam

Abstract

The purpose of this chapter is to introduce and study some new ideal convergence sequence spaces ${}_F\mathcal{S}^{\mathcal{J}_\theta}(\mathcal{T})$, ${}_F\mathcal{S}_0^{\mathcal{J}_\theta}(\mathcal{T})$ and ${}_F\mathcal{S}_\infty^{\mathcal{J}_\theta}(\mathcal{T})$ on a fuzzy real number F defined by a compact operator $\mathcal{T}$. We investigate algebraic properties like linearity, solidness and monotinicity with some important examples. Further, we also analyze closedness of the subspace and inclusion relations on the said spaces.

Keywords: Ideal, J{convergence, J{Cauchy, Fuzzy number, Lacunary sequence, Compact operator

1. Introduction

The concepts of fuzzy sets were initiated by Zadeh [1], since then it has become an active area of researchers. Matloka [2] initiated the notion of ordinary convergence of a sequence of fuzzy real numbers and studied convergent and bounded sequences of fuzzy numbers and some of their properties, and proved that every convergent sequence of fuzzy numbers is bounded. Nanda [3] investigated some basic properties for these sequences and showed that the set of all convergent sequences of fuzzy real numbers form a complete metric space. Alaba and Norahun [4] studied fuzzy Ideals and fuzzy filters of pseudocomplemented semilattices Moreover, Nuray and Savas [5] extended the notion of convergence of the sequence of fuzzy real numbers to the notion of statistical convergence.

Fast [6] introduced the theory of statistical convergence. After that, and under different names, statistical convergence has been discussed in the ergodic theory, Fourier analysis and number theory. Furthermore, it was examined from the sequence space point of view and linked with summability theory. Esi and Acikgoz [7] examined almost λ-statistical convergence of fuzzy numbers. Kostyrko et al. [8] introduced ideal $\mathcal{J}-$convergence which is based on the natural density of the subsets of positive integers. Kumar and Kumar [9] extended the theory of ideal convergence to apply to sequences of fuzzy numbers. Khan et al. [10–12] studied the notion of $\mathcal{J}$-convergence in intuitionistic fuzzy normed spaces. Subsequently, Hazarika [3] studied the concept of lacunary ideal convergent sequence of fuzzy real numbers. Where a lacunary sequence is an increasing integer sequence $\theta = (k_r)$ such that $k_0 = 0$ and $h_r = k_r - k_{r-1} \to \infty$ as $r \to \infty$. The intervals are determined by θ and defined by $I_r = (k_{r-1}, k_r]$.

We outline the present work as follows. In Section 2, we recall some basic definitions related to the fuzzy number, ideal convergent, monotonic sequence and compact operator. In Section 3, we introduce the spaces of fuzzy valued lacunary ideal convergence of sequence with the help of a compact operator and prove our maim results. In Section 4, we state the conclusion of this chapter.

2. Preliminaries

In this section, we recall some basic notion, definitions and lemma that are required for the following sections.

Definition 2.1. A mapping $F : \mathbb{R} \mapsto \lambda(= [0,1])$ is a real fuzzy number on the set $\mathbb{R}$ associating real number s with its grade of membership $F(s)$. Let $\mathcal{C}$ denote the set of all closed and bounded intervals $F = [f_1, f_2]$ on the real line $\mathbb{R}$. For $G = [g_1, g_2]$ in $\mathcal{C}$, one define $F \leq G$ if and only if $f_1 \leq f_2$ and $g_1 \leq g_2$. Determine a metric ρ on $\mathcal{C}$ by

$$\rho(F,G) = \max\{|f_1 - g_1|, |f_2 - g_2|\}. \quad (1)$$

It can be easily seen that "$\leq$" is a partial order on $\mathcal{C}$ and $(\mathcal{C}, \rho)$ is a complete metric space. The absolute value $|F|$ of $F \in \mathbb{R}(\lambda)$ is defined by

$$|F|(s) = \begin{cases} \max\{F(s), F(-s)\}, & \text{if } s > 0 \\ 0, & \text{if } s < 0. \end{cases}$$

Suppose $\overline{\rho} : \mathbb{R}(\lambda) \times \mathbb{R}(\lambda) \mapsto \mathbb{R}$ be determined as

$$\overline{\rho}(F,G) = \sup \rho(F,G).$$

Hence, $\overline{\rho}$ defines a metric on $\mathbb{R}(\lambda)$. The multiplicative and additive identity in $\mathbb{R}(\lambda)$ are denoted by $\overline{1}$ and $\overline{0}$, respectively.

Definition 2.2. A family of subsets $\mathcal{J}$ of the power set $P(\mathbb{N})$ of the natural number $\mathbb{N}$ is known as an ideal if and only if the following conditions are satisfied [8]

i. $\emptyset \in \mathcal{J}$,

ii. for every $\mathcal{A}_1, \mathcal{A}_2 \in \mathcal{J}$ one obtain $\mathcal{A}_1 \cup \mathcal{A}_2 \in \mathcal{J}$,

iii. for every $\mathcal{A}_1 \in \mathcal{J}$ and every $\mathcal{A}_2 \subseteq \mathcal{A}_1$ one obtain $\mathcal{A}_2 \in \mathcal{J}$.

An ideal $\mathcal{J}$ is known as non– trivial if $\mathcal{J} \neq P(\mathbb{N})$ and non– trivial ideal is said to be an admissible if $\mathcal{J} \supseteq \{\{n\} : n \in \mathbb{N}\}$.

Lemma 1. *If ideal $\mathcal{J}$ is maximal, then for every $\mathcal{A} \subset \mathbb{N}$ we have either $\mathcal{A} \in \mathcal{J}$ or $\mathbb{N} \backslash \mathcal{A} \in \mathcal{J}$* [8].

Definition 2.3. A family of subsets $\mathcal{H}$ of the power set $P(\mathbb{N})$ of the natural number $\mathbb{N}$ is known as filter in if and only if following condition are satisfied [8].

i. $\emptyset \notin \mathcal{H}$,

ii. for every $\mathcal{A}_1, \mathcal{A}_2 \in \mathcal{H}$ one have $\mathcal{A}_1 \cap \mathcal{A}_1 \in \mathcal{H}$,

iii. for every $\mathcal{A}_1 \in \mathcal{H}$ and $\mathcal{A}_2 \supseteq \mathcal{A}_1$ one have $\mathcal{A}_2 \in \mathcal{H}$.

Remark 1. Filter associated with the ideal $\mathcal{J}$ is defined by the family of sets

$$\mathcal{H}(\mathcal{J}) = \{K \subset \mathbb{N} : \exists \quad \mathcal{A} \in \mathcal{J} : K = \mathbb{N} \backslash \mathcal{A}\}$$

Definition 2.4. A sequence (F_k) of fuzzy real numbers is known as $\mathcal{J}-$ convergent to fuzzy real numbers F_0 if for each $\varepsilon > 0$, the set [9].

$$\{k \in \mathbb{N} : \overline{\rho}(F_k, F_0) \geq \varepsilon\} \in \mathcal{J}.$$

Definition 2.5. A sequence (F_k) is known as $\mathcal{J}-$ null if there exists a fuzzy real numbers $\overline{0}$ such that for each $\varepsilon > 0$ [9],

$$\{k \in \mathbb{N} : \overline{\rho}(F_k, \overline{0}) \geq \varepsilon\} \in \mathcal{J}.$$

Definition 2.6. A sequence (F_k) of fuzzy real numbers is known as $\mathcal{J}$-Cauchy if there exists a subsequence (F_{l_ε}) of (F_k) in such a way that for every $\varepsilon > 0$ [13],

$$\{k \in \mathbb{N} : \overline{\rho}(F_k, F_{l_\varepsilon}) \geq \varepsilon\} \in \mathcal{J}.$$

Definition 2.7. A sequence (F_k) is known as $\mathcal{J}-$ bounded if there exists a fuzzy real numbers $\mathcal{M} > 0$ so that, the set [9].

$$\{k \in \mathbb{N} : \overline{\rho}(F_k, \overline{0}) > \mathcal{M}\} \in \mathcal{J}.$$

Definition 2.8. Suppose $K = \{k_i \in \mathbb{N} : k_1 < k_2 < \dots\} \subseteq \mathbb{N}$ and $\mathbb{E}$ be a sequence space. A $K-$ step space of E is a sequence space [12].

$$\Lambda_K^{\mathbb{E}} = \{(x_{k_i}) \in \omega : (x_k) \in \mathbb{E}\}.$$

The canonical pre-image of a sequence $(x_{k_i}) \in \lambda_K^{\mathbb{E}}$ is a sequence $(y_k) \in \omega$ defined as follows:

$$y_k = \begin{cases} x_k, & \text{if } n \in K \\ 0, & \text{otherwise.} \end{cases}$$

y is in canonical pre-image of $\Lambda_K^{\mathbb{E}}$ if y is canonical pre-image of some element $x \in \Lambda_K^{\mathbb{E}}$.

Definition 2.9. A sequence space $\mathbb{E}$ is known as monotone, if it is contains the canonical pre-images of it is step space [12].

That is, if for all infinite $K \subseteq \mathbb{N}$ and $(x_k) \in \mathbb{E}$ the sequence $(\alpha_k x_k)$, where $\alpha_k = 1$ for $k \in K$ and $\alpha_k = 0$ otherwise, belongs to $\mathbb{E}$.

Definition 2.10. [12] A sequence space $\mathbb{E}$ is known as convergent free, if $(x_k) \in \mathbb{E}$ whenever $(y_k) \in \mathbb{E}$ and $(y_k) = 0$ implies that $(x_k) = 0$ for all $k \in \mathbb{N}$.

Lemma 2.1. Every solid space is monotone [12].

Definition 2.11. Suppose U and V are normed spaces. An operator $\mathcal{T} : U \to V$ is known as compact linear operator if [12].

1. $\mathcal{T}$ is linear

2. $\mathcal{T}$ maps every bounded sequence (x_k) in U onto a sequence $\mathcal{T}(x_k)$ in V which has a convergent subsequence.

3. Main results

In this section, we introduce the spaces of fuzzy valued lacunary ideal convergence of sequence with the help of a compact operator and investigate some topological and algebraic properties on these spaces. We denote ω^F the class of all sequences of fuzzy real numbers and $\mathcal{J}$ be an admissible ideal of the subset of the natural numbers $\mathbb{N}$.

$$ {}_F\mathcal{S}^{\mathcal{J}_\theta}(\mathcal{T}) = \left\{ F = (F_k) \in \omega^F : \{ r \in \mathbb{N} : h_r^{-1} \sum_{k \in J_r} \overline{\rho}(\mathcal{T}(F_k), F_0) \geq \varepsilon \} \in \mathcal{J} \text{ for some } F_0 \in \mathbb{R}(\lambda) \right\}, $$

$$ {}_F\mathcal{S}_0^{\mathcal{J}_\theta}(\mathcal{T}) = \left\{ F = (F_k) \in \omega^F : \{ r \in \mathbb{N} : h_r^{-1} \sum_{k \in J_r} \overline{\rho}(\mathcal{T}(F_k), \overline{0}) \geq \varepsilon \} \in \mathcal{J} \right\}, $$

$$ {}_F\mathcal{S}_\infty^{\mathcal{J}_\theta}(\mathcal{T}) = \left\{ F = (F_k) \in \omega^F : \exists \ \mathcal{M} > 0 : \{ r \in \mathbb{N} : h_r^{-1} \sum_{k \in J_r} \overline{\rho}(\mathcal{T}(F_k), \overline{0}) \geq \mathcal{M} \} \in \mathcal{J} \right\}, $$

$$ {}_F\mathcal{S}_\infty^{\theta}(\mathcal{T}) = \left\{ F = (F_k) \in \omega^F : \sup_r h_r^{-1} \sum_{k \in J_r} \overline{\rho}(\mathcal{T}(F_k), \overline{0}) < \infty \right\}. $$

Theorem 3.1. The sequence spaces ${}_F\mathcal{S}^{\mathcal{J}_\theta}(\mathcal{T})$, ${}_F\mathcal{S}_0^{\mathcal{J}_\theta}(\mathcal{T})$ and ${}_F\mathcal{S}_\infty^{\mathcal{J}}(\mathcal{T})$ are linear spaces.

Proof. Suppose α and β be scalars, and assume that $F = (F_k)$, $G = (g_k) \in {}_F\mathcal{S}^{\mathcal{J}_\theta}(\mathcal{T})$. Since F_k, $G_k \in {}_F\mathcal{S}^{\mathcal{J}_\theta}(\mathcal{T})$. Then for a given $\varepsilon > 0$, there exists $F_1, F_2 \in \mathbb{C}$ in such a manner that

$$ \left\{ r \in \mathbb{N} : h_r^{-1} \sum_{k \in J_r} \overline{\rho}(\mathcal{T}(F_k), F_1) \geq \frac{\varepsilon}{2} \right\} \in \mathcal{J}. $$

and

$$ \left\{ r \in \mathbb{N} : h_r^{-1} \sum_{k \in J_r} \overline{\rho}(\mathcal{T}(F_k), F_2) \geq \frac{\varepsilon}{2} \right\} \in \mathcal{J}. $$

Now, let

$$ A_1 = \left\{ r \in \mathbb{N} : h_r^{-1} \sum_{k \in J_r} \overline{\rho}(\mathcal{T}(F_k), F_1) < \frac{t}{2|\alpha|} \right\} \in \mathcal{H}(\mathcal{J}). $$

$$ A_2 = \left\{ r \in \mathbb{N} : h_r^{-1} \sum_{k \in J_r} \overline{\rho}(\mathcal{T}(F_k), F_2) < \frac{t}{2|\beta|} \right\} \in \mathcal{H}(\mathcal{J}). $$

be such that $A_1^c, A_2^c \in \mathcal{J}$. Therefore, the set

$$ \begin{aligned} A_3 &= \left\{ r \in \mathbb{N} : h_r^{-1} \sum_{k \in J_r} \overline{\rho}(\mathcal{T}(\alpha(F_k) + \beta(G_k)), (\alpha L_1 + \beta L_2)) < \varepsilon \right\} \\ &\supseteq \left\{ r \in \mathbb{N} : h_r^{-1} \sum_{k \in J_r} \overline{\rho}(\mathcal{T}(F_k), L_1) < \frac{\varepsilon}{2|\alpha|} \right\} \cap \left\{ r \in \mathbb{N} : h_r^{-1} \sum_{k \in J_r} \overline{\rho}(\mathcal{T}(F_k), L_1) < \frac{\varepsilon}{2|\beta|} \right\}. \end{aligned} \tag{2} $$

Thus, the sets on right hand side of (2) belong to $\mathcal{H}(\mathcal{J})$. Therefore A_3^c belongs to $\mathcal{J}$. Therefore, $\alpha(F_k) + \beta(G_k) \in {}_F\mathcal{S}^{\mathcal{J}_\theta}(\mathcal{T})$. Hence ${}_F\mathcal{S}^{\mathcal{J}_\theta}(\mathcal{T})$ is linear space.

In similar manner, one can easily prove that ${}_F\mathcal{S}_0^{\mathcal{J}_\theta}(\mathcal{T})$ and ${}_F\mathcal{S}_\infty^{\mathcal{J}}(\mathcal{T})$ are linear spaces. □

Example 3.1. Suppose $\mathcal{J} = \mathcal{J}_\rho = \{ \mathcal{B} \subseteq \mathbb{N} : \rho(\mathcal{B}) = 0 \}$, where $\rho(\mathcal{B})$ denotes the asymptotic density of $\mathcal{B}$. In this case ${}_F\mathcal{S}^{\mathcal{J}_\rho}(\mathcal{T}) = {}_F\mathcal{S}_\theta(\mathcal{T})$, where

$$_F\mathcal{S}_\theta(T) = \left\{ F_k \in \omega^F : \rho\left(\left\{ r \in \mathbb{N} : h_r^{-1} \sum_{k \in J_r} \overline{\rho}(\mathcal{T}(F_k), F_0) \geq \varepsilon \right\}\right) = 0 \text{ for some } F_0 \in \mathbb{R}(\gamma) \right\}.$$

Example 3.2. Suppose $\mathcal{J} = \mathcal{J}_f = \{\mathcal{B} \subseteq \mathbb{N} : \mathcal{B} \text{ is finite}\}$. $\mathcal{J}_f$ is an admissible ideal in $\mathbb{N}$ and $_F\mathcal{S}^{\mathcal{J}_f}(\mathcal{T}) = {_F\mathcal{S}^\theta}(\mathcal{T})$.

Theorem 3.2. The spaces $_F\mathcal{S}^{\mathcal{J}_\theta}(\mathcal{T})$ and $_F\mathcal{S}_0^{\mathcal{J}_\theta}(\mathcal{T})$ are not convergent free.

Proof. For the proof of the theorem, we consider the subsequent example. □

Example 3.3. Suppose $\mathcal{J} = \mathcal{J}_\delta$ and $\mathcal{T}(F_k) = F_k$.

Consider the sequence $F_k \in {_F\mathcal{S}_0^{\mathcal{J}_\theta}}(\mathcal{T}) \subset {_F\mathcal{S}^{\mathcal{J}_\theta}}(\mathcal{T})$ as:

For $k \neq i^2, i \in \mathbb{N}$

$$F_k(s) = \begin{cases} 1, & \text{if } 0 \leq s \leq k^{-1} \\ 0, & \text{otherwise.} \end{cases}$$

For $k = i^2, i \in \mathbb{N}, F_k(s) = \overline{0}$. Therefore

$$F_k = \begin{cases} [0, 0], & \text{if } k^2 = i \\ \left[0, k^{-1}\right], & k \neq i^2. \end{cases}$$

Hence, $F_k \to \overline{0}$ *as* $k \to \infty$. Thus $F_k \in {_F\mathcal{S}_0^{\mathcal{J}_\theta}}(\mathcal{T}) \subset {_F\mathcal{S}^{\mathcal{J}_\theta}}(\mathcal{T})$. Let G_k be sequence in such a way that, for $k = i^2, i \in \mathbb{N}, F_k(s) = \overline{0}$. Therefore, one obtain

$$G_k(s) = \begin{cases} 1, & \text{if } 0 \leq s \leq k \\ 0, & \text{otherwise.} \end{cases}$$

For $k = i^2, i \in \mathbb{N}, F_k(s) = \overline{0}$. Therefore

$$G_k = \begin{cases} [0, 0], & \text{if } k^2 = i \\ [0, k], & k \neq i^2. \end{cases}$$

It can be easily seen that $G_k \notin {_F\mathcal{S}_0^{\mathcal{J}_\theta}}(\mathcal{T}) \subset {_F\mathcal{S}^{\mathcal{J}_\theta}}(\mathcal{T})$.

Hence $_F\mathcal{S}^{\mathcal{J}_\theta}(\mathcal{T})$ and $_F\mathcal{S}_0^{\mathcal{J}_\theta}(\mathcal{T})$ are not convergence free.

Theorem 3.3. The sequence $F = (F_k) \in {_F\mathcal{S}_\infty^\theta}(\mathcal{T})$ is $\mathcal{J}-$ convergent if and only if for every $\varepsilon > 0$, there exists $N_\varepsilon \in \mathbb{N}$ in such a way that

$$\left\{ r \in \mathbb{N} : h_r^{-1} \sum_{k \in J_r} \overline{\rho}(\mathcal{T}(F_k), \mathcal{T}(F_{N_\varepsilon})) < \varepsilon \right\} \in \mathcal{H}(\mathcal{J}). \tag{3}$$

Proof. Suppose that $F = (F_k) \in {_F\mathcal{S}_\infty^\theta}(\mathcal{T})$ Let $F_0 = \mathcal{J} - \lim F$. Then for every $\varepsilon > 0$, the set

$$B_\varepsilon = \left\{ r \in \mathbb{N} : h_r^{-1} \sum_{k \in J_r} \overline{\rho}(\mathcal{T}(F_k), F_0) < \frac{\varepsilon}{2} \right\} \in \mathcal{H}(\mathcal{J}).$$

Fix an $N_\varepsilon \in B_\varepsilon$. Then, we have

$$\overline{\rho}(\mathcal{T}(F_k),\mathcal{T}(F_{N_\varepsilon}))\leq\overline{\rho}(\mathcal{T}(F_k),F_0)$$
$$+\overline{\rho}(\mathcal{T}(F_{N_\varepsilon}),F_0)<\frac{\varepsilon}{2}+\frac{\varepsilon}{2}=\varepsilon.$$

Which holds for all $N_\varepsilon\in B_\varepsilon$. Hence

$$\left\{r\in\mathbb{N}:h_r^{-1}\sum_{k\in J_r}\overline{\rho}(\mathcal{T}(F_k),\mathcal{T}(F_{N_\varepsilon}))<\varepsilon\right\}\in\mathcal{H}(\mathcal{J}).$$

On Contrary, assume that

$$\left\{r\in\mathbb{N}:h_r^{-1}\sum_{k\in J_r}\overline{\rho}(\mathcal{T}(F_k),\mathcal{T}(F_{N_\varepsilon}))<\varepsilon\right\}\in\mathcal{H}(\mathcal{J}).$$

That is

$$\left\{r\in\mathbb{N}:h_r^{-1}\sum_{k\in J_r}\overline{\rho}(\mathcal{T}(F_k),\mathcal{T}(F_{N_\varepsilon}))<\varepsilon\right\}\in\mathcal{H}(\mathcal{J})\ \text{ for all }\ \varepsilon>0.$$

Then, the set

$$C_\varepsilon=\{k\in\mathbb{N}:\mathcal{T}(F_k)\in[\mathcal{T}(F_{N_\varepsilon})-\varepsilon,\mathcal{T}(F_{N_\varepsilon})+\varepsilon]\}\in\mathcal{H}(\mathcal{J})\ \text{ for all }\ \varepsilon>0.$$

Let $I_\varepsilon=[\mathcal{T}(F_{N_\varepsilon})-\varepsilon,\mathcal{T}(F_{N_\varepsilon})+\varepsilon]$. If we fix an $\varepsilon>0$, then we have $C_\varepsilon\in\mathcal{H}(\mathcal{J})$ as well as $C_{\frac{\varepsilon}{2}}\in\mathcal{H}(\mathcal{J})$. Hence $C_{\frac{\varepsilon}{2}}\cap C_\varepsilon\in\mathcal{H}(\mathcal{J})$. This implies that $I=I_\varepsilon\cap I_{\frac{\varepsilon}{2}}\neq\emptyset$. That is

$$\{k\in\mathbb{N}:\mathcal{T}(F_k)\in I\}\in\mathcal{H}(\mathcal{J}).$$

That is

$$\text{diam }\ I\leq\text{diam }\ I_\varepsilon.$$

Where the diam of I denotes the length of interval I. Continuing in this way, by induction, we get the sequence of closed intervals.

$$I_\varepsilon=\mathcal{I}_0\supseteq\mathcal{I}_1\supseteq\ \dots\ \supseteq\mathcal{I}_k\supseteq\ \dots$$

With the property that

$$\text{diam }\ \mathcal{I}_k\leq\frac{1}{2}\ \text{ diam }\ \mathcal{I}_{k-1}\ \text{ for }\ (k=2,3,4,\ \dots)$$

and

$$\{k\in\mathbb{N}:\mathcal{T}(F_k)\in\mathcal{I}_k\}\in\mathcal{H}(\mathcal{J})\ \text{ for }\ (k=2,3,4,\ \dots).$$

Then there exists a $\xi\in\cap\mathcal{I}_k$ where $k\in\mathbb{N}$ in such a way that $\xi=\mathcal{J}-\lim T(F_k)$. Therefore, the result holds. □

Theorem 3.4. *The inclusions* ${}_FS_0^{\mathcal{J}_\theta}(\mathcal{T})\subset{}_FS^{\mathcal{J}_\theta}(\mathcal{T})\subset{}_FS_\infty^{\mathcal{J}_\theta}(\mathcal{T})$ *hold.*

Proof. Let $F = (F_k) \in {}_F\mathcal{S}^{\mathcal{J}_\theta}(\mathcal{T})$. Then there exists a number $F_0 \in \mathbb{R}$ such that

$$\mathcal{J} - \lim_{k\to\infty} \overline{\rho}(\mathcal{T}(F_k), F_0) = 0.$$

That is, the set

$$\left\{ r \in \mathbb{N} : h_r^{-1} \sum_{k \in J_r} \overline{\rho}(\mathcal{T}(F_k), F_0) \geq \varepsilon \right\} \in \mathcal{J}.$$

Here

$$\begin{aligned} h_r^{-1} \sum_{k \in J_r} \overline{\rho}(\mathcal{T}(F_k), \overline{0}) &= h_r^{-1} \sum_{k \in I_r} \overline{\rho}(\mathcal{T}(F_k), -F_0 + F_0) \\ &\leq h_r^{-1} \sum_{k \in I_r} \overline{\rho}(\mathcal{T}(F_k) - F_0) + h_r^{-1} \sum_{k \in J_r} \overline{\rho}(\mathcal{T}(F_k), F_0) \end{aligned} \tag{4}$$

On the both sides, taking the supremum over k of the above equation, we obtain $F_k \in {}_F\mathcal{S}^{\mathcal{J}_\theta}(\mathcal{T})$. Therefore inclusion holds.

$$\left\{ r \in \mathbb{N} : h_r^{-1} \sum_{k \in J_r} \overline{\rho}(\mathcal{T}(F_k^p), a_p) \geq \varepsilon \right\} \in \mathcal{J}.$$

Then, it proves that ${}_F\mathcal{S}_0^{\mathcal{J}_\theta}(\mathcal{T}) \subset {}_F\mathcal{S}^{\mathcal{J}_\theta}(\mathcal{T})$. Hence ${}_F\mathcal{S}_0^{\mathcal{J}_\theta}(\mathcal{T}) \subset {}_F\mathcal{S}^{\mathcal{J}_\theta}(\mathcal{T}) \subset {}_F\mathcal{S}_\infty^{\mathcal{J}_\theta}(\mathcal{T})$. □

Theorem 3.5. The space ${}_F\mathcal{S}^{\mathcal{J}_\theta}(T)$ is neither normal nor monotone if $\mathcal{J}$ is not maximal ideal.

Example 3.4. Suppose fuzzy number

$$G_k(s) = \begin{cases} \dfrac{1+s}{2}, & \text{if } \ -1 \leq s \leq 1 \\ \dfrac{3-s}{2}, & \text{if } \ 1 \leq s \leq 3 \\ 0, & \text{otherwise.} \end{cases}$$

Then $G_k(s) \in {}_F\mathcal{S}^{\mathcal{J}_\theta}(\mathcal{T})$. Applying Lemma 1, there exists a subset K of $\mathbb{N}$ in such a way that $K \notin \mathcal{J}$ and $\mathbb{N} \backslash K \notin \mathcal{J}$. Determine a sequence $G = (G_k)$ as

$$G_k = \begin{cases} F_k, & k \in K \\ 0, & \text{otherwise.} \end{cases}$$

Therefore (G_k) belong to the canonical pre- image of the K- step spaces of ${}_F\mathcal{S}^{\mathcal{J}_\theta}(\mathcal{T})$. But $(G_k) \notin {}_F\mathcal{S}^{\mathcal{J}_\theta}(\mathcal{T})$. Hence ${}_F\mathcal{S}^{\mathcal{J}_\theta}(\mathcal{T})$ is not monotone. Hence, by Lemma (2.1) ${}_F\mathcal{S}^{\mathcal{J}_\theta}(\mathcal{T})$ is not normal.

Theorem 3.6. The sequence space ${}_F\mathcal{S}_0^{\mathcal{J}_\theta}(\mathcal{T})$ is solid and monotone.

Proof. Suppose $F = (F_k) \in {}_F\mathcal{S}_0^{\mathcal{J}_\theta}(\mathcal{T})$, then for $\varepsilon > 0$, the set

$$\left\{ r \in \mathbb{N} : h_r^{-1} \sum_{k \in J_r} \overline{\rho}(\mathcal{T}(F_k), F_0) \geq \varepsilon \right\} \in \mathcal{J}. \tag{5}$$

Suppose that sequence of scalars (α_k) with the property $|\alpha_k| \leq 1 \ \forall \ k \in \mathbb{N}$. Therefore

$$\begin{aligned} &\overline{\rho}(\mathcal{T}(\alpha_k F_k), F_0) = \overline{\rho}(\alpha_k \mathcal{T}(F_k), F_0) \\ &\leq |\alpha_k| \overline{\rho}(\mathcal{T}(F_k), F_0) \ \text{ for all } \ k \in \mathbb{N}. \end{aligned} \tag{6}$$

Hence, from the Eq. (5) and above inequality, one obtain

$$\left\{ r \in \mathbb{N} : h_r^{-1} \sum_{k \in J_r} \overline{\rho}(\mathcal{T}(\alpha_k F_k), F_0) \geq \varepsilon \right\}$$

$$\subseteq \left\{ r \in \mathbb{N} : h_r^{-1} \sum_{k \in J_r} \overline{\rho}(\mathcal{T}(F_k), F_0) \geq \varepsilon \right\} \in \mathcal{J}.$$

Implies

$$\left\{ r \in \mathbb{N} : h_r^{-1} \sum_{k \in J_r} \overline{\rho}(\mathcal{T}(\alpha_k F_k), F_0) \geq \varepsilon \right\} \in \mathcal{J}.$$

Therefore, $(\alpha_k F_k) \in {}_F\mathcal{S}_0^{\mathcal{J}_\theta}(\mathcal{T})$. Then ${}_F\mathcal{S}_0^{\mathcal{J}_\theta}(\mathcal{T})$ is solid and monotone by lemma 2.1.

□

Theorem 3.7. The sequence space ${}_F\mathcal{S}^{\mathcal{J}_\theta}(\mathcal{T})$ is closed subspace of ${}_F\mathcal{S}_\infty^\theta(\mathcal{T})$.

Proof. Suppose (F_k^q) be a Cauchy sequence in ${}_F\mathcal{S}^{\mathcal{J}_\theta}(\mathcal{T})$. Then, $F_k^q \to F$ in ${}_F\mathcal{S}_\infty(\mathcal{T})$ as $q \to \infty$. Since $(F_k^q) \in {}_F\mathcal{S}^{\mathcal{J}_\theta}(\mathcal{T})$, then for each $\varepsilon > 0$ there exists a_q such that converges to a.

$$\mathcal{J} - \lim F = a.$$

Since (F_k^q) be a Cauchy sequence in ${}_F\mathcal{S}^{I_\theta}(\mathcal{T})$. Then for each $\varepsilon > 0$ there exists $n_0 \in \mathbb{N}$, such that

$$\sup_k \overline{\rho}(\mathcal{T}(F_k^q), \mathcal{T}(F_k^s)) < \frac{\varepsilon}{3} \ \text{ for all } \ q, s \geq n_0.$$

For a given $\varepsilon > 0$,

$$B_{q,s} = \left\{ r \in \mathbb{N} : h_r^{-1} \sum_{k \in J_r} \overline{\rho}(\mathcal{T}(F_k^q), \mathcal{T}(F_k^s)) < \frac{\varepsilon}{3} \right\}.$$

Now,

$$B_q = \left\{ r \in \mathbb{N} : h_r^{-1} \sum_{k \in J_r} \overline{\rho}(\mathcal{T}(F_k^p), a_q) < \frac{\varepsilon}{3} \right\}$$

and

$$B_s = \left\{ r \in \mathbb{N} : h_r^{-1} \sum_{k \in I_r} \overline{\rho}(\mathcal{T}(F_k^s), a_s) < \frac{\varepsilon}{3} \right\}.$$

Then $B^c_{q,s}, B^c_q, B^c_s \in \mathcal{J}$. Let $B^c = B^c_{q,s} \cup B^c_q \cup B^c_s$. Where

$$B = \left\{ r \in \mathbb{N} : h_r^{-1} \sum_{k \in J_r} \overline{\rho}(a_q, a_s) < \varepsilon \right\} \text{ then } B^c \in \mathcal{J}.$$

Assume $n_0 \in B^c$. Then, for every $q, s \geq n_0$ it has

$$\begin{aligned} \overline{\rho}(a_q, a_s) &\leq \overline{\rho}(\mathcal{T}(F^q_k), a_q) + \overline{\rho}(\mathcal{T}(F^s_k), a_s) + \overline{\rho}(\mathcal{T}(F^q_k), \mathcal{T}(F^s_k)) \\ &< \frac{\varepsilon}{3} + \frac{\varepsilon}{3} + \frac{\varepsilon}{3} = \varepsilon. \end{aligned}$$

Hence, a_q is a Cauchy sequence of scalars in $\mathbb{C}$, so there exists scalar $a \in \mathbb{C}$ in such a way that $(a_q) \to a$ as $q \to \infty$. For this step, let $0 < \alpha < 1$ be given. Therefore it proved that whenever

$$U = \left\{ r \in \mathbb{N} : h_r^{-1} \sum_{k \in J_r} \overline{\rho}(\mathcal{T}(F), a) < \alpha \right\} \text{ then } U^c \in \mathcal{J}. \quad (7)$$

Since $(F^q_k) \to F$, there exists $q_0 \in \mathbb{N}$ so that

$$P = \left\{ r \in \mathbb{N} : h_r^{-1} \sum_{k \in J_r} \overline{\rho}(\mathcal{T}(F^{q_0}), \mathcal{T}(F)) < \frac{\alpha}{3} \right\}.$$

implies that $P^c \in \mathcal{J}$. The number, q_o can be chosen together with Eq. (7), it have

$$Q = \left\{ r \in \mathbb{N} : h_r^{-1} \sum_{k \in J_r} \overline{\rho}(a_{q_0}, a) < \frac{\alpha}{3} \right\}.$$

Which implies that $Q^c \in \mathcal{J}$. Since

$$\left\{ r \in \mathbb{N} : h_r^{-1} \sum_{k \in J_r} \overline{\rho}(\mathcal{T}(F^{q_0}_k), a_{q_0}) \geq \frac{\alpha}{3} \right\} \in \mathcal{J}.$$

Then it has a subset S such that $S^c \in \mathcal{J}$, where

$$S = \left\{ r \in \mathbb{N} : h_r^{-1} \sum_{k \in J_r} \overline{\rho}(\mathcal{T}(F^{q_0}_k), a_{q_0}) < \frac{\alpha}{3} \right\}.$$

Suppose that $U^c = P^c \cup Q^c \cup S^c$. Then, for every $k \in U^c$ it has

$$\begin{aligned} \left\{ r \in \mathbb{N} : h_r^{-1} \sum_{k \in J_r} \overline{\rho}(\mathcal{T}(F), a) < \alpha \right\} &\supseteq \left\{ r \in \mathbb{N} : h_r^{-1} \sum_{k \in I_r} \overline{\rho}(\mathcal{T}(F^{q_0}), (F)) < \frac{\alpha}{3} \right\} \cap \\ &\left\{ r \in \mathbb{N} : h_r^{-1} \sum_{k \in J_r} \overline{\rho}(a_{q_0}, a) < \frac{\alpha}{3} \right\} \cap \left\{ r \in \mathbb{N} : h_r^{-1} \sum_{k \in J_r} \overline{\rho}(\mathcal{T}(F^{q_0}_k), a_{q_0}) < \frac{\alpha}{3} \right\}. \end{aligned} \quad (8)$$

The right hand side of Eq. (8) belongs to $\mathcal{H}(\mathcal{J})$. Hence, the sets on the left hand side of Eq. (8) belong to $\mathcal{H}(\mathcal{J})$. Therefore its complement belongs to $\mathcal{J}$. Thus, $\mathcal{J} - \lim \overline{\rho}(a_{q_0}, a) = 0$. □

In the following example to prove that ${}_F\mathcal{S}^{\mathcal{J}_\theta}(\mathcal{T})$ is closed subspace of ${}_F\mathcal{S}^{\theta}_{\infty}(\mathcal{T})$.

Example 3.5. Suppose that sequence of fuzzy number determine by

$$F_k(s) = \begin{cases} 2^{-1}(1+s), & \text{for } s \in [-1, 1] \\ 2^{-1}(3-s), & \text{for } s \in [1, 3] \\ 0, & \text{otherwise.} \end{cases}$$

Hence, $\overline{\rho}(F_k, \overline{0}) = \sup \rho(F_k, \overline{0})$. Therefore $(F_k) \in {}_F\mathcal{S}^{\mathcal{J}_\theta}(\mathcal{T})$ and $L = \overline{0}$. So, it can be easily seen that ${}_F\mathcal{S}^{\mathcal{J}_\theta}(\mathcal{T})$ is closed subspace of ${}_F\mathcal{S}^{\theta}_{\infty}(\mathcal{T})$.

4. Conclusion

The spaces of fuzzy valued lacunary ideal convergence of sequence with the help of a compact operator and investigate algebraic and topological properties together with some examples on the given spaces. We proved that the new introduced sequence spaces are linear. Some spaces are convergent free and we also proved that space ${}_F\mathcal{S}^{\mathcal{J}_\theta}(\mathcal{T})$ is closed subspace of ${}_F\mathcal{S}^{\theta}_{\infty}(\mathcal{T})$. These new spaces and results provide new tools to help the authors for further research and to solve the engineering problems.

Author details

Vakeel A. Khan[1*], Mobeen Ahmad[1] and Masood Alam[2]

1 Department of Mathematics, Aligarh Muslim University, Aligarh, 202002, India

2 Department of Mathematics and IT Center for Preparatory Studies, Sultan Qaboos University, P.O. Box 162, PC 123, Al Khoud Muscat, Sultanate of Oman

*Address all correspondence to: vakhanmaths@gmail.com

References

[1] Esi A, Mehmet Acikgoz, On almost λ-statistical convergence of fuzzy numbers, Acta Scientiarum.Technology, 2014, vol. 36(1), 129-133.

[2] Alaba BA, Norahun WZ, Fuzzy Ideals and Fuzzy Filters of Pseudocomplemented Semilattices, Advances in Fuzzy Systems, 2019, Vol. (2019), 1-13.

[3] Hazarika B, Fuzzy real valued lacunary I-convergent sequences, Applied Mathematics Letters, 2012, 25 (3):466-470.

[4] Nuray F, Savas E, Statistical convergence of sequences of fuzzy numbers, Mathematica Slovaca, 1995, 45(3):269-273.

[5] Fast H, Sur la convergence statistique Colloquium Math., 1951, (2), 241-244.

[6] Zadeh LA, Fuzzy sets. Information and control, 1965, 8(3):338-353.

[7] Matloka M, Sequences of fuzzy numbers, Busefal, 1986, 28(1):28-37.

[8] Kostyrko P, Wilczynski W, Salat T, I-convergence, Real Analysis Ex-change, 2000, 26(2):669-686.

[9] Nanda S, On sequences of fuzzy numbers, Fuzzy sets and systems, 1989, 33(1):123-126.

[10] Khan VA, Ahmad M, Fatima F, Khna MF, On some results in intuitionistic fuzzy ideal convergence double sequence spaces, Advances in Difference Equations, 2019, vol.(2019): 1-10.

[11] Khan VA, Kara EE, Altaf H, Ahmad M, Intuitionistic fuzzy I-convergent Fibonacci difference sequence spaces, Journal of Inequalities and Applications, 2019, Vol.(2019)(1), 1-7

[12] Khan VA, Rababah RKA, Esi A, Sameera Ameen Ali Abdullah, and Kamal Mohammed Ali Suleiman Alshlool, Some new spaces of ideal convergent double sequences by using compact operator, Journal of Applied Sciences, 2017, 17(9):467-474.

[13] Kumar V, Kumar K, On the ideal convergence of sequences of fuzzy numbers, Information Sciences, 2008, 178(24):4670-4678.

Chapter 4

Towards a Fuzzy Context Logic

Hedda Schmidtke

Abstract

A key step towards trustworthy, reliable and explainable, AI is bridging the gap between the quantitative domain of sensor-actuator systems and the qualitative domain of intelligent systems reasoning. Fuzzy logic is a well-known formalism suitable for aiming at this gap, featuring a quantitative mechanism that at the same time adheres to logical principles. Context logic is a two-layered logical language originally aimed at pervasive computing systems for reasoning about and within context, i.e., changing logical environments. Both logical languages are linguistically motivated. This chapter uncovers the close connection between the two logical languages presenting two new results. First, a proof is presented that context logic with a lattice semantics can be understood as an extension of fuzzy logic. Second, a fuzzification for context logic is proposed. The resulting language, which can be understood as a two-layered fuzzy logic or as a fuzzified context logic, expands both fields in a novel manner.

Keywords: intelligent systems, fuzzy logic, context logic, context

1. Introduction

Fuzzy logic has been employed successfully in intelligent systems, sensor-actuator systems, expert systems, and machine learning techniques for more than 50 years [1]. Being a tool for inference at both the logical and the sensor-actuator systems level its use for reliable and explainable autonomous systems has become a focus of recent research [2–5]. One key building block for this has been a growing understanding of fuzzy logic semantics over the past 20 years [6] and the position this family of logics assumes within the field of logics in general. In particular, the connection to residuated lattices plays an important role for novel perspectives [7, 8]. One such new perspective is the connection to context logic, which is developed in this chapter.

Context logic was introduced in [9–11] as a logic for representing context-dependency and context phenomena in pervasive computing systems. Recent developments in context logic focus on a logical actuator control mechanism [12–14]. This chapter presents the logic with a fuzzy logic lattice semantics highlighting the close relation between the two formalisms and the close relation between context logic and the sensory and machine learning components of intelligent sensor actuator systems (ISAS), such as robotics and autonomous vehicles. We show that context logic can be understood as a fuzzy logic since it can be given an algebraic semantics like that of fuzzy logic as based upon lattice structures.

2. Fuzzy logic and context logic

We briefly review the basics of how fuzzy logic handles quantitative information and contrast this with the approach chosen in context logic. Here, it may appear we go into basic aspects at a greater depth than what may seem necessary. However, to bring the two logics together, establishing the common ground conceptually is a critical first step.

Fuzzy logic [15] was developed as a linguistically motivated logic that was to be more akin to how human beings reason with uncertain information and how experts analyze alternatives and act upon them [16]. Its main cognitive motivation was that human beings are able to relay, for instance, control information without the use of numerical values. In fact, human language outside scientific and technical contexts rarely employs quantities to express relations regarding a scale, amounts, or probabilities. We prefer to say, e.g., *rarely* rather than giving an estimate about a concrete percentage, or give a color term, such as *yellow*, instead of providing RGB values and we reason with such information. We "compute with words" [17]. One reason for this is the inherent uncertainty of perceptual or sensory information and the presence of intersubjective differences. Rules we receive or provide verbally benefit from this vagueness, as they have a wide applicability, allow a concise formulation, and allow for intersubjective differences: two people may disagree whether a certain fruit is yellow or rather a light orange, but they will agree that to at least some degree, something that has a light orange color is yellow. A rule given by an expert to a novice, such as "if a fruit is yellow, then it is ripe," is easy to understand for a human being, and accordingly fuzzy expert systems, fuzzy sensor-actuator systems, and the output of some fuzzy learning systems, can be understood and verified by human beings better than purely numerical systems that operate with numerical equations.

In natural language, human beings convey information about continuous sensory domains, such as color or height, by use of adjectives. The phenomena of vagueness, uncertainty, and context-dependency are the main challenges for formalization from a linguistic point of view [18]. Adjectives can be used in several different ways. The main categories are:

Positive: Anne is tall (for her age).

Comparative: Anne is taller than Betty.

Equative: Ann is as tall as Betty.

Superlative: Ann is the tallest (girl on the team).

While the comparative and equative use are most easily mapped to a corresponding ordering and equivalence relation for the dimension in question (here: height), the positive and superlative can change their applicability depending on context. If we talk about children, 1.50 m (5 ft) may be tall. If we talk about the average European female adult, this is comparatively small. Likewise, the superlative changes with the context: Ann may be the smallest person in the room and still be called the tallest while the current topic is her team. Context logic is interesting from a cognitive science perspective as it enables the modeling of such influence of the context.

From a cognitive science point of view, fuzzy logic is an interesting formalism as it addresses issues of vagueness and uncertainty that appear especially in the semantics of adjectives. But it is also one of only few approaches bridging logical reasoning and machine learning [19].

Fuzzy logic goes beyond multi-valued logics [20] by proposing semantics for approximate reasoning. In particular, [15, p.424] proposes to "[view] the process of inference [...] as the solution of a system of relational assignment equations." This emphasizes the connection to both sensor-actuator systems and classical methods of system modeling and evaluation with recent advances reaching from explainable machine learning [5] to advanced uncertainty mechanisms for ontology design [21]. Combining the two languages promises to make the full expressiveness of natural language adjectives available for modeling, reasoning, and explanation in ISAS design.

3. Fuzzy logic as a logical language

While the linguistic background facilitates usability of fuzzy logic, it is easier to see logical connections with respect to a more restrictive and conventional logic syntax. We therefore use a simple propositional logical language as a classical background language in this chapter. We adopt the following syntax for the set of all formulae $\mathcal{L}_F$ based on a set of variables $\mathcal{V}_F$ and a set of predicate symbols $\mathcal{P}_F$.

For $\mathrm{P} \in \mathcal{P}_\mathrm{F}$ and $\mathrm{x} \in \mathcal{V}_\mathrm{F}$, $\mathrm{P(x)}$ is an (atomic) formula.
For any formula $\phi \in \mathcal{L}_\mathrm{F}$, $\neg\phi$ is a formula.
For any formulae $\phi, \psi \in \mathcal{L}_\mathrm{F}$, $\phi \wedge \psi$, $\phi \vee \psi$, and $\phi \rightarrow \psi$ are formula.

Using this syntax, we can formalize a proposition similar to the above example as:

$$\mathrm{Yellow(color)} \rightarrow \mathrm{Ripe(ripeness)}.$$

We can use the usual semantics for predicate logics to interpret this sentence based on a structure (U, I, i_V, i_P). Here U is the universe of discourse, which needs to contain in this example: the referents for the constants, i.e., concrete colors, e.g., as RGB values, and degrees of ripeness, e.g., as sets of tuples containing percentage of sugar and other substances indicative of degrees of ripeness. The term interpretation function $i_V : \mathcal{V}_F \rightarrow U$ maps the variable symbols *ripeness* and *color* to elements from U, distinct measurement values in a measurement value space. Predicate symbols are interpreted by the function $i_P : \mathcal{V}_P \rightarrow 2^U$ mapping out regions in U. The classical formula interpretation function $I : \mathcal{L}_F \rightarrow \{0, 1\}$ maps formulae to values in $\{0, 1\}$.

3.1 Interpretation of predicates based on fuzzy sets

A fundamental point where fuzzy logic differs from classical predicate logic is in the interpretation of the predicates and predication: classical logic considers $I(\mathrm{Yellow(color)})$ as true iff $i_V(\mathrm{color}) \in i_P(\mathrm{Yellow})$, realizing predication by set membership ($\in$). Fuzzy logic, in contrast, interprets predicate symbols such as Yellow with fuzzy sets $\mu_P : U \rightarrow [0, 1]$, e.g., $\mu_{\mathrm{Yellow}} : U \rightarrow [0, 1]$, i.e., as functions into $[0, 1]$. It then can replace the classical membership function $\in$ (of type $U \times 2^U \rightarrow \{0, 1\}$), with a fuzzy set membership function $\mu : U \times (U \rightarrow [0, 1]) \rightarrow [0, 1]$ that simply applies the fuzzy set membership function: $\mu(u, \mu_\mathrm{P}) \mapsto \mu_\mathrm{P}(u)$. Being based on fuzzy sets μ_P, formulae $\mathcal{L}_F$ in fuzzy logic can then be interpreted with a fuzzy semantics using a suitable function $I : \mathcal{L}_F \rightarrow [0, 1]$ for complex formulae.

3.2 Interpretation of connectives based on *t*-norms

To evaluate complex formulae, fuzzy logic requires extended semantics for the propositional connectives that can handle arbitrary values in $[0, 1]$, while remaining true to the classical interpretation in the cases $\{0, 1\}$. A general strategy in fuzzy logic is to allow different semantics to take the place of the classical semantics for propositional connectives ($\neg, \wedge, \vee, \rightarrow$), in particular, as t-norms (functions $t : [0,1] \times [0,1] \rightarrow [0,1]$) with corresponding t-conorms (functions $s : [0,1] \times [0,1] \rightarrow [0,1]$), and their residuals ($r : [0,1] \times [0,1] \rightarrow [0,1]$), respectively [6]. These functions are described and discussed in more detail below. A t-norm based semantics interprets the logical language we defined above in the following way:

$$\text{For } Q \in \mathcal{P}_F \text{ and } x \in \mathcal{V}_F\text{: } I(Q(x)) = \mu(i_V(x), i_P(Q)) = \mu_Q(i_V(x))$$
$$\text{For any formula } \phi \in \mathcal{L}_F\text{: } I(\neg\phi) = 1 - I(\phi)$$
$$\text{For any formulae } \phi, \psi \in \mathcal{L}_F\text{: } I(\phi \wedge \psi) = t(I(\phi), I(\psi))$$
$$I(\phi \vee \psi) = s(I(\phi), I(\psi))$$
$$I(\phi \rightarrow \psi) = r(I(\phi), I(\psi))$$

3.3 Properties of *t*-norms

If the semantics for $\wedge$ are based on a t-norm, this guarantees that important semantic properties of the classical conjunction are retained. A t-norm $[0,1]^2 \rightarrow [0,1]$ is a commutative (1), associative (2), and monotone function (3), with a neutral element 1 (4).

$$t(x,y) = t(y,x) \tag{1}$$
$$t(t(x,y),z) = t(x,t(y,z)) \tag{2}$$
$$\text{If } x \preceq y \text{ then } t(x,z) \preceq t(y,z) \tag{3}$$
$$t(1,x) = x \tag{4}$$

Examples are the minimum t-norm (5), used in Gödel logics, and the product t-norm (6), used in probability theory:

$$t_{\text{min}}(a,b) = \min(a,b) \tag{5}$$
$$t_{\text{prod}}(a,b) = a * b \tag{6}$$

The corresponding t-conorms, denoted by the symbol s and accordingly also called s-norms, can be obtained by applying De Morgan's laws assuming the semantics of negation of a value t to be $1 - t$. Their neutral element is 0.

$$s(a,b) = 1 - t(1-a, 1-b) \tag{7}$$

The corresponding s-norms for the above example t-norms are then s_{min}, the minimum s-norm (8), and the product s-norm s_{prod} (9):

$$s_{\text{min}}(a,b) = \max(a,b) \tag{8}$$
$$s_{\text{prod}}(a,b) = a + b - a * b \tag{9}$$

There are several ways to interpret the implication and different approaches are suitable for different purposes (cf. [22], for a detailed overlook and comparison). As with other operators, fuzzy implication should be conservative for values in $\{0,1\}$. A widely used notion is the *left-residual* [23]:

$$r(a,b) = \sup\{z \in [0,1] | t(a,z) \preceq b\} \tag{10}$$

The relation between the residual and the t-norm/s-norm are covered by two additional axioms, continuity (11) and pre-linearity (12):

$$t(x,y) \preceq z \text{ iff } x \preceq r(y,z) \tag{11}$$

$$s(r(x,y), r(y,s)) = 1 \tag{12}$$

For the above two t-norms $t_{\min}, t_{\text{prod}}$ the following are corresponding residuals:

$$r_{\min}(a,b) = \begin{cases} 1 \text{ iff } a \leq b \\ b \text{ otherwise.} \end{cases} \tag{13}$$

$$r_{\text{prod}}(a,b) = \begin{cases} 1 \text{ iff } a \leq b \\ b/a \text{ otherwise.} \end{cases} \tag{14}$$

3.4 Generalized *t*-norms: the set-theoretic lattice

The most widely used examples of functions μ_P map elements of U to values in $[0, 1]$, with, e.g., the minimum or product t-norm. However, fuzzy logic can be given a generalized t-norm semantics based on residuated lattices, i.e., other lattice structures $(L, \preceq)$ instead of $([0, 1], \preceq)$. A particularly interesting residuated lattice for the purposes of comparison with context logic is $L = (2^B, \subseteq)$, where B is a given base set and $U = B$. Given this structure, we can define interpretation functions $i_V : \mathcal{V}_F \rightarrow B$ for variable symbols as before. But we can now interpret predicates not with classical $[0, 1]$-fuzzy sets but with generalized L-fuzzy sets $\mu_P : B \rightarrow 2^B$, so that $I : \mathcal{L}_F \rightarrow 2^B$ for formulae:

$$I(P(x)) = \mu_P(i_V(x)), \text{with } \mu_P : U \rightarrow 2^B \tag{15}$$

$$I(\neg\phi) = 2^B - I(\phi) \tag{16}$$

$$I(\phi \wedge \psi) = I(\phi) \cap I(\psi) \tag{17}$$

$$I(\phi \vee \psi) = I(\phi) \cup I(\psi) \tag{18}$$

$$I(\phi \rightarrow \psi) = I(\psi) \cup (2^B - I(\phi)) \tag{19}$$

The intuition behind this is to map elements $x \in U$ to, e.g., sets of points, i.e., spatial regions or temporal or sensory values intervals. Instead of saying x is P to a degree of 0.5, for instance, we could thus distinguish x as in a specific area of space, time, or sensor value space. E.g., we can assign a function μ_{Yellow} to map measured RGB colors x to sets that form a filter around the color #FFFF00. Measuring an orange x and a lime y we could determine they are yellow to the same degree as $\mu_{\text{Yellow}}(x)$ and $\mu_{\text{Yellow}}(y)$ yielding the same large region around the core value #FFFF00. We could say x is as yellow as y is yellow, since with $I(\text{Yellow}(x)) = I(\text{Yellow}(y))$ holds $I(\text{Yellow}(x) \leftrightarrow \text{Yellow}(y)) = 2^B$. This would be the same result as with classical fuzzy sets, but we would be able to additionally avoid comparing incompatible contexts, e.g.: while a red apple z may be as aubergine as an orange is yellow with classical fuzzy sets, the set theoretic interpretation yields $I(\text{Yellow}(x) \leftrightarrow \text{Aubergine}(z)) \subseteq I(\text{Yellow}(x) \leftrightarrow \text{Yellow}(y))$, as the regions for $I(\text{Yellow}(x))$ and $I(\text{Aubergine}(z))$ overlap but are distinct. In contrast to the strictly ordered $[0, 1]$, the partially ordered 2^B thus allows higher expressiveness.

Partial orders and corresponding lattice structures are at the heart of the semantics for context logic, and the two languages can on this basis be combined in a natural manner.

4. An overview of context logic

We now specify the context logic language and describe a semantics similarly in terms of a predicate logical language, which in turn can be related to lattice structures and thus fuzzy logical semantics.

4.1 Contextualization in context logic

Context logic has only one type of basic entity, *context variables*, and a single partial order relation $\sqsubseteq$ (*part of* or *sub-context*): the city of London, for instance, is a sub-context of England, and March 2017 is a sub-context of the year 2017:

$$\text{London} \sqsubseteq \text{England}$$

$$\text{March2017} \sqsubseteq \text{Year2017}$$

The language provides three term operators $\sqcap$ (intersection), $\sqcup$ (sum), and $\sim$ (complement).

Since any pre-order can be expressed as a sub-relation of a partial order relation, and be extended to a partial order relation over its equivalence classes, the single sub-context relation together with the $\sqcap$ operators allows the specification of arbitrarily many different partial order relations [24]. More accurately we may, for instance, want to say that the city of London is a *spatial* sub-context or a sub-region of England, and that March 2017 is a *temporal* sub-context or a sub-interval of the year 2017.

$$\text{London} \sqcap \text{Space} \sqsubseteq \text{England}$$

$$\text{March2017} \sqcap \text{Time} \sqsubseteq \text{Year2017}$$

This and the following examples feature one simple spatial sub-context and one temporal sub-context relation. We can in the same manner however express, for instance, directional relations [25], temporal ordering relations (bi-directionally branching), and class hierarchies [9]. Ordering relations between thematic values, such as expressed by the comparative use of adjectives (Section 2) can also be added in the same way. The main purpose of the language is to facilitate expressing the common partial order core of all these theories, including the tractable fragments of these theories in a unified syntax.

A syntactic shorthand reflects – linguistically speaking – a topicalized adverbial position:

$$c : [a \sqsubseteq b] \overset{def}{\Leftrightarrow} c \sqcap a \sqsubseteq b$$

$$\text{Space} : [\text{London} \sqsubseteq \text{England}]$$

$$\text{Time} : [\text{March2017} \sqsubseteq \text{Year2017}]$$

Spatially, London is a sub-context of England. Temporally, March 2017 is a sub-context of the year 2017. For entities such as cities or months, this may seem redundant. But contexts, such as a birthday party, which have both temporal and spatial extent can thus be located temporally within one context and spatially within another:

$$\text{Space} : [\text{John'sBirthday} \sqsubseteq \text{London}]$$

$$\text{Time} : [\text{John'sBirthday} \sqsubseteq \text{March2017}]$$

We can also reflect that speakers may choose to topicalize the other way around [26], as the last two sentences are logically equivalent to the following:

$$\text{John'sBirthday} : [\text{Space} \sqsubseteq \text{London}]$$
$$\text{John'sBirthday} : [\text{Time} \sqsubseteq \text{March2017}]$$

or, leveraging the propositional second layer,

$$\text{John'sBirthday} : ([\text{Space} \sqsubseteq \text{London}] \wedge [\text{Time} \sqsubseteq \text{March2017}])$$

where, for any propositional junctor $\circ \in \{\wedge, \vee, \rightarrow\}$:

$$c : (\phi_1 \circ \phi_2) \overset{def}{\Leftrightarrow} c : \phi_1 \circ c : \phi_2 \qquad \text{and} \qquad c : \neg\phi \overset{def}{\Leftrightarrow} \neg c : \phi$$

Regarding John's birthday party: the location is in London, the time is in March 2017. Moreover, we can allow contexts to be stacked or combined, in order to express more complex contextualization:

$$\text{MarySays} : \text{John'sBirthday} : [\text{Time} \sqsubseteq \text{March2017}]$$
$$\text{TomSays} : \text{John'sBirthday} : [\text{Time} \sqsubseteq \text{August2017}]$$

Similarly to how we would express conflicting opinions in natural language, we can equivalently state:

$$\text{John'sBirthday} \sqcap \text{Time} : ([\text{MarySays} \sqsubseteq \text{March2017}] \wedge [\text{TomSays} \sqsubseteq \text{August2017}])$$
$$d : c : [a \sqsubseteq b] \equiv d : [c \sqcap a \sqsubseteq b] \equiv d \sqcap c \sqcap a \sqsubseteq b$$

Regarding John's birthday party and the time, Mary says in March 2017 and Tom says in August 2017. Context logic thus allows to reflect colloquial contextualizations well, but also to represent conflicting information.

4.2 Context logic as a logical language

Context logic thus employs two syntactic layers: the term layer with the term operators $\sqcap$, $\sqcup$, $\sim$ and the propositional layer with the logic connectives $(\wedge, \vee, \neg, \rightarrow)$. Context terms $\mathcal{T}_C$ are defined over a set of variables $\mathcal{V}_C$:[1]

Any context variable $v \in \mathcal{V}_C$ and the special symbols $\top$ and $\bot$ are atomic context terms.

If c is a context term, then $\sim c$ is a context term.

If c and d are context terms then $c \sqcap d$ and $c \sqcup d$ are context terms.

Context formulae $\mathcal{L}_C$ are defined as follows:

If c and d are context terms then $c \sqsubseteq d$ is an atomic context formula.

[1] We leave out brackets as possible applying the following precedence: $\sim$, $\sqcap$, $\sqcup$, $\sqsubseteq$, $:$, $\neg$, $\wedge$, $\vee$, $\rightarrow$, $\leftrightarrow$. The scope of quantifiers is to be read as maximal.

If ϕ is a context formula, then $\neg\phi$ is a context formula.

If ϕ and ψ are context formulae then $\phi \wedge \psi$, $\phi \vee \psi$ and $\phi \rightarrow \psi$ are context formulae.

We further define:

$$c = d \overset{def}{\Leftrightarrow} [c \sqsubseteq d] \wedge [d \sqsubseteq c] \tag{20}$$

$$c \sqsubset d \overset{def}{\Leftrightarrow} [c \sqsubseteq d] \wedge \neg[d \sqsubseteq c] \tag{21}$$

Different variant semantics have been proposed [10, 11, 26]. The different approaches slightly differ in the resulting semantics, but all three employ a lattice structure for specifying the meanings of context terms, assigning a partial order to give a semantics to $\sqsubseteq$. Here, we give a semantics by mapping the language to a predicate logic with a single binary predicate P, describing a pre-order relation, to give the fundamental $\sqsubseteq$ its semantics. We use a function $\tau_{\mathrm{CL}}^{\mathrm{PL}} : \mathcal{L}_C \times \mathcal{V}_P \rightarrow \mathcal{L}_P$, where $\mathcal{L}_C$ is the set of context logic formulae, $\mathcal{V}_P$ is a vocabulary of predicate logic variables, and $\mathcal{L}_P$ is the set of predicate logic formulae. We also employ $\mathcal{V}_C$, the set of variables, as the set of constants for $\mathcal{L}_P$, and require $\mathcal{V}_P \cap \mathcal{V}_C = \emptyset$:

$$\tau_{\mathrm{CL}}^{\mathrm{PL}}(\top \sqsubseteq \top, m) = \top \tag{22}$$

$$\tau_{\mathrm{CL}}^{\mathrm{PL}}(\top \sqsubseteq \bot, m) = \bot \tag{23}$$

$$\tau_{\mathrm{CL}}^{\mathrm{PL}}(\top \sqsubseteq v, m) = P(m, v), \text{ for } v \in \mathcal{V}_C \tag{24}$$

$$\tau_{\mathrm{CL}}^{\mathrm{PL}}(\top \sqsubseteq \sim c, m) = \tau_{\mathrm{CL}}^{\mathrm{PL}}(c \sqsubseteq \bot, m) \text{ for } c \in \mathcal{T}_C \tag{25}$$

$$\tau_{\mathrm{CL}}^{\mathrm{PL}}(\top \sqsubseteq c \sqcap d, m) = \tau_{\mathrm{CL}}^{\mathrm{PL}}(\top \sqsubseteq c, m) \wedge \tau_{\mathrm{CL}}^{\mathrm{PL}}(\top \sqsubseteq d, m) \tag{26}$$

$$\begin{aligned} &\tau_{\mathrm{CL}}^{\mathrm{PL}}(\top \sqsubseteq c \sqcup d, m) = \\ &\quad \forall m', P(m', m) : \exists m'', P(m', m) : \tau_{\mathrm{CL}}^{\mathrm{PL}}(\top \sqsubseteq c, m'') \vee \tau_{\mathrm{CL}}^{\mathrm{PL}}(\top \sqsubseteq d, m'') \\ &\quad \text{where } m' \text{ and } m'' \text{ are new variables.} \end{aligned} \tag{27}$$

$$\begin{aligned} &\tau_{\mathrm{CL}}^{\mathrm{PL}}(c \sqsubseteq d, m) = \forall m', P(m', m) : \tau_{\mathrm{CL}}^{\mathrm{PL}}(\top \sqsubseteq c, m') \rightarrow \tau_{\mathrm{CL}}^{\mathrm{PL}}(\top \sqsubseteq d, m') \\ &\quad \text{where } m' \text{ is a new variable.} \end{aligned} \tag{28}$$

$$\tau_{\mathrm{CL}}^{\mathrm{PL}}(\neg\phi, m) = \neg\tau_{\mathrm{CL}}^{\mathrm{PL}}(\phi, m) \tag{29}$$

$$\tau_{\mathrm{CL}}^{\mathrm{PL}}(\phi \wedge \psi, m) = \tau_{\mathrm{CL}}^{\mathrm{PL}}(\phi, m) \wedge \tau_{\mathrm{CL}}^{\mathrm{PL}}(\psi, m) \tag{30}$$

$$\tau_{\mathrm{CL}}^{\mathrm{PL}}(\phi \vee \psi, m) = \tau_{\mathrm{CL}}^{\mathrm{PL}}(\phi, m) \vee \tau_{\mathrm{CL}}^{\mathrm{PL}}(\psi, m) \tag{31}$$

$$\tau_{\mathrm{CL}}^{\mathrm{PL}}(\phi \rightarrow \psi, m) = \tau_{\mathrm{CL}}^{\mathrm{PL}}(\phi, m) \rightarrow \tau_{\mathrm{CL}}^{\mathrm{PL}}(\psi, m) \tag{32}$$

We note that although we introduce new variables m', m'' in (27) and (28), each new variable is only used together with the variable last introduced – m' with m, m'' with m' but not with m –, not with any other variables introduced before. This means, we can alternate between two variables and reuse m after m', i.e., that $\mathcal{V}_P = \{m, m'\}$. We also note, that the context variables $v \in \mathcal{V}_C$ are constants with respect to the predicate logic and that they only appear in the second position of P in (24). This property allows us to reformulate any binary expression $P(m, v)$ for $v \in \mathcal{V}_C$ using a different monadic predicate P_v for each $v \in \mathcal{V}_C$, and write $P_v(m)$ instead of $P(m, v)$.

Consequently, the fragment of predicate logic required in application of $\tau_{\mathrm{CL}}^{\mathrm{PL}}$ alone is in the two-variable fragment known to be decidable. Moreover, the variables, such as m and m', only occur together in the *atomic guard*, as $P(m', m)$, suggesting that the language as defined so far is in the so-called *guarded fragment* GF [27] defined as [cited after 28, p.1664f]:

Every atomic formula belongs to GF.

GF is closed under $\neg, \wedge, \vee, \rightarrow, \leftrightarrow$.

If x, y are tuples of variables, $\alpha(x, y)$ is an atomic formula, $\psi(x,y)$ is in GF, and free ($\psi \subseteq$ free $(\alpha) = \{x,y\}$, where free (ϕ) is the set of the free variables of ϕ, then the formulae

$$\exists y : \alpha(x,y) \wedge \psi(x,y)$$
$$\forall y : \alpha(x,y) \rightarrow \psi(x,y)$$

belong to GF.

In order to obtain the reasoning capabilities, however, we would need to add pre-order axioms for P, so as to be able to specify $\sqsubseteq$ as a partial order relation:

$$\forall x,y,z : P(x,y) \wedge P(y,z) \rightarrow P(x,z) \tag{33}$$
$$\forall x : P(x,x) \tag{34}$$

and we see that transitivity (13) cannot be axiomatized in the two-variable fragment, as it requires three variables. Fortunately, [28, 29] have shown that for GF^2 + PG – the guarded fragment limited to two variables and a single binary pre-order that can only appear in the guard – is in 2-EXPTIME. Moreover, this result is a loose upper bound, since the language under inspection here can be expressed using the transitive binary relation P in only one direction – namely from wholes to parts –, using otherwise only the monadic predicates $P_v, v \in \mathcal{V}_C$, placing the translation of context logic with the axioms for $\sqsubseteq$ into the class $\mathrm{MGF}^2 + \overleftarrow{TG}$, the two-variable monadic guarded fragment with one-way transitive guards, which is decidable and whose satisfiability problem is in EXPSPACE [28].

In addition to the pre-order axioms, we can also add a localized guarded variant of the so-called *weak supplementation principle* [30, Ch. 3] for $\sqsubseteq$ ensuring a minimal homogeneity constraint over $v_1, v_2 \in \mathcal{V}_C$:[2]

$$\begin{aligned}\forall x : &(\forall x', P(x',x) : P(x',v_1) \rightarrow P(x',v_2)) \\ &\wedge (\exists x', P(x',x) : P(x',v_2) \wedge \neg P(x',v_1)) \\ &\rightarrow \left(\exists x', P(x',x) : P(x',v_2) \wedge \neg \exists x'', P\left(x'',x'\right) : P\left(x'',v_1\right)\right).\end{aligned} \tag{35}$$

The principle says that, if for any x all its parts x' that are in v_1 are also in v_2, but there is a part x' that is in v_2 but not in v_1 (paraphrasing: v_1 is a proper part of v_2), then there is a part x' of v_2 that has no parts in v_1 (i.e.: x' does not overlap v_1), i.e., is completely outside of v_1. Axiom 35 ensures that the entities described by $v_1, v_2 \in \mathcal{V}_C$ do not have, e.g., singular points that are not entities themselves in the domain under inspection. This axiom is required for proving several of the lattice laws. Note that we thus characterize a weak supplementation principle only for $\sqsubseteq$, that

[2] The interested reader may find a brief discussion on mereological and ontological properties in Section 4.5.

we, however, cannot formulate a weak supplementation principle for P without leaving the guarded fragment.

In order to do this, however, we have to employ $v_1, v_2 \in \mathcal{V}_C$ as schema variables, i.e., we have to formally see this actually not as one axiom but $|\mathcal{V}_C^2|$ axioms. This means that for infinite $\mathcal{V}_C$, the axiomatization becomes infinite. For practical, finite knowledge bases, $\mathcal{V}_C$ will be finite. If an infinite vocabulary $\mathcal{V}_C$ is employed, a practical realization would be to use a unification mechanism suitable for the particular language $\mathcal{V}_C$ employed.

Intuitively, the meaning of $a \sqsubseteq b$ is that all parts of a are part of b. The reading thus corresponds to a universal quantification, and the properties expressed by contexts in this statement describe homogenous properties inherited from wholes to their parts. Correspondingly, $\neg[a \sqsubseteq b]$ expresses an existential quantification, stating that not all parts of a are parts of b, which means that there is a part of a that is not part of b, or that does not have property b. We can thus express heterogeneity.

The complement $\sim c$, is interpreted with respect to the pseudo-0-element $\bot$: the atomic formula $\top \sqsubseteq\ \sim a$ is interpreted as equivalent to $a \sqsubseteq \bot$, meaning that no part is in a, implying universal quantification. There are thus two types of negation $\neg$ on the logical level and $\sim$ on the context level. $\bot$ is a pseudo-element, it disappears in the translation when applying (28). We do not need to assume that an empty element exists:

$$\begin{aligned}
\tau_{\mathrm{CL}}^{\mathrm{PL}}(c \sqsubseteq \bot, m) &\equiv \forall m', P(m', m) : \tau_{\mathrm{CL}}^{\mathrm{PL}}(\top \sqsubseteq c, m') \rightarrow \tau_{\mathrm{CL}}^{\mathrm{PL}}(\top \sqsubseteq \bot, m') \\
&\equiv \forall m', P(m', m) : \tau_{\mathrm{CL}}^{\mathrm{PL}}(\top \sqsubseteq c, m') \rightarrow \bot \\
&\equiv \forall m', P(m', m) : \neg\tau_{\mathrm{CL}}^{\mathrm{PL}}(\top \sqsubseteq c, m') \\
&\equiv \neg\exists m', P(m', m) : \tau_{\mathrm{CL}}^{\mathrm{PL}}(\top \sqsubseteq c, m')
\end{aligned} \tag{36}$$

A crucial consequence of adopting weak supplementation (35) is (2). It says that if all parts m'' of a part m' have a part m''' that is part of a, this is equivalent to m' being part of a:

$$\forall m'', P(m'', m') : \exists m''', P(m''', m'') : P(m''', a) \equiv P(m', a) \tag{37}$$

Proof ($\Dashv$): this holds immediately with the reflexivity (34) and transitivity (33) of P: if $P(m', a)$ then all parts m'' of m' fulfill $P(m'', a)$ by transitivity, and therefore there is a part m''' of m'', namely m'' itself, by reflexivity, so that $P(m''', a)$.

Proof ($\vDash$): we prove the reverse direction by contradiction, applying (35). Assume $\forall m'', P(m'', m') : \exists m''', P(m''', m'') : P(m''', a)$ and not $P(m', a)$, i.e., that there is an m_1'' that has $m''', P(m''', a)$ but not $P(m_1'', a)$. Then by (35) there has to be a part m_2'' of m' that does not have a part m''' where $P(m''', a)$. But this is prevented by the premise $\forall m'', P(m'', m') : \exists m''', P(m''', m'') : P(m''', a)$.

It can be shown (Section 4.4) that the definition of $\tau_{\mathrm{CL}}^{\mathrm{PL}}$ together with the two pre-order axioms and the local guarded variant of the weak supplementation principle is sufficient to characterize context terms as spanning a bounded lattice. We note that with a different axiomatization other types of lattice structures could be realized for different application domains.

4.3 A fuzzy logic perspective on context logic

This section shows context logic as specified above is a two-layered language with a generalized t-norm-based fuzzy logic at the term level and a classical

$\{0,1\}$-based semantics at the formula level. From there it is a small step to also add a $[0,1]$-based multivalued semantics to the formula level, so as to obtain a full two-layered fuzzy logic in Section 5.

To see that the context terms $\mathcal{T}_C$ can be viewed as a generalized t-norm, we set the intersection $\sqcap$, the meet operation of the lattice, as the monoid operation and the term $\top$ as the identity element of the monoid. The monoid properties associativity and identity element are fulfilled by any lattice (see Section 4.4, (44) and (46)). For the generalized fuzzy logic semantics, the lattice meet-operation $\sqcap$ will be shown to fulfill the properties of a t-norm, the join-operation $\sqcup$, those of the corresponding s-norm. Both are required to be commutative (1), associative (2), and support an identity element (4) and monotonicity (3) (for the full proofs see Section 4.4). We prove monotonicity for $\sqcap$ (38) and $\sqcup$ (39):

$$\begin{aligned}
&\tau_{\mathrm{CL}}^{\mathrm{PL}}([a\sqsubseteq c]\wedge[b\sqsubseteq d]\rightarrow[a\sqcap b\sqsubseteq c\sqcap d],m)\equiv\\
&\qquad(\forall m',P(m',m):P(m',a)\rightarrow P(m',c))\\
&\qquad\wedge(\forall m',P(m',m):P(m',b)\rightarrow P(m',d))\\
&\qquad\rightarrow(\forall m',P(m',m):P(m',a)\wedge P(m',b)\rightarrow P(m',c)\wedge P(m',d))
\end{aligned}\tag{38}$$

$$\begin{aligned}
&\tau_{\mathrm{CL}}^{\mathrm{PL}}([a\sqsubseteq c]\wedge[b\sqsubseteq d]\rightarrow[a\sqcup b\sqsubseteq c\sqcup d],m)\equiv\\
&\qquad(\forall m',P(m',m):P(m',a)\rightarrow P(m',c))\\
&\qquad\wedge(\forall m',P(m',m):P(m',b)\rightarrow P(m',d))\\
&\qquad\rightarrow\forall m',P(m',m):\\
&\qquad\quad(\forall m'',P(m'',m'):\exists m''',P(m''',m''):P(m''',a)\vee P(m''',b))\\
&\qquad\quad\rightarrow(\forall m'',P(m'',m'):\exists m''',P(m''',m''):P(m''',c)\vee P(m''',d))
\end{aligned}\tag{39}$$

Proof (3): if every m' that is part of a is in c and every m' that is part of b is in d, then every m' that is part of a and b is also in both c and d. Proof (4): we see that it follows from this condition also that any m''' that exists as part of any m'' in a or b must also be part of c or d in m''.

The generalized De Morgan law connects t-norms with s-norms (7). It follows for the translations of $\sqcap$ and $\sqcup$ directly from the De Morgan laws in predicate logic.

$$\begin{aligned}
&\tau_{\mathrm{CL}}^{\mathrm{PL}}(a\sqcup b={\sim}({\sim}a\sqcap{\sim}b),m)\equiv\\
&\qquad\forall m',P(m',m):\ (\forall m'',P(m'',m'):\exists m''',P(m''',m''):\\
&\qquad\qquad P(m''',a)\vee P(m''',b))\\
&\qquad\leftrightarrow\neg\exists m'',P(m'',m'):(\neg\exists m''',P(m''',m''):P(m''',a))\\
&\qquad\qquad\wedge(\neg\exists m''',P(m''',m''):P(m''',b))
\end{aligned}\tag{40}$$

The residual can then be derived from its characterization:

$$r(a,b)=\sup\{z|t(a,z)\preccurlyeq b\}.$$

The operation $\Rightarrow$ with the definition

$$a\Rightarrow b\overset{def}{\Leftrightarrow}{\sim}a\sqcup b\tag{41}$$

has the required property $t(a,z) \preccurlyeq b$ (with $\sqcap$ the t-norm and $\sqsubseteq$, the lattice partial order $\preccurlyeq$).[3]

$$\begin{aligned}\tau^{\mathrm{PL}}_{\mathrm{CL}}(a \sqcap (\sim a \sqcup b) \sqsubseteq b, m) \equiv \\ \forall m', P(m',m) : \ (\forall m'', P(m'',m') : \exists m''', P(m''',m'') : \\ (\neg \exists m^{iv}, P(m^{iv},m''') : P(m^{iv},a)) \\ \vee P(m''',b)) \wedge P(m',a) \rightarrow P(m',b)\end{aligned} \tag{42}$$

We prove that for any $m', P(m',m)$:

$$\begin{aligned}\forall m'', P(m'',m') : \ \exists m''', P(m''',m'') : \\ (\neg \exists m^{iv}, P(m^{iv},m''') : P(m^{iv},a) \vee P(m''',b)) \wedge P(m',a) \\ \vDash P(m',b)\end{aligned}$$

and the term $\sim a \sqcup b$ expresses the maximal element local to m with this property.

Proof: assume the antecedent is true, then because of transitivity of P (33) and the conjunct $P(m',a)$, there can be no m''' part of m' for which all parts m^{iv}, including m''' itself fulfill $\neg P(m^{iv},a)$. Therefore the second disjunct $P(m''',b)$ must be true. But if we know that for all m'' with $P(m'',m')$ exists m''', so that $P(m''',b)$, we know by (2), a consequence of the localized guarded variant of the weak supplementation principle (35), that $P(m',b)$. To see that it is maximal, assume there is m'_1 outside of $\sim a \sqcup b$ and $P(m'_1,a)$ and $P(m'_1,b)$. To be outside of $\sim a \sqcup b$, there would have to be an $m'', P(m'',m'_1)$ so that for all $m''', P(m''',m'')$ there is m^{iv}, so that $P(m^{iv},m''')$ and $P(m^{iv},a)$ and $\neg P(m''',b)$, but this cannot be, because $P(m''',m'_1)$ and by the assumption $P(m'_1,b)$, thus by transitivity (33) $P(m''',b)$.

This result indicates that, at least with respect to the supplementation property expressed through (35), $\sim a \sqcup b$ fulfills the characterization of a residual. We can also show continuity (54) and pre-linearity (55) (Section 4.4).

We are thus justified to say that context logic terms have a generalized t-norm semantics and we can give a t-norm-based semantics to context logic.

We obtain: a t-norm-based classical semantics for context logic is a structure $(I, i, a, L_T, \preccurlyeq, 1_T, t, s)$, where the terms are interpreted by $i : \mathcal{T}_C \rightarrow L_T$ together with the function $a : \mathcal{V}_C \rightarrow L_T$ assigning context terms and variables, respectively, to elements of a lattice L_T, and the formulae, by the classical interpretation function $I : \mathcal{L}_C \rightarrow \{0,1\}$:

---

[3] To understand the meaning of $a \Rightarrow b$, we can translate

$$\begin{aligned}&\tau^{\mathrm{PL}}_{\mathrm{CL}}(T \sqsubseteq \ \sim a \sqcup b, x) \\ &\equiv \forall m', P(m',m) : \exists m'', P(m'',m') : (\neg \exists m''', P(m''',m'') : P(m''',a)) \vee P(m'',b) \\ &\equiv \forall m', P(m',m) : (\neg \forall m'', P(m'',m') : \exists m''', P(m''',m'') : P(m''',a)) \\ &\qquad \vee \exists m'', P(m'',m') : P(m'',b) \\ &\equiv \forall m', P(m',m) : (\forall m'', P(m'',m') : \exists m''', P(m''',m'') : P(m''',a)) \\ &\qquad \rightarrow \exists m'', P(m'',m') : P(m'',b) \\ &\equiv \forall m', P(m',m) : P(m',a) \rightarrow \exists m'', P(m'',m') : P(m'',b) \\ &\equiv \forall m', P(m',m) : P(m',a) \rightarrow P(m',b),\end{aligned}$$

that is, for every part m' of m holds that: if m' is inside a it is inside b. Here the last two steps follow by (2) a consequence of (A9) .

$$i(v) = a(v), \text{for } v \in V_C$$
$$i(\sim c) = 1_T - i(c),$$

With $1_T - i(c)$the pseudo-complement of the lattice

$$i(c \sqcap d) = t(i(c), i(d))$$
$$i(c \sqcup d) = s(i(c), i(d))$$
$$I(c \sqsubseteq d) = 1 \text{ iff } i(c) \preccurlyeq i(d)$$
$$I(\neg\phi) = 1 - I(\phi)$$
$$I(\phi \wedge \psi) = \min(I(\phi), I(\psi))$$
$$I(\phi \vee \psi) = \max(I(\phi), I(\psi))$$

It only remains to show that the context term operators indeed support the lattice requirements.

4.4 Proof: context logic with local, guarded weak supplementation characterizes a bounded lattice

For the purpose of completeness, the proofs are listed here in detail. However, the results are part of basic, fundamental lattice theory and no novelty is claimed.

We prove that $\sqcap$ and $\sqcup$ fulfill the laws for a bounded lattice. We start by showing that $\sqcap$ fulfills the laws of a semilattice: $\sqcap$ is idempotent (7), associative (8), commutative (9), and has $\top$ as its neutral element (46).

$$a \sqcap a = a \tag{43}$$
$$a \sqcap (b \sqcap c) = (a \sqcap b) \sqcap c \tag{44}$$
$$a \sqcap b = b \sqcap a \tag{45}$$
$$a \sqcap \top = a \tag{46}$$

These properties hold, since $\sqcap$ directly translates into $\wedge$:

$$\tau_{\text{CL}}^{\text{PL}}(\top \sqsubseteq c \sqcap d, m) = \tau_{\text{CL}}^{\text{PL}}(\top \sqsubseteq c, m) \wedge \tau_{\text{CL}}^{\text{PL}}(\top \sqsubseteq d, m)$$

We show the translations:

$$\tau_{\text{CL}}^{\text{PL}}(a \sqcap a = a, m) \equiv \forall P(m', m) : P(m', a) \wedge P(m', a) \leftrightarrow P(m', a)$$
$$\tau_{\text{CL}}^{\text{PL}}(a \sqcap (b \sqcap c) = (a \sqcap b) \sqcap c, m) \equiv \forall P(m', m) : P(m', a) \wedge (P(m', b) \wedge P(m', c))$$
$$\leftrightarrow (P(m', a) \wedge P(m', b)) \wedge P(m', c)$$
$$\tau_{\text{CL}}^{\text{PL}}(a \sqcap b = b \sqcap a, m) \equiv \forall P(m', m) : P(m', a) \wedge P(m', b)$$
$$\leftrightarrow P(m', b) \wedge P(m', a)$$
$$\tau_{\text{CL}}^{\text{PL}}(a \sqcap \top = a, m) \equiv \forall P(m', m) : P(m', a) \wedge \top \leftrightarrow P(m', a)$$

We can see that all translations of properties are tautologies and follow directly from the properties of $\wedge$. The semantics of $\sqcup$ requires a closer look. We first note that a basic requirement of extensionality holds:

$$\begin{aligned} \tau_{\text{CL}}^{\text{PL}}(\top \sqsubseteq a, m) &\equiv P(m, a) \equiv \forall m', P(m', m) : P(m', a) \\ &\equiv \forall P(m', m) : \tau_{\text{CL}}^{\text{PL}}(\top \sqsubseteq a, m') \end{aligned} \tag{47}$$

The property (47) holds because $P(m,a)$ entails $P(m',a)$ for all $P(m',m)$ because of transitivity of P. Also, for all $m', P(m',m)$: $P(m',a)$ entails $P(m,a)$, since P is reflexive.

We can now prove the semilattice laws for $\sqcup$.

$$a \sqcup a = a \tag{48}$$

$$a \sqcup (b \sqcup c) = (a \sqcup b) \sqcup c \tag{49}$$

$$a \sqcup b = b \sqcup a \tag{50}$$

$$a \sqcup \bot = a \tag{51}$$

When we translate idempotency (48):

$$\begin{aligned} &\tau_{\mathrm{CL}}^{\mathrm{PL}}(a \sqcup a = a, m) \\ \equiv\ & \forall m', P(m',m) : (\forall m'', P(m'',m') : \exists m''', P(m''',m'') : \\ & \quad P(m''',a) \vee P(m''',a)) \leftrightarrow P(m',a) \\ \equiv\ & \forall m', P(m',m) : (\forall m'', P(m'',m') : \exists m''', P(m''',m'') : \\ & \quad P(m''',a)) \leftrightarrow P(m',a) \end{aligned}$$

we see that the translation of $\sqcup$ provides one direction of the proof. With (2), a consequence of weak supplementation, we obtain the other direction.

The other laws follow in a similar manner. We show associativity (49):

$$\begin{aligned} &\tau_{\mathrm{CL}}^{\mathrm{PL}}(a \sqcup (b \sqcup c) = (a \sqcup b) \sqcup c, m) \equiv \forall m', P(m',m) : \\ &\forall m'', P(m'',m') : \exists m''', P(m''',m'') : P(m''',a) \vee \forall m^{iv}, P(m^{iv},m''') : \\ &\quad \exists m^{v}, P(m^{v},m^{iv}) : P(m^{v},b) \vee P(m^{v},c) \\ &\leftrightarrow \forall m'', P(m'',m') : \exists m''', P(m''',m'') : (\forall m^{iv}, P(m^{iv},m''') \\ &\quad \exists m^{v}, P(m^{v},m^{iv}) : P(m^{v},a) \vee P(m^{v},b) \vee P(m''',c) \end{aligned}$$

By proving the following for any m' from which the above then follows directly via the associativity and commutativity of $\vee$:

$$\begin{aligned} &\forall m'', P(m'',m') : \exists m''', P(m''',m'') : P(m''',a) \vee \forall m^{iv}, P(m^{iv},m''') : \\ &\quad \exists m^{v}, P(m^{v},m^{iv}) : P(m^{v},b) \vee P(m^{v},c) \\ \equiv\ & \forall m'', P(m'',m') : \exists m''', P(m''',m'') : P(m''',a) \vee P(m''',b) \vee P(m''',c) \end{aligned}$$

We prove in two steps.

Proof ($\models$): assume we choose an arbitrary $m'', P(m'',m')$. The antecedent says that if there is an $m''', P(m''',m'')$ so that $P(m''',a)$ or there is $P(m''',m'')$ so that for all its parts m^{iv}, we can find m^{v}, so that $P(m^{v},b)$ or $P(m^{v},c)$. If there is an $m''', P(m''',a)$, the consequent holds. If there is no such $m''', P(m''',a)$, there must be an m''', so that all its parts m^{iv} have a part m^{v} in b or c. Since each such m^{v} is also a part of m'', we can conclude that for all $m'', P(m'',m')$ there is an m''' – namely, the m^{v} we identified –, so that $P(m''',b) \vee P(m''',c)$.

Proof ($\rightleftharpoons$): assume we have for each $m'', P(m'',m')$: $\exists m''', P(m''',m'')$: $P(m''',a) \vee P(m''',b) \vee P(m''',c)$ and the consequent is false. In this case, there must be an m_1'', so that $P(m''',a)$ must be false for all $m''', P(m''',m_1'')$ and that there is a part of any such m''' so that all its subparts are neither in b nor in c. By the premise however, we know that m_1'' must have a part m'''_1 so that $P(m'''_1,b) \vee P(m'''_1,c)$. But since P is transitive we know that for all parts m_1^{iv} of m'''_1 holds either $P(m_1^{iv},b)$ or

$P(m_1^{iv}, c)$. By reflexivity we moreover know that each m_1^{iv} has a part, namely itself, for which $P(m_1^{iv}, b)$ or $P(m_1^{iv}, c)$ hold.

Applying this result twice via the associativity and commutativity of $\vee$, we can conclude (49) must hold:

$$\begin{aligned}
&\forall m'', P(m'', m') : \exists m''', P(m''', m'') : P(m''', a) \vee \forall m^{iv}, P(m^{iv}, m''') : \\
&\quad \exists m^{v}, P(m^{v}, m^{iv}) : P(m^{v}, b) \vee P(m^{v}, c) \\
&\equiv \forall m'', P(m'', m') : \exists m''', P(m''', m'') : P(m''', a) \vee P(m''', b) \vee P(m''', c) \\
&\equiv \forall m'', P(m'', m') : \exists m''', P(m''', m'') : (\forall m^{iv}, P(m^{iv}, m''') \\
&\quad \exists m^{v}, P(m^{v}, m^{iv}) : P(m^{v} a \vee P(m^{v} b) \vee P(m''', c)
\end{aligned}$$

Theorem 14 holds immediately given the definition of the translation for $\sqcup$ and the commutativity of $\vee$:

$$\begin{aligned}
&\tau_{\text{CL}}^{\text{PL}}(a \sqcup b = b \sqcup a) \equiv \\
&(\forall m', P(m', m) : \ \forall m'', P(m'', m') : \exists m''', P(m''', m'') : \\
&\quad P(m''', a) \vee P(m''', b)) \\
&\leftrightarrow (\forall m', P(m', m) : \ \forall m'', P(m'', m') : \exists m''', P(m''', m'') : \\
&\quad P(m''', b) \vee P(m''', a))
\end{aligned}$$

Proving the neutral element property (51) requires (35).

$$\begin{aligned}
&\tau_{\text{CL}}^{\text{PL}}(a \sqcup \bot = a, m) \\
&\equiv \forall m', P(m', m) : (\forall m'', P(m'', m') : \exists m''', P(m''', m'') : \\
&\quad \tau_{\text{CL}}^{\text{PL}}(\top \sqsubseteq a, m''') \vee \tau_{\text{CL}}^{\text{PL}}(\top \sqsubseteq \bot, m''')) \\
&\quad \leftrightarrow \tau_{\text{CL}}^{\text{PL}}(\top \sqsubseteq a, m') \\
&\equiv \forall m', P(m', m) : \ (\forall m'', P(m'', m') : \exists m''', P(m''', m'') : P(m''', a) \vee \bot) \\
&\quad \leftrightarrow P(m', a) \\
&\equiv \forall m', P(m', m) : \ (\forall m''', P(m'', m) : \exists m''', P(m''', m'') : P(m''', a)) \\
&\quad \leftrightarrow P(m', a)
\end{aligned}$$

The proof follows immediately by (2).

In summary, we needed (35) for proving idempotency (48) and the neutral element (51). Associativity (49) and commutativity (50) were proven without using (35).

We have thus shown that $\sqcap$ and $\sqcup$ each create a semilattice structure over the $x \in \mathcal{V}_C$. When we prove the absorption laws, we see that the absorption law (52) can be proven without requiring (35), while the proof for the absorption law (53) uses it:

$$a \sqcap (a \sqcup b) = a \tag{52}$$

$$a \sqcup (a \sqcap b) = a \tag{53}$$

For (52):

$$\begin{aligned}
&\tau_{\text{CL}}^{\text{PL}}(a \sqcap (a \sqcup b) = a, m) \\
&\equiv \forall m', P(m', m) : (P(m', a) \wedge \forall m'', P(m'', m') : \\
&\quad (\exists m''', P(m''', m'') : P(m''', a)) \vee (\exists m''', P(m''', m'') : P(m''', b))) \\
&\leftrightarrow P(m', a)
\end{aligned}$$

we show that for any m':

$$\begin{aligned} P(m',a) \wedge \forall m'', P(m'',m') : (\exists m''', P(m''',m'') : P(m''',a)) \\ \vee(\exists m''', P(m''',m'') : P(m''',b)) \\ \equiv P(m',a) \end{aligned}$$

Proof (⫤): this holds because of transitivity (33) and reflexivity (34) of P. If $P(m',a)$, is true, we know $\forall m'', P(m'',m') : (\exists m''', P(m''',m'') : P(m''',a))$ is also true and thus also the disjunct. Therefore, the whole consequent must be true.

Proof (⊨): here we already know $P(m',a)$ in the antecedent, so the consequent cannot be false.

We prove (53):

$$\begin{aligned} &\tau_{\mathrm{CL}}^{\mathrm{PL}}(a \sqcup (a \sqcap b) = a, m) \\ \equiv\;& \forall m', P(m',m) : (\forall m'', P(m'',m') : (\exists m''', P(m''',m'') : P(m''',a)) \\ &\qquad \vee(\exists m''', P(m''',m'') : P(m''',a) \wedge P(m''',b))) \\ &\leftrightarrow P(m',a) \end{aligned}$$

by showing for any m':

$$\begin{aligned} &\forall m'', P(m'',m') : (\exists m''', P(m''',m'') : P(m''',a)) \\ &\qquad \vee(\exists m''', P(m''',m'') : P(m''',a) \wedge P(m''',b)) \equiv P(m',a) \end{aligned}$$

Proof (⊨): $\forall m'', P(m'',m') : (\exists m''', P(m''',m'') : P(m''',a)) \vee (\exists m''', P(m''',m'') : P(m''',a) \wedge P(m''',b))$ is true iff $\forall m'', P(m'',m') : (\exists m''', P(m''',m'') : P(m''',a))$ is true, and this entails $P(m',a)$ by (2). Proof (⫤): this holds by transitivity and reflexivity.

The relation between the residual and the t-norm were covered by two additional axioms above: continuity (11) and pre-linearity (12):

$$x \sqcap y \sqsubseteq z \text{ iff } x \sqsubseteq (y \Rightarrow z) \tag{54}$$

$$(x \Rightarrow y) \sqcup (y \Rightarrow x) = \top \tag{55}$$

We prove continuity (54) by translation using $\tau_{\mathrm{CL}}^{\mathrm{PL}}$.

$$\begin{aligned} &\tau_{\mathrm{CL}}^{\mathrm{PL}}(x \sqcap y \sqsubseteq z, m) \equiv \tau_{\mathrm{CL}}^{\mathrm{PL}}(x \sqsubseteq (y \Rightarrow z), m) \text{ translates into :} \\ &\forall m', P(m',m) : P(m',x) \wedge P(m',y) \rightarrow P(m',z) \\ \equiv\;& \forall m', P(m',x) \rightarrow \forall m'', P(m'',m') : \exists m''', P(m''',m'') : \\ &\quad (\neg\exists m^{iv}, P(m^{iv},m''') : P(m^{iv},y)) \vee P(m''',z) \end{aligned}$$

Proof (⊨): assume the antecedent holds, and $P(m',x)$ for some m'. Then, for $\forall m'', P(m''',m'') : \ldots P(m''',z)$ to be false, there must be an $m''_1, P(m''_1,m')$, so that $\forall m''', P(m''',m''_1) : (\exists m^{iv}, P(m^{iv},m''') : P(m^{iv},y))$ and $\forall m''', P(m''',m''_1) : \neg P(m''',z)$. However, if $\forall m''', P(m''',m''_1) : (\exists m^{iv}, P(m^{iv},m''') : P(m^{iv},y))$ holds then by (2), $P(m''_1,y)$ and by transitivity also $P(m''_1,x)$ and by the assumption thus $P(m''_1,z)$, which cannot hold since all parts m''' of m''_1 including by reflexivity m''_1 itself fulfill $\neg P(m''',z)$.

Proof (⫤): assume the antecedent $\forall m', P(m',x) : \ldots P(m''',z)$ holds. For $\forall m', P(m',m) : P(m',x) \wedge P(m',y) \rightarrow P(m',z)$ to be false, there must be $m'_1, P(m'_1,m)$, so that $P(m'_1,x)$ and $P(m'_1,y)$ must hold, but $P(m'_1,z)$ must be false. But then we also know that $\exists m''', P(m''',m'') : (\neg\exists m^{iv}, P(m^{iv},m''') : P(m^{iv},y))$ cannot hold for any

$m'', P(m'', m_1')$. Thus, $\exists m''', P(m''', m'') : P(m''', z)$ must hold for all m'', and thus by (2) $P(m_1', z)$.

We prove pre-linearity (55):

$$\tau_{CL}^{PL}((x \Rightarrow y) \sqcup (y \Rightarrow x) = \top, m) \equiv$$
$$\forall m', P(m', m) : [\exists m'', P(m'', m') : \forall m''', P(m''', m'') : \exists m^{iv}, P(m^{iv}, m''') :$$
$$(\neg \exists m^v, P(m^v, m^{iv}) : P(m^v, x)) \vee P(m^{iv}, y)]$$
$$\vee [\exists m'', P(m'', m') : \forall m''', P(m''', m'') : \exists m^{iv}, P(m^{iv}, m''') :$$
$$(\neg \exists m^v, P(m^v, m^{iv}) : P(m^v, y)) \vee P(m^{iv}, x)]$$

by showing for any$m', P(m', m)$:

$$\neg [\exists m'', P(m'', m') : \forall m''', P(m''', m'') : \exists m^{iv}, P(m^{iv}, m''') :$$
$$(\neg \exists m^v, P(m^v, m^{iv}) : P(m^v, x)) \vee P(m^{iv}, y)]$$
$$\models [\exists m'', P(m'', m') : \forall m''', P(m''', m'') : \exists m^{iv}, P(m^{iv}, m''') :$$
$$(\neg \exists m^v, P(m^v, m^{iv}) : P(m^v, y)) \vee P(m^{iv}, x)]$$

Proof: we obtain for the antecedent:

$$\forall m'', P(m'', m') : \exists m''', P(m''', m'') : \forall m^{iv}, P(m^{iv}, m''') : (\exists m^v, P(m^v, m^{iv}) : P(m^v, x)) \wedge \neg P(m^{iv}, y)$$

Since this holds for all $m'', P(m'', m')$ it also holds for m' itself, i.e., it follows that:

$$\exists m''', P(m''', m') : \forall m^{iv}, P(m^{iv}, m''') : (\exists m^v, P(m^v, m^{iv}) : P(m^v, x)) \wedge \neg P(m^{iv}, y)$$

We rename the variables to better show the structure:

$$\exists m'', P(m'', m') : \forall m''', P(m''', m'') : \neg P(m''', y) \wedge (\exists m^{iv}, P(m^{iv}, m''') : P(m^{iv}, x))$$
$$\equiv \exists m'', P(m'', m') : (\forall m''', P(m''', m'') : \neg P(m''', y)) \wedge$$
$$(\forall m''', P(m''', m'') : (\exists m^{iv}, P(m^{iv}, m''') : P(m^{iv}, x)))$$

and by (2):

$$\equiv \exists m'', P(m'', m') : (\neg \exists m''', P(m''', m'') : P(m''', y)) \wedge P(m'', x).$$

We now know that m' has a part m'' that is in x and none of its parts is in y. With $P(m'', x)$, however we also know by transitivity of P that all parts $m''', P(m''', m'')$ fulfill $P(m''', x)$, and thus by reflexivity of P that there is an $m^{iv}, P(m^{iv}, m''')$, namely $m^{iv} = m'''$, for each m''', which fulfills $P(m^{iv}, x)$. Moreover, since all parts $m^v, P(m^v, m^{iv})$ are by transitivity also parts of m'', we know that $\neg \exists m^v, P(m^v, m^{iv}) : P(m^{iv}, y)$ and thus:

$$\exists m'', P(m'', m') : \forall m''', P(m''', m'') : \exists m^{iv}, P(m^{iv}, m''') : (\neg \exists m^v, P(m^v, m^{iv}) : P(m^v, y)) \wedge P(m^{iv}, x),$$

which entails the consequent.

4.5 A note on mereological and ontological status

The mereologically interested reader may notice that adding even the weakened variant of the weak supplementation principle is sufficient to collapse context logic term structures to a single level by (2). The reason for this is that the weak supplementation principle considerably strengthens the expressiveness of negation, which given the principle always ensures the existence of a fully negative individual. This is the case, although our system mereologically speaking is an MM system, i.e., supports M1-M4 [30] only, with M4 acting as an axiom schema.

We may note also, that we need not ensure product (M5) or sum (M6) to exists, nor do we need or posit a universal $\top$ or null object $\bot$ to exist. The symbols $\top, \bot, \sqcap, \sqcup, -$ are, so to speak, "syntactic sugar" only. The assumed mereology thus is slightly weaker than MM and ontologically careful and minimalistic. For a deeper discussion, cf. [30, 31].

4.6 Example: set-theoretical model

To make the discussion more concrete, we briefly sketch a set-theoretical interpretation. An example of a suitable model is the set-theoretic lattice, assuming the set of all subsets of a base universe as the universe for the interpretation of the translation $\tau_{\mathrm{CL}}^{\mathrm{PL}}(\phi, m)$ of a context formula ϕ, and mapping t to $\cap$ (17), s to $\cup$ (18), and the residual r according to (19). Note that, within this interpretation, the variables m, m', etc., as well as the constants a, b, c, etc. of the translation $\tau_{\mathrm{CL}}^{\mathrm{PL}}(\phi, m)$ range over sets, not elements, of the base universe. With a set-theoretical model $(I, i, a, U, \subseteq, B, \cap, \cup)$, where $U \subseteq 2^B$ for B, the base set, we get:

$$
\begin{aligned}
i(v) &= a(v) \text{ for } v \in \mathcal{V}_C \\
i(\sim c) &= 1_T - i(c) = B - i(c) \\
i(c \sqcap d) &= t(i(c), i(d)) = i(c) \cap i(d) \\
i(c \sqcup d) &= s(i(c), i(d)) = i(c) \cup i(d) \\
I(c \sqsubseteq d) &= 1 \text{ iff } i(c) \subseteq i(d) \\
I(\neg\phi) &= 1 - I(\phi) \\
I(\phi \wedge \psi) &= \min\left(I(\phi), I(\psi)\right) \\
I(\phi \vee \psi) &= \max\left(I(\phi), I(\psi)\right)
\end{aligned}
$$

We can show that, if the canonical interpretation $(I, i, a, U, \subseteq, B, \cap, \cup)$ is a model for formula ϕ, then there is a corresponding predicate logic model for $\tau_{\mathrm{CL}}^{\mathrm{PL}}(\phi, m)$ with interpretations for m, m' from $U \subseteq 2^B - \{\varnothing\}$, interpreting P as $\subseteq$ and individuals $v \in V_C$ using an assignment function $a : \mathcal{V}_C \to 2^B$. While we allow the constants $v \in V_C$ to be empty, the variables used to describe their extension cannot.

The pre-order axioms for P obviously hold for $\subseteq$. Also, the weak supplementation principle holds for $\subseteq$:

$$
\begin{aligned}
\forall x: \quad & (\forall x' \subseteq x : x' \subseteq a \to x' \subseteq b) \\
& \wedge (\exists x' \subseteq x : x' \subseteq b \wedge \neg x' \subseteq a) \\
& \to \left(\exists x' \subseteq x : x' \subseteq b \wedge \neg \exists x'' \subseteq x' : x'' \subseteq a\right).
\end{aligned}
$$

Proof: assume a set $x' \subseteq x$ supports that $x' \subseteq a$ implies $x' \subseteq b$, and there is a set $x_1' \subseteq x$ supporting $x_1' \subseteq b$ but not $x_1' \subseteq a$. We can then construct $x_2' \subseteq x$ as $x_2' = x_1' - a$, which supports that $x_2' \subseteq b$ and none of its subsets $x_2'' \subseteq x_2'$ supports $x_2'' \subseteq a$.

We prove that $\cap, \cup, -$ over non-empty sets m, m', m'' fulfill the characteristic properties for translations for $\sqcap$, $\sqcup$, $\sim$, respectively:

$$m \subseteq a \cap b \quad \equiv m \subseteq a \wedge m \subseteq b \tag{56}$$

$$m \subseteq a \cup b \quad \equiv \forall m' \subseteq m : \exists m'' \subseteq m' : m'' \subseteq a \vee m'' \subseteq b \tag{57}$$

$$m \subseteq \sim a \quad \equiv \forall m' \subseteq m : \neg \exists m'' \subseteq m' : m'' \subseteq a \tag{58}$$

The case of (20) is immediately clear. For (21), we look at the definition of $\cup$ in terms of elements $P \in B$, which we call *points*:

$$\begin{aligned} m \subseteq a \cup b &\equiv \forall P \in m : P \in a \vee P \in b \\ &\equiv \forall m' \subseteq m : \exists P \in m' : P \in a \vee P \in b \\ &\equiv \forall m' \subseteq m : \exists m'' \subseteq m' : m'' \subseteq a \vee m'' \subseteq b \end{aligned}$$

Proof ($\models$): if the points in m are in a or in b in the first step, then, since the m' are non-empty, it follows that each $m' \subseteq m$ has a point either in a or in b. In the second step, if there is a point P in each m', that is in a or in b, then there is a set $m'' \subseteq m'$, namely the singleton containing P, for which $m'' \subseteq a$ or $m'' \subseteq b$ must hold.

Proof ($\dashv$): assume that for every m', there is a non-empty $m'' \subseteq m'$, with $m'' \subseteq a$ or $m'' \subseteq b$, then, since m'' non-empty, it must have a point $P \in a$ or $P \in b$. Since this holds for all non-empty sets m', including all singleton sets, which have only one element, it must hold for all points $P \in m$.

For (22), we similarly look at the definition of $-$ in terms of elements of m, i.e., points:

$$\begin{aligned} m \subseteq \sim a &\equiv \forall P \in m : \neg P \in a \\ &\equiv \forall m' \subseteq m : \neg \exists P \in m' : P \in a \\ &\equiv \forall m' \subseteq m : \neg \exists m'' \subseteq m' : m' \subseteq a \end{aligned}$$

Proof ($\models$): the property carries over to all parts m' of m in the second step. The third step follows, because any set $m'' \subseteq m'$ must be non-empty, and if it contains a point, $m'' \subseteq a$ cannot be true, since no point in m' is in a and $\subseteq$ is transitive.

Proof ($\dashv$): as in the proof for $\sqcup$, we can argue over singleton sets. If for all sets m', no subset m'' is subset of a, then this also holds for the singletons, and thus no set m' has a point P in a, but this again holds also for singleton sets $m' \subseteq m$, and thus all points of m are outside a.

We have thus seen that the set-thoretical standard model is a concrete example of a structure for interpreting context terms and formulae.

5. A fuzzy context logic

The key to the proposed fuzzy context logic is to additionally provide a fuzzy interpretation for the atomic formulae, via the symbol $\sqsubseteq$. To do that, we need a residual that takes two elements from the context lattice and produces a fuzzy value in $[0, 1]$. Then we can apply one of the well-known standard fuzzy semantics to the formula level.

The fuzzy semantics is defined by two lattices: a bounded lattice $(L_T, \preccurlyeq_T, t_T, s_T, 1_T)$ for the term level, and another bounded lattice $(L_F, \preccurlyeq_F, s_F, t_F, r_F, 1_F)$, where $L_F = [0, 1]$ for the formula level together with the interpretation functions $a : \mathcal{V}_C \to L_T$ for context variables, $i : \mathcal{T}_C \to L_T$ for terms, and $I : \mathcal{L}_C \to L_F$ for formulae. We interpret the terms as before based on L_T:

$$i(v) = a(v), \text{for } v \in V_C$$
$$i(\sim c) = 1_T - i(c),$$

with the $1_T - i(c)$term-level pseudo-complement

$$i(c \sqcap d) = t_T(i(c), i(d))$$
$$i(c \sqcup d) = s_T(i(c), i(d)).$$

We will need to characterize a fuzzified variant of $\preccurlyeq_T$ to obtain atomic formulae that can have a value outside of $\{0, 1\}$:

$$I(c \sqsubseteq d) = \preccurlyeq_{TF}(i(c), i(d).$$

On this basis, the interpretation of formulae can then follow one of the standard models of fuzzy logic in L_F:

$$I(\neg\phi) = 1_F - I(\phi)$$
$$I(\phi \wedge \psi) = t_F(I(\phi), I(\psi))$$
$$I(\phi \vee \psi) = s_F(I(\phi), I(\psi))$$

The key is to provide a function $\preccurlyeq_{TF} : L_T \times L_T \rightarrow L_F$ for connecting the fuzzy term and formula layers. As usual, we want the relation to be conservative with respect to the classical partial order relation on the classical cases:

$$\preccurlyeq_{TF}(x, y) = 1_F \text{iff } x \preccurlyeq_T y.$$
$$\preccurlyeq_{TF}(x, x) = 1_F \text{holds for all } x \in L_T.$$
$$\text{If} \preccurlyeq_{TF}(x, y) = 1_F \text{ and} \preccurlyeq_{TF}(y, x) = 1_F \text{for } x, y \in L_T \text{ then } x = y.$$
$$\text{If} \preccurlyeq_{TF}(x, y) = 1_F \text{ and } \preccurlyeq_{TF}(y, z) = 1_F \text{ then also} \preccurlyeq_{TF}(x, z) = 1_F.$$

What is a good choice depends on both L_T and L_F, and given a particular choice, different functions may support this weak restriction. A candidate for spatial applications for $L_T = 2^B$ for a base set B and $L_F = [0, 1]$ is a fuzzified variant of the qualitative granular relation systems proposed in [32]. Here, several types of granular relations between regions are distinguished based on an absolute ranking of sizes, such as the largest circle a spatial region is contained in, or its diameter, or the length of an interval. Complementing topological notions, such as *part-of* or *overlap*, granular relations can be defined [32]:

- Two regions are *adjacent* iff they overlap but only in a part smaller than grain-size.
- Two regions are *spatially indistinguishable* iff they differ only in a part smaller than grain-size.
- Two regions *relevantly overlap* iff they overlap in a part larger than grain-size and differ in a part larger than grain-size.

We can generalize this notion using a $[0, 1]$ perspective instead of a discrete partitioning of the space of possible overlap-relations. For the example of a set-theoretical model, we could proceed, e.g., to find a fuzzification of $\subseteq$ into a function $\subseteq_{TF}$ mapping to $[0, 1]$ by assessing the largest difference between two arguments x, y in comparison to the diameter of x. The intervals $(14, 46]$ and

$[14, 46)$, for instance differ only in boundary points. The intervals $x = (11, 34)$ and $y = (12, 36)$ overlap in $x \sqcap y = (12, 34)$. With the overlap $|x \sqcap y| = |(12, 34)| = 12$ and $|x| = |(11, 34)| = 13$, this is an overlap of $|x \sqcap y|/|x| = 12/13 = 92\%$.

Generally, we can employ a granularity function $\gamma : L_T \to \mathbb{R}^+$ to compute a mapping from entities of L_T to $\mathbb{R}^+$. Based on this, we can use a suitable function $r : L_T \to L_F$ to make the transition between the term layer and the formula layer in such a way that it also connects appropriately to the basic properties of the residual r_F, e.g., by employing r_F itself:

$$\preccurlyeq_{TF}(x, y) = r_F(\gamma(x), \gamma(x \sqcap y)).$$

We obtain a fully specified family of fuzzy context logics. Note that with $r_F = r_{\text{prod}}$, we receive the conditional probability:

$$r_{\text{prod}}(\gamma(x), \gamma(x \sqcap y)) = \frac{\gamma(x \sqcap y)}{\gamma(x)} = \begin{cases} 1 & \text{if } x \sqsubseteq y \\ \dfrac{\gamma(x \sqcap y)}{\gamma(x)} & \text{otherwise} \end{cases}$$

For $r_F = r_{\min}$ we obtain:

$$r_{\min}(\gamma(x), \gamma(x \sqcap y)) = \begin{cases} 1 & \text{iff } x \sqsubseteq y \\ \gamma(x \sqcap y) & \text{otherwise.} \end{cases}$$

Among the potential applications, a two-layered fuzzy logic can help to reason about fuzzy logic systems. The base logic being decidable for the classical semantics, we can, at least for the classical case, make absolute guarantees for a given system. We can prove whether a given fuzzy system, e.g., the output of a machine learning mechanism, such as an ANFIS, together with a description of possible situations in the domain and desirable properties yields a tautology, thus proving that the system has the desirable properties under all possible circumstances. If we are interested in gaining an understanding of systems that are not tautological in this sense, so as to obtain, e.g., degrees of possibility of failure under certain circumstances, more advanced fuzzy proof methods are required.

6. Conclusions

This chapter illustrated that the lwo-layered logic context logic and fuzzy logic can be combined in a meaningful way. We first mapped both logics to a predicate logical background language, so as to highllight their commonalities and differences and to obtain a background compatible with both. In both cases, we discussed a common set-theoretical model that can be used to interpret the background language. We formally proved that the lattice-based generalized t-norms of fuzzy logic provide a suitable semantics for the term-layer of context logic. To do this, we expressed context logic in terms of a single pre-order relation that additionally supports the weak supplementation principle and showed that, with this translation providing semantics, context logic fulfills the properties of a residuated lattice. We also derived that the language is decidable in EXPSPACE.

The formula-layer of context logic could then additionally be imbued with a $[0, 1]$-based fuzzification. Proposals for adding either the product t-norm or the minimum t-norm for the formula layer on top of the lattice-based generalized t-norm of the context term layer were suggested, and a mechanism for combining this with granularity to further expand expressiveness was discussed.

Acknowledgements

This work was financially supported by the Hanse-Wissenschaftskolleg, Delmenhorst, Germany; and received infrastructural support from the University of Bremen, Germany.

Author details

Hedda Schmidtke
Independent Scientist, Schwerin, Germany

*Address all correspondence to: schmidtke@acm.org

References

[1] Dubois D, Prade H. The emergence of fuzzy sets: A historical perspective. In: *Fuzzy logic in its 50th year*, pages 3–19. Springer. 2016

[2] M. Mendler, J. Aguado, B. Bodin, P. Roop, and R. von Hanxleden. Logic meets algebra: Compositional timing analysis for synchronous reactive multithreading. In *Models, Mindsets, Meta: The What, the How, and the Why Not?*, pages 45–67. Springer, 2019.

[3] B. M. Keneni, D. Kaur, A. Al Bataineh, V. K. Devabhaktuni, A. Y. Javaid, J. D. Zaientz, and R. P. Marinier. Evolving rule-based explainable artificial intelligence for unmanned aerial vehicles. *IEEE Access*, 7:17001–17016, 2019.

[4] E. Lughofer. Model explanation and interpretation concepts for stimulating advanced human-machine interaction with "expert-in-the-loop". In *Human and Machine Learning*, pages 177–221. Springer, 2018.

[5] B. Ruprecht, W. Wu, M. A. Islam, D. Anderson, J. Keller, G. Scott, C. Davis, F. Petry, P. Elmore, K. Nock, et al. Possibilistic clustering enabled neuro fuzzy logic. In *2020 IEEE International Conference on Fuzzy Systems (FUZZ-IEEE)*, pages 1–8. IEEE, 2020.

[6] P. Hájek. *Metamathematics of fuzzy logic*, volume 4 of *Trends in Logic*. Springer Science & Business Media, 1998.

[7] S. Rasouli, Z. Zarin, and A. Hassankhani. Characterization of a new subquasivariety of residuated lattice. *Journal of applied logics-The IfCoLog journal of logics and their applications*, 5 (1):33–63, 2018.

[8] D. Dubois, H. Prade, and A. Rico. Graded cubes of opposition and possibility theory with fuzzy events. *International Journal of Approximate Reasoning*, 84:168–185, 2017.

[9] H. R. Schmidtke and W. Woo. Towards ontology-based formal verification methods for context aware systems. In H. Tokuda, M. Beigl, A. Brush, A. Friday, and Y. Tobe, editors, *Pervasive 2009*, pages 309–326. Springer, 2009.

[10] H. R. Schmidtke, D. Hong, and W. Woo. Reasoning about models of context: A context-oriented logical language for knowledge-based context-aware applications. *Revue d'Intelligence Artificielle*, 22(5):589–608, 2008.

[11] D. Hong, H. R. Schmidtke, and W. Woo. Linking context modelling and contextual reasoning. In A. Kofod-Petersen, J. Cassens, D. B. Leake, and S. Schulz, editors, *4th International Workshop on Modeling and Reasoning in Context (MRC)*, pages 37–48. Roskilde University, 2007.

[12] H. R. Schmidtke. A survey on verification strategies for intelligent transportation systems. *Journal of Reliable Intelligent Environments*, 4(4): 211–224, 2018.

[13] H. R. Schmidtke. Textmap: A general purpose visualization system. *Cognitive Systems Research*, 59:27–36, 2020.

[14] H. R. Schmidtke. Logical rotation with the Activation Bit Vector Machine. *Procedia Computer Science*, 169:568–577, 2020.

[15] L. A. Zadeh. Fuzzy logic and approximate reasoning. *Synthese*, 30(3): 407–428, 1975.

[16] L. A. Zadeh. Fuzzy logic. *Computer*, 21(4):83–93, 1988.

[17] L. A. Zadeh. Fuzzy logic= computing with words. *IEEE*

transactions on fuzzy systems, 4(2):103–111, 1996.

[18] M. Bierwisch. The semantics of gradation. *Dimensional adjectives*, 71:261, 1989.

[19] J.-S. R. Jang. ANFIS: adaptive-network-based fuzzy inference system. *Systems, Man and Cybernetics, IEEE Transactions on*, 23(3):665–685, 1993.

[20] S. Kundu and J. Chen. Fuzzy logic or Lukasiewicz logic: A clarification. In Z. W. Ras′ and M. Zemankova, editors, *Methodologies for Intelligent Systems: 8th International Symposium, ISMIS '94 Charlotte, North Carolina, USA, October 16–19, 1994 Proceedings*, pages 56–64. Springer Berlin Heidelberg, Berlin, Heidelberg, 1994.

[21] F. Bobillo and U. Straccia. Generalizing type-2 fuzzy ontologies and type-2 fuzzy description logics. *International Journal of Approximate Reasoning*, 87:40–66, 2017.

[22] *M. mas*, M. Monserrat, J. Torrens, and E. Trillas. A survey on fuzzy implication functions. *IEEE Transactions on fuzzy systems*, 15(6):1107–1121, 2007.

[23] F. Esteva and L. Godo. Monoidal t-norm based logic: towards a logic for left-continuous t-norms. *Fuzzy sets and systems*, 124(3):271–288, 2001.

[24] H. R. Schmidtke. Contextual reasoning in context-aware systems. In *Workshop Proceedings of the 8th International Conference on Intelligent Environments*, pages 82–93. IOS Press, 2012.

[25] H. R. Schmidtke and M. Beigl. Distributed spatial reasoning for wireless sensor networks. In *Modeling and Using Context*, pages 264–277. Springer, 2011.

[26] H. R. Schmidtke, H. Yu, P. Masomo, A. Kinai, and A. Shema. Contextual reasoning in an intelligent electronic patient leaflet system. In P. Brézillon and A. Gonzalez, editors, *Context in Computing: A Cross-disciplinary Approach for Modeling the Real World*, pages 557–573. Springer, 2014.

[27] H. Andréka, I. Németi, and J. van Benthem. Modal languages and bounded fragments of predicate logic. *Journal of Philosophical Logic*, 27(3):217–274, 1998.

[28] E. Kieron′ski. On the complexity of the two-variable guarded fragment with transitive guards. *Information and Computation*, 204(11):1663–1703, 2006.

[29] E. Kieronski and L. Tendera. Finite Satisfiability of the Two-Variable Guarded Fragment with Transitive Guards and Related Variants. *ArXiv e-prints*, November 2016.

[30] R. Casati and A. C. Varzi. *Parts and Places: the Structure of Spatial Representations*. MIT Press, 1999.

[31] P. Simons. *Parts – A Study in Ontology*. Oxford University Press, Oxford, 1987.

[32] H. R. Schmidtke and W. Woo. A size-based qualitative approach to the representation of spatial granularity. In M. M. Veloso, editor, *Twentieth International Joint Conference on Artificial Intelligence*, pages 563–568, 2007.

Section 2

Decision Making

Chapter 5

Modified Expression to Evaluate the Correlation Coefficient of Dual Hesitant Fuzzy Sets and Its Application to Multi-Attribute Decision Making

Akanksha Singh

Abstract

The main objective of this paper is to understand all the existing correlation coefficients (CoCfs) to determine the relation and dependency between two variables of the fuzzy sets and its extensions for solving decision-making (DM) problems. To study the weighted CoCfs between two variables the environment chosen here is dual hesitant fuzzy set (DHFS) which is a generalization of a fuzzy set which considers the hesitant value of both the membership and non-membership elements of a set. Although there exists CoCfs for DHFS but a detailed mathematical analysis suggests that there exists some shortcomings in the existing CoCfs for DHFS. Thus, an attempt has been made to properly understand the root cause of the posed limitation in the weighted CoCfs for DHFS and hence, modified weighted CoCfs for DHFS has been proposed for solving DHFS multi-attribute decision making (MADM) problems i.e., DM problems in which rating value of each alternative over each criterion is represented by a DHFS in the real-life. Also, to validate the proposed expressions of weighted CoCfs for solving DHFS MADM problems, an existing real-life problem is evaluated and a systematic comparison of the solution is presented for clarification.

Keywords: decision-making, dual hesitant fuzzy set, correlation coefficient, multi-attribute decision-making

1. Introduction

Decision-making is a process which has a wide range of real-life applications which requires a great precision for desirable outcomes. Real-world applications like supply chain management, marketing management, healthcare, telecommunication, finance, energy, banking, forestry, pattern recognition, investment, personnel selection etc., has a set of data which includes information with both certainties and uncertainties. The study of uncertainties can be handled well by fuzzy sets [1] and its extensions [1–7], thus measures of decision-making helps in removing and controlling the existing constraints or uncertainties, it increases productivity, helps in better coordination etc.

IntechOpen

To rank fuzzy sets and its generalizations there exists various ranking measures like distance measures, similarity measures, score function, accuracy function, certainty function, divergence measure, CoCfs etc. Although in literature there exist expressions to evaluate the CoCf between fuzzy sets and many of its extensions as proposed by several researchers like, the CoCf between two fuzzy sets [8], the CoCf between two intuitionistic fuzzy sets [9–17], the CoCf between two interval-valued intuitionistic fuzzy sets [18], the CoCf between two Pythagorean fuzzy sets [19, 20], the CoCf between two intuitionistic multiplicative sets [21], the CoCf between two hesitant fuzzy sets [22–27], the CoCf between two dual hesitant fuzzy sets [28–30] etc.. Ye [29] proposed an expressions for evaluating the weighted CoCfs between two DHFSs and solved a real-life problem (finding the best investment company) where the uncertainty is represented as a DHFS. However, after a deep study, it is observed that some mathematical incorrect assumptions are considered in the existing weighted CoCf and hence it is scientifically incorrect to apply existing weighted CoCf in real-life MADM problems for DHFSs in its present form. This limitation is a real motivation to modify the CoCfs for DHFSs which would be applicable for the evaluation of the real-life problems. Considering the existing weighted CoCf [29] for solving DHFSs MADM problems as a base, a modified weighted CoCf for DHFSs is proposed and using the modified expressions, the exact results of the real-life problem, considered in the existing paper [29] have been obtained.

The paper is organized as follows. Section 2. Preliminaries. Section 3. A brief review of the existing CoCf of DHFSs is presented here. Section 3.1. Gaps in the existing weighted CoCf for DHFSs. Section 3.2. Mathematical incorrect assumptions. Section 4. It proposes the modified CoCf for DHFSs. Section 5. Origin of the proposed CoCf for the DHFSs is discussed here. Section 6. It presents the exact solution to the existing real-life problem. Section 7. Advantages of modified CoCf for DHFSs. Section 8. Discussion and Concludes the presented paper.

2. Preliminaries

This section states some requisites concerned with the DHFSs and the correlation coefficients while applying in the real-life application during DM process.

Definition 2.1 [31] A set $\tilde{A} = \{\langle x, \mu_{\tilde{A}}(x)\rangle | x \in X, 0 \le \mu_{\tilde{A}}(x) \le 1\}$, defined on the universal set XX, is said to be an fuzzy set (FS), where $\mu_{\tilde{A}}(x)$ represents the degree of membership of the element x xin $\tilde{A}$.

Definition 2.2 [1] A set $\tilde{A} = \{\langle x, \mu_{\tilde{A}}(x), \nu_{\tilde{A}}(x)\rangle | x \in X, 0 \le \mu_{\tilde{A}}(x) \le 1, 0 \le \nu_{\tilde{A}}(x) \le 1, \mu_{\tilde{A}}(x) + \nu_{\tilde{A}}(x) \le 1\}$, defined on the universal set X, is said to be an intuitionistic fuzzy set (IFS), where, $\mu_{\tilde{A}}(x)$ and $\nu_{\tilde{A}}(x)$ represents the degree of membership and degree of non-membership respectively of the element x in $\tilde{A}$. The pair $\langle \mu_{\tilde{A}}, \nu_{\tilde{A}}\rangle$ is called an intuitionistic fuzzy number (IFN) or an intuitionistic fuzzy (IFV), where, $\mu_{\tilde{A}} \in [0,1], \nu_{\tilde{A}} \in [0,1], \mu_{\tilde{A}} + \nu_{\tilde{A}} \le 1$.

Definition 2.3 [29] Let X be an initial universe of objects. A set $\tilde{A}$ on X defined as $\tilde{A} = \{\langle x, \mu_{\tilde{A}}(x^{(s)})\rangle | x \in X\}$ is called a hesitant fuzzy set (HFS), where $\mu_{\tilde{A}}(x^{(s)})$ is a mapping defined by $\mu_{\tilde{A}}(x^{(s)}) : X \to [0,1]$ here, $\mu_{\tilde{A}}(x^{(s)})$ is a set of some different values in [0,1] and $'s'$ represent the number of possible membership degrees of the element $x \in X$ to $\tilde{A}$. For convenience, we call $\mu_{\tilde{A}}(x^{(s)})$ as a hesitant fuzzy element (HFE).

Definition 2.4 [29] Let X be an initial universe of objects. A set $\tilde{A}$ on X then for a given HFE $\mu_{\tilde{A}}(x^{(s)})$, its lower and upper bounds are defined as $\mu_{\tilde{A}}^{-}(x^{(s)}) = \min \mu_{\tilde{A}}(x^{(s)})$ and $\mu_{\tilde{A}}^{+}(x^{(s)}) = \max \mu_{\tilde{A}}(x^{(s)})$, respectively, where 's' represent the number of possible membership degrees of the element $x \in X$ to $\tilde{A}$.

Definition 2.5 [29] Let X be an initial universe of objects. A set $\tilde{A}$ on X then for a given HFE $\mu_{\tilde{A}}(x^{(s)})$, $A_{env}(\mu_{\tilde{A}}(x^{(s)}))$ is called the envelope of $\mu_{\tilde{A}}(x^{(s)})$ which is denoted as $\left(\mu_{\tilde{A}}^{-}, 1-\mu_{\tilde{A}}^{+}\right)$, with the lower bound $\mu_{\tilde{A}}^{-}$ and upper bound $\mu_{\tilde{A}}^{+}$. Also, $A_{env}(\mu_{\tilde{A}}(x^{(s)}))$ establishes the relation between HFS and IFS i.e., $A_{env}(\mu_{\tilde{A}}(x^{(s)})) = \{\langle x, \mu_{\tilde{A}}(x^{(s)}), \nu_{\tilde{A}}(x^{(s)})\rangle\}$, where $\mu_{\tilde{A}}(x^{(s)}) = \mu_{\tilde{A}}^{-}$ and $\nu_{\tilde{A}}(x^{(s)}) = 1-\mu_{\tilde{A}}^{+}$.

Definition 2.6 [29] For a HFE $\mu_{\tilde{A}}$, $s(\mu_{\tilde{A}}) = \left(\frac{1}{l(\mu_{\tilde{A}})}\right)\sum_{\gamma \in \mu_{\tilde{A}}} \gamma$ is called the score function of $\mu_{\tilde{A}}$, where $l(\mu_{\tilde{A}})$ is the number of the values in $\mu_{\tilde{A}}$. For any two HFEs $\mu_{\tilde{A}_1}$ and $\mu_{\tilde{A}_2}$, the comparison between two HFEs is done as follows:

i. If $s\left(\mu_{\tilde{A}_1}\right) > s\left(\mu_{\tilde{A}_2}\right)$, then $\mu_{\tilde{A}_1} > \mu_{\tilde{A}_2}$.

ii. If $s\left(\mu_{\tilde{A}_1}\right) = s\left(\mu_{\tilde{A}_2}\right)$, then $\mu_{\tilde{A}_1} = \mu_{\tilde{A}_2}$.

Let $\mu_{\tilde{A}_1}$ and $\mu_{\tilde{A}_2}$ be two HFEs such that $l\left(\mu_{\tilde{A}_1}\right) \neq l\left(\mu_{\tilde{A}_2}\right)$. For convenience, let $l = \max\left\{l\left(\mu_{\tilde{A}_1}\right), l\left(\mu_{\tilde{A}_2}\right)\right\}$, then while comparing them, the shorter one is extended by adding the same value till both are of same length. The selection of the value to be added is dependent on the decision makers risk preferences. For example (adopted from 28), let $\mu_{\tilde{A}_1} = \{0.1, 0.2, 0.3\}$, $\mu_{\tilde{A}_2} = \{0.4, 0.5\}$ and $l\left(\mu_{\tilde{A}_1}\right) > l\left(\mu_{\tilde{A}_2}\right)$, then for the correct arithmetic operations $\mu_{\tilde{A}_2}$ must be extended to $\mu_{\tilde{A}_2}{}'$, i.e. either $\mu_{\tilde{A}_2}{}' = \{0.4, 0.5, 0.5\}$ as an optimist or $\mu_{\tilde{A}_2}{}' = \{0.4, 0.4, 0.5\}$ as a pessimist depending on the risk taking factor of the decision-maker though their results would vary definitely.

Definition 2.7 [29] A set $\tilde{A}$ on X defined as $\tilde{A} = \{\langle x, \mu_{\tilde{A}}(x^{(s)}), \nu_{\tilde{A}}(x^{(t)})\rangle | x \in X\}$ is called a DHFS, where, $\mu_{\tilde{A}}(x^{(s)}), \nu_{\tilde{A}}(x^{(t)})$ is a mapping defined by $\mu_{\tilde{A}}(x^{(s)}), \nu_{\tilde{A}}(x^{(t)}) : X \to [0,1]$, here $\mu_{\tilde{A}}(x^{(s)}), \nu_{\tilde{A}}(x^{(t)})$ is a set of some different values in [0,1], 's' represent the number of possible membership degrees and 't' represent the number of possible non membership degrees of the element $x \in X$ to $\tilde{A}$. For convenience, we call $d = \{\mu_{\tilde{A}}(x^{(s)}), \nu_{\tilde{A}}(x^{(t)})\}$ as a dual hesitant fuzzy element (DHFE).

Definition 2.8 [29] Let $d_1 = \left\{\mu_{\tilde{A}_1}, \nu_{\tilde{A}_1}\right\}$ and $d_2 = \left\{\mu_{\tilde{A}_2}, \nu_{\tilde{A}_2}\right\}$ be any two DHFEs, then the score function for DHFSs $d_i (i = 1, 2)$ is defined as

$s(d_i) = \left(\frac{1}{l\left(\mu_{\tilde{A}_i}\right)}\right)\sum_{\gamma \in \mu_{\tilde{A}_i}} \gamma - \left(\frac{1}{m\left(\nu_{\tilde{A}_i}\right)}\right)\sum_{\eta \in \nu_{\tilde{A}_i}} \eta \ (i = 1, 2)$ and the accuracy function for DHFSs $d_i (i = 1, 2)$ is defined as $p(d_i) = \left(\frac{1}{l\left(\mu_{\tilde{A}_i}\right)}\right)\sum_{\gamma \in \mu_{\tilde{A}_i}} \gamma + \left(\frac{1}{m\left(\nu_{\tilde{A}_i}\right)}\right)\sum_{\eta \in \nu_{\tilde{A}_i}} \eta \ (i = 1, 2)$

where $l\left(\mu_{\tilde{A}_i}\right)$ and $m\left(\nu_{\tilde{A}_i}\right)$ are the number of the values in $\mu_{\tilde{A}_i}$ and $\nu_{\tilde{A}_i}$ respectively. For any two DHFEs d_1 and d_2, the comparison between two DHFEs is done as follows:

i. If $s(d_1) > s(d_2)$, then $d_1 > d_2$.

ii. If $s(d_1) = s(d_2)$, then check the accuracy function of DHFSs

1. If $s(d_1) > s(d_2)$, then $d_1 > d_2$.

2. If $s(d_1) = s(d_2)$, then $d_1 = d_2$.

Definition 2.9 [29] Correlation coefficient of HFSs.

The values in HFEs are generally not in order, so they are arranged in descending order i.e., for HFE $\mu_{\tilde{A}}$, let $\sigma : (1, 2, \ldots, n) \to (1, 2, \ldots, n)$ be such that $\mu_{\tilde{A}\sigma(j)} \geq \mu_{\tilde{A}\sigma(j+1)}$ for $j = 1, 2, \ldots, n-1$ and $\mu_{\tilde{A}\sigma(j)}$ be the jth largest value in $\mu_{\tilde{A}}$.

Definition 2.9.1 Let $X = \{x_1, x_2, \ldots, x_n\}$ be an initial universe of objects and a set $\tilde{A}$ on X defined as $\tilde{A} = \{\langle x, \mu_{\tilde{A}}(x^{(s)})\rangle | x \in X\}$ be a HFS, then the information energy of $\tilde{A}$ is defined as $E_{HFS}(\tilde{A}) = \sum_{i=1}^{n}\left(\frac{1}{l_i}\sum_{j=1}^{l_i}\mu^2_{\tilde{A}\sigma(j)}(x_i)\right)$, where $l_i = l(\mu_{\tilde{A}}(x_i))$ denotes the total number of membership values in $\mu_{\tilde{A}}(x_i)$, $x_i \in X$.

Definition 2.9.2 Let $X = \{x_1, x_2, \ldots, x_n\}$ be a universal set and a set $\tilde{A} = \{\langle x, \mu_{\tilde{A}}(x^{(s)})\rangle | x \in X\}$ and $\tilde{B} = \{\langle x, \mu_{\tilde{B}}(x^{(s)})\rangle | x \in X\}$ be any two HFSs on X, then the correlation between $\tilde{A}$ and $\tilde{B}$ is defined by $C_{HFS}(\tilde{A}, \tilde{B}) = \sum_{i=1}^{n}\left(\frac{1}{l_i}\sum_{j=1}^{l_i}\mu_{\tilde{A}\sigma(j)}(x_i)\, \mu_{\tilde{B}\sigma(j)}(x_i)\right)$ where $l_i = \max\{l(\mu_{\tilde{A}}(x_i)), l(\mu_{\tilde{B}}(x_i))\}$ for each $x_i \in X$. Also, when $l(\mu_{\tilde{A}}(x_i)) \neq l(\mu_{\tilde{B}}(x_i))$, then they can be made equal by adding number of membership values in HFE which has least number of membership values in it. This can be done by adding the smallest membership values to make the lengths of both HFE $\tilde{A}$ and $\tilde{B}$ equal i.e. $l(\mu_{\tilde{A}}(x_i)) = l(\mu_{\tilde{B}}(x_i))$. For example, $\tilde{A} = \{\langle 0.3, 0.6, 0.8\rangle\}$ and $\tilde{B} = \{\langle 0.5, 0.4\rangle\}$, be any two HFSs and their lengths are not equal therefore it can be made equal as $\tilde{A} = \{\langle 0.3, 0.6, 0.8\rangle\}$ and $\tilde{B} = \{\langle 0.5, 0.4, 0.4\rangle\}$ respectively.

Definition 2.9.3 Let $X = \{x_1, x_2, \ldots, x_n\}$ be a universal set and a set $\tilde{A} = \{\langle x, \mu_{\tilde{A}}(x^{(s)})\rangle | x \in X\}$ and $\tilde{B} = \{\langle x, \mu_{\tilde{B}}(x^{(s)})\rangle | x \in X\}$ be any two HFSs on X, then the correlation coefficient between $\tilde{A}$ and $\tilde{B}$ is defined by

$$\rho_{HFS}(\tilde{A}, \tilde{B}) = \frac{C_{HFS}(\tilde{A}, \tilde{B})}{\sqrt{E_{HFS}(\tilde{A})}\sqrt{E_{HFS}(\tilde{B})}} = \frac{\sum_{i=1}^{n}\left(\frac{1}{l_i}\sum_{j=1}^{l_i}\mu_{\tilde{A}\sigma(j)}(x_i)\, \mu_{\tilde{B}\sigma(j)}(x_i)\right)}{\sqrt{\sum_{i=1}^{n}\left(\frac{1}{l_i}\sum_{j=1}^{l_i}\mu^2_{\tilde{A}\sigma(j)}(x_i)\right)}\sqrt{\sum_{i=1}^{n}\left(\frac{1}{l_i}\sum_{j=1}^{l_i}\mu^2_{\tilde{B}\sigma(j)}(x_i)\right)}}.$$

Definition 2.10 [29] Correlation coefficient of DHFSs.

The values in DHFEs are generally not in order, so they are arranged in descending order i.e., for DHFE $d = \{\mu_{\tilde{A}}, \nu_{\tilde{A}}\}$, let $\sigma : (1, 2, \ldots, n) \to (1, 2, \ldots, n)$ be such that $\mu_{\tilde{A}\sigma(s)} \geq \mu_{\tilde{A}\sigma(s+1)}$ for $s = 1, 2, \ldots, n-1$, and $\mu_{\tilde{A}\sigma(s)}$ be the sth largest value in $\mu_{\tilde{A}}$; let $\delta : (1, 2, \ldots, m) \to (1, 2, \ldots, m)$ be such that $\nu_{\tilde{A}\delta(t)} \geq \nu_{\tilde{A}\delta(t+1)}$ for $t = 1, 2, \ldots, m-1$, and $\nu_{\tilde{A}\delta(t)}$ be the tth largest value in $\nu_{\tilde{A}}$.

Definition 2.10.1 Let $X = \{x_1, x_2, \ldots, x_n\}$ be an initial universe of objects and a set $\tilde{A}$ on X defined as $\tilde{A} = \{\langle x, \mu_{\tilde{A}}(x^{(s)}, \nu_{\tilde{A}}(x^{(t)},))\rangle | x \in X\}$ be a DHFS, then the information energy of $\tilde{A}$ is defined as $E_{DHFS}(\tilde{A}) = \sum_{i=1}^{n}\left(\frac{1}{k_i}\sum_{s=1}^{k_i}\mu^2_{\tilde{A}\sigma(s)}(x_i) + \frac{1}{l_i}\sum_{t=1}^{l_i}\nu^2_{\tilde{A}\sigma(t)}(x_i)\right)$, where $k_i = k(\mu_{\tilde{A}}(x_i))$ denotes the total number of membership values in $\mu_{\tilde{A}}(x_i)$ and $l_i = l(\nu_{\tilde{A}}(x_i))$ denotes the total number of non-membership values in $\nu_{\tilde{A}}(x_i)$ respectively.

Definition 2.10.2 Let $X = \{x_1, x_2, \dots, x_n\}$ be a universal set and a set $\tilde{A} = \{\langle x, \mu_{\tilde{A}}(x^{(s)}, \nu_{\tilde{A}}(x^{(t)},))\rangle | x \in X\}$ and $\tilde{B} = \{\langle x, \mu_{\tilde{B}}(x^{(s)}, \nu_{\tilde{B}}(x^{(t)},))\rangle | x \in X\}$ be any two DHFSs on X, then the correlation between $\tilde{A}$ and $\tilde{B}$ is defined by $C_{DHFS}(\tilde{A}, \tilde{B}) = \sum_{i=1}^{n}\left(\frac{1}{k_i}\sum_{s=1}^{k_i}\mu_{\tilde{A}\sigma(s)}(x_i)\,\mu_{\tilde{B}\sigma(s)}(x_i) + \frac{1}{l_i}\sum_{t=1}^{l_i}\nu_{\tilde{A}\delta(t)}(x_i)\,\nu_{\tilde{B}\delta(t)}(x_i)\right)$ where $k_i = \max\{k(\mu_{\tilde{A}}(x_i)), k(\mu_{\tilde{B}}(x_i))\}$ $l_i = \max\{l(\nu_{\tilde{A}}(x_i)), l(\nu_{\tilde{B}}(x_i))\}$ for each $x_i \in X$. Also, when $k(\mu_{\tilde{A}}(x_i)) \neq k(\mu_{\tilde{B}}(x_i))$ or $l(\nu_{\tilde{A}}(x_i)) \neq l(\nu_{\tilde{B}}(x_i))$, then they can be made equal by adding some elements in DHFE which has least number of elements in it. This can be done by adding the smallest membership values or smallest non-membership values to make the lengths of both DHFE $\tilde{A}$ and $\tilde{B}$ equal i.e. $k(\mu_{\tilde{A}}(x_i)) = k(\mu_{\tilde{B}}(x_i))$ or $l(\nu_{\tilde{A}}(x_i)) = l(\nu_{\tilde{B}}(x_i))$. For example, $\tilde{A} = \{\langle\{0.3, 0.8\}, \{0.2, 0.5\}\rangle\}$ and $\tilde{B} = \{\langle\{0.1, 0.7\}, \{0.8, 0.9, 0.4\}\rangle\}$, be any two DHFSs and their lengths are not equal therefore it can be made equal as $\tilde{A} = \{\langle\{0.3, 0.8\}, \{0.2, 0, 5, 0.2\}\rangle\}$ and $\tilde{B} = \{\langle\{0.1, 0.7\}, \{0.8, 0.9, 0.4\}\rangle\}$ respectively.

Definition 2.10.3 Let $X = \{x_1, x_2, \dots, x_n\}$ be a universal set and a set $\tilde{A} = \{\langle x, \mu_{\tilde{A}}(x^{(s)}, \nu_{\tilde{A}}(x^{(t)},))\rangle | x \in X\}$ and $\tilde{B} = \{\langle x, \mu_{\tilde{B}}(x^{(s)}, \nu_{\tilde{B}}(x^{(t)},))\rangle | x \in X\}$ be any two DHFSs on X, then the correlation coefficient between $\tilde{A}$ and $\tilde{B}$ is defined by $\rho_{DHFS}(\tilde{A}, \tilde{B}) = \frac{C_{DHFS}(\tilde{A}, \tilde{B})}{\sqrt{E_{DHFS}(\tilde{A})}\sqrt{E_{DHFS}(\tilde{B})}}$

$$= \frac{\sum_{i=1}^{n}\left(\frac{1}{k_i}\sum_{s=1}^{k_i}\mu_{\tilde{A}\sigma(s)}(x_i)\,\mu_{\tilde{B}\sigma(s)}(x_i) + \frac{1}{l_i}\sum_{t=1}^{l_i}\nu_{\tilde{A}\delta(t)}(x_i)\,\nu_{\tilde{B}\delta(t)}(x_i)\right)}{\sqrt{\sum_{i=1}^{n}\left(\frac{1}{k_i}\sum_{s=1}^{k_i}\mu^2_{\tilde{A}\sigma(s)}(x_i) + \frac{1}{l_i}\sum_{t=1}^{l_i}\nu^2_{\tilde{A}\sigma(t)}(x_i)\right)}\sqrt{\sum_{i=1}^{n}\left(\frac{1}{k_i}\sum_{s=1}^{k_i}\mu^2_{\tilde{B}\sigma(s)}(x_i) + \frac{1}{l_i}\sum_{t=1}^{l_i}\nu^2_{\tilde{B}\sigma(t)}(x_i)\right)}}.$$

3. Brief review of the existing CoCf between two DHFSs

In the existing literature [29] it is claimed that, there does not exist any expression to evaluate the CoCf between two DHFSs, so to fill this gap, the expression (1) is proposed to evaluate the weighted CoCf between two DHFSs $A = \{\langle h_{A\sigma(s)}(x_i), g_{A\sigma(t)}(x_i)\rangle\}$ and $B = \{\langle h_{B\sigma(s)}(x_i), g_{B\sigma(t)}(x_i)\rangle\}$, where $i = 1, 2, \dots, n$, and s, t represents the number of values in $h_{A\sigma(s)}$ and $g_{A\sigma(t)}$ respectively.

$$\rho_{WDHFS}(A, B) = \frac{\sum_{i=1}^{n} w_i\left(\frac{1}{k_i}\sum_{s=1}^{k_i}(h_{A\sigma(s)}(x_i)\,h_{B\sigma(s)}(x_i)) + \frac{1}{l_i}\sum_{t=1}^{l_i}(g_{A\sigma(t)}(x_i)\,g_{B\sigma(t)}(x_i))\right)}{\sqrt{\sum_{i=1}^{n} w_i\left(\frac{1}{k_i}\sum_{s=1}^{k_i}(h^2_{A\sigma(s)}(x_i)) + \frac{1}{l_i}\sum_{t=1}^{l_i}(g^2_{A\sigma(t)}(x_i))\right)}\sqrt{\sum_{i=1}^{n} w_i\left(\frac{1}{k_i}\sum_{s=1}^{k_i}(h^2_{B\sigma(s)}(x_i)) + \frac{1}{l_i}\sum_{t=1}^{l_i}(g^2_{B\sigma(t)}(x_i))\right)}} \tag{1}$$

where,

i. w_i represents the normalized weight ($w_i \geq 0$, and $\sum_{i=1}^{n} w_i = 1$) of the ith element.

ii. n represents the number of elements.

iii. $h_{A\sigma(s)}(x_i)$ and $g_{A\sigma(t)}(x_i)$ are two sets of some values in $[0, 1]$. Out of these two, $h_{A\sigma(s)}(x_i)$ represents the set of all the possible membership degree and $g_{A\sigma(t)}(x_i)$ represents the set of all the possible non-membership degree.

iv. k_i represents the number of values in $h_{A\sigma(s)}(x_i)$.

v. l_i represents the number of values in $g_{A\sigma(t)}(x_i)$.

Its claimed that if $w_i = \frac{1}{n}$ for all i then the expression (1) will be transformed into expression (2).

$$\rho_{DHFS}(A,B) = \frac{\sum_{i=1}^{n}\left(\frac{1}{k_i}\sum_{s=1}^{k_i}\left(h_{A\sigma(s)}(x_i)\,h_{B\sigma(s)}(x_i)\right) + \frac{1}{l_i}\sum_{t=1}^{l_i}\left(g_{A\sigma(t)}(x_i)\,g_{B\sigma(t)}(x_i)\right)\right)}{\sqrt{\sum_{i=1}^{n}\left(\frac{1}{k_i}\sum_{s=1}^{k_i}\left(h^2_{A\sigma(s)}(x_i)\right) + \frac{1}{l_i}\sum_{t=1}^{l_i}\left(g^2_{A\sigma(t)}(x_i)\right)\right)}\sqrt{\sum_{i=1}^{n}\left(\frac{1}{k_i}\sum_{s=1}^{k_i}\left(h^2_{B\sigma(s)}(x_i)\right) + \frac{1}{l_i}\sum_{t=1}^{l_i}\left(g^2_{B\sigma(t)}(x_i)\right)\right)}}. \tag{2}$$

3.1 Gaps in the existing weighted CoCf for DHFSs

In this paper, it is claimed that the existing CoCf (1) [29] is not valid in its present form. To prove that this claim is valid, there is a need to discuss the origin of the expressions (1). Therefore, the same is discussed in this section.

It can be easily verified that the expression (1) can be obtained mathematically in the following manner:

$$\sum_{i=1}^{n} w_i\left(\frac{1}{k_i}\sum_{s=1}^{k_i} h_{A\sigma(s)}(x_i)h_{B\sigma(s)}(x_i) + \frac{1}{l_i}\sum_{t=1}^{l_i} g_{A\sigma(t)}(x_i)\,g_{B\sigma(t)}(x_i)\right)$$
$$= \sum_{i=1}^{n}\left(\sum_{s=1}^{k_i} w_i\frac{1}{k_i}\left(h_{A\sigma(s)}(x_i)h_{B\sigma(s)}(x_i)\right)\right) + \sum_{i=1}^{n}\left(\sum_{t=1}^{l_i} w_i\frac{1}{l_i}\left(g_{A\sigma(t)}(x_i)\,g_{B\sigma(t)}(x_i)\right)\right)$$
$$= \left(\left(\sum_{i=1}^{n}\sum_{s=1}^{k_i}\frac{\sqrt{w_i}}{\sqrt{k_i}}h_{A\sigma(s)}(x_i)\right)\times\left(\sum_{i=1}^{n}\sum_{s=1}^{k_i}\frac{\sqrt{w_i}}{\sqrt{k_i}}h_{B\sigma(s)}(x_i)\right)\right)$$
$$+\left(\left(\sum_{i=1}^{n}\sum_{t=1}^{l_i}\frac{\sqrt{w_i}}{\sqrt{l_i}}g_{A\sigma(t)}(x_i)\right)\times\left(\sum_{i=1}^{n}\sum_{t=1}^{l_i}\frac{\sqrt{w_i}}{\sqrt{l_i}}g_{B\sigma(t)}(x_i)\right)\right)$$

Assuming, $X_1 = \sum_{i=1}^{n}\sum_{s=1}^{k_i}\frac{\sqrt{w_i}}{\sqrt{k_i}}h_{A\sigma(s)}(x_i)$, $Y_1 = \sum_{i=1}^{n}\sum_{s=1}^{k_i}\frac{\sqrt{w_i}}{\sqrt{k_i}}h_{B\sigma(s)}(x_i)$,

$X_2 = \sum_{i=1}^{n}\sum_{t=1}^{l_i}\frac{\sqrt{w_i}}{\sqrt{l_i}}g_{A\sigma(t)}(x_i)$ and $Y_2 = \sum_{i=1}^{n}\sum_{t=1}^{l_i}\frac{\sqrt{w_i}}{\sqrt{l_i}}g_{B\sigma(t)}(x_i)$.

$$\sum_{i=1}^{n} w_i\left(\frac{1}{k_i}\sum_{s=1}^{k_i} h_{A\sigma(s)}(x_i)h_{B\sigma(s)}(x_i) + \frac{1}{l_i}\sum_{t=1}^{l_i} g_{A\sigma(t)}(x_i)\,g_{B\sigma(t)}(x_i)\right)$$
$$= (X_1Y_1 + X_2Y_2) \le \sqrt{X_1^2 + X_2^2}\sqrt{Y_1^2 + Y_2^2}$$

$$\le\left(\sqrt{\left(\sum_{i=1}^{n}\sum_{s=1}^{k_i}\frac{\sqrt{w_i}}{\sqrt{k_i}}h_{A\sigma(s)}(x_i)\right)^2 + \left(\sum_{i=1}^{n}\sum_{t=1}^{l_i}\frac{\sqrt{w_i}}{\sqrt{l_i}}g_{A\sigma(t)}(x_i)\right)^2}\right)\times$$

$$\left(\sqrt{\left(\sum_{i=1}^{n}\sum_{s=1}^{k_i}\frac{\sqrt{w_i}}{\sqrt{k_i}}h_{B\sigma(s)}(x_i)\right)^2 + \left(\sum_{i=1}^{n}\sum_{t=1}^{l_i}\frac{\sqrt{w_i}}{\sqrt{l_i}}g_{B\sigma(t)}(x_i)\right)^{2.}}\right)$$

$$\leq \left(\left(\sqrt{\sum_{i=1}^{n}(\sqrt{w_i})^2\left(\left(\sum_{s=1}^{k_i}\frac{h_{A\sigma(s)}(x_i)}{\sqrt{k_i}}\right)^2+\left(\sum_{t=1}^{l_i}\frac{g_{A\sigma(t)}(x_i)}{\sqrt{l_i}}\right)^2\right)}\right)\right)\times$$

$$\left(\sqrt{\sum_{i=1}^{n}(\sqrt{w_i})^2\left(\left(\sum_{s=1}^{k_i}\frac{h_{B\sigma(s)}(x_i)}{\sqrt{k_i}}\right)^2+\left(\sum_{t=1}^{l_i}\frac{g_{B\sigma(t)}(x_i)}{\sqrt{l_i}}\right)^2\right)}\right)$$

$$\leq \left(\left(\sqrt{\sum_{i=1}^{n}w_i\left(\left(\sum_{s=1}^{k_i}\frac{h_{A\sigma(s)}(x_i)}{\sqrt{k_i}}\right)^2+\left(\sum_{t=1}^{l_i}\frac{g_{A\sigma(t)}(x_i)}{\sqrt{l_i}}\right)^2\right)}\right)\right)\times$$

$$\left(\sqrt{\sum_{i=1}^{n}w_i\left(\left(\sum_{s=1}^{k_i}\frac{h_{B\sigma(s)}(x_i)}{\sqrt{k_i}}\right)^2+\left(\sum_{t=1}^{l_i}\frac{g_{B\sigma(t)}(x_i)}{\sqrt{l_i}}\right)^2\right)}\right)$$

$$\leq \left(\left(\sqrt{\sum_{i=1}^{n}w_i\left(\frac{1}{k_i}\sum_{s=1}^{k_i}h^2_{A\sigma(s)}(x_i)+\frac{1}{l_i}\sum_{t=1}^{l_i}g^2_{A\sigma(t)}(x_i)\right)}\right)\right)\times$$

$$\left(\sqrt{\sum_{i=1}^{n}w_i\left(\frac{1}{k_i}\sum_{s=1}^{k_i}h^2_{B\sigma(s)}(x_i)+\frac{1}{l_i}\sum_{t=1}^{l_i}g^2_{B\sigma(t)}(x_i)\right)}\right)$$

$$\Rightarrow \sum_{i=1}^{n}w_i\left(\frac{1}{k_i}\sum_{s=1}^{k_i}h_{A\sigma(s)}(x_i)h_{B\sigma(s)}(x_i)+\frac{1}{l_i}\sum_{t=1}^{l_i}g_{A\sigma(t)}(x_i)g_{B\sigma(t)}(x_i)\right)$$

$$\leq \left\{\begin{array}{c}\left(\left(\sqrt{\sum_{i=1}^{n}w_i\left(\frac{1}{k_i}\sum_{s=1}^{k_i}h^2_{A\sigma(s)}(x_i)+\frac{1}{l_i}\sum_{t=1}^{l_i}g^2_{A\sigma(t)}(x_i)\right)}\right)\right)\times \\ \left(\sqrt{\sum_{i=1}^{n}w_i\left(\frac{1}{k_i}\sum_{s=1}^{k_i}h^2_{B\sigma(s)}(x_i)+\frac{1}{l_i}\sum_{t=1}^{l_i}g^2_{B\sigma(t)}(x_i)\right)}\right)\end{array}\right\}$$

$$\Rightarrow \frac{\sum_{i=1}^{n}w_i\left(\frac{1}{k_i}\sum_{s=1}^{k_i}h_{A\sigma(s)}(x_i)h_{B\sigma(s)}(x_i)+\frac{1}{l_i}\sum_{t=1}^{l_i}g_{A\sigma(t)}(x_i)g_{B\sigma(t)}(x_i)\right)}{\left\{\left(\left(\sqrt{\sum_{i=1}^{n}w_i\left(\frac{1}{k_i}\sum_{s=1}^{k_i}h^2_{A\sigma(s)}(x_i)+\frac{1}{l_i}\sum_{t=1}^{l_i}g^2_{A\sigma(t)}(x_i)\right)}\right)\right)\times\left(\sqrt{\sum_{i=1}^{n}w_i\left(\frac{1}{k_i}\sum_{s=1}^{k_i}h^2_{B\sigma(s)}(x_i)+\frac{1}{l_i}\sum_{t=1}^{l_i}g^2_{B\sigma(t)}(x_i)\right)}\right)\right\}}$$

$$\leq 1.$$

3.2 Mathematical incorrect assumptions

In this section, the mathematical incorrect assumptions, considered in existing literature [29] to obtain the expressions (1) have been discussed.

It can be easily verified from Section 3.1 that to obtain the expressions (1) it have been assumed that,

i. $\sum_{i=1}^{n}\left(\sum_{s=1}^{k_i}\frac{w_i}{k_i}h_{A\sigma(s)}(x_i)h_{B\sigma(s)}(x_i)\right)=\left(\sum_{i=1}^{n}\sum_{s=1}^{k_i}\frac{\sqrt{w_i}}{\sqrt{k_i}}h_{A\sigma(s)}(x_i)\right)\times\left(\sum_{i=1}^{n}\sum_{s=1}^{k_i}\frac{\sqrt{w_i}}{\sqrt{k_i}}h_{B\sigma(s)}(x_i)\right),$

ii. $\sum_{i=1}^{n}\left(\sum_{t=1}^{l_i}\frac{w_i}{l_i}g_{A\sigma(t)}(x_i)g_{B\sigma(t)}(x_i)\right)=\left(\sum_{i=1}^{n}\sum_{s=1}^{l_i}\frac{\sqrt{w_i}}{\sqrt{l_i}}g_{A\sigma(t)}(x_i)\right)\times\left(\sum_{i=1}^{n}\sum_{s=1}^{l_i}\frac{\sqrt{w_i}}{\sqrt{l_i}}g_{B\sigma(t)}(x_i)\right)$

iii. $\left(\sum_{s=1}^{k}\frac{1}{\sqrt{k}}h_{A\sigma(s)}(x_1)\right)^2=\frac{1}{k}\sum_{s=1}^{k}h^2_{A\sigma(s)}(x_1)$

iv. $\left(\sum_{t=1}^{l}\frac{1}{\sqrt{l}}g_{A\sigma(t)}(x_1)\right)^2=\frac{1}{l}\sum_{t=1}^{l}g^2_{A\sigma(t)}(x_1)$

v. $\left(\sum_{s=1}^{k} \frac{1}{\sqrt{k}} h_{B\sigma(s)}(x_1)\right)^2 = \frac{1}{k}\sum_{s=1}^{k} h^2_{B\sigma(s)}(x_1)$

vi. $\left(\sum_{t=1}^{l} \frac{1}{\sqrt{l}} g_{B\sigma(t)}(x_1)\right)^2 = \frac{1}{l}\sum_{t=1}^{l} g^2_{B\sigma(t)}(x_1)$.

Let us consider an example,
Example 1: Let

$$A = \left\{ \begin{array}{c} \langle x_1, \{0.1, 0.2, 0.5\}, \{0.3\}\rangle, \langle x_2, \{0.2, 0.4, 0.6\}, \{0.4, 0.5, 0.8\}\rangle, \\ \langle x_3, \{0.1, 0.2, 0.4\}, \{0.6, 0.8, 0.9\}\rangle, \langle x_4, \{\{0.2, 0.4, 0.1\}, \{0.8, 0.9, 0.6\}\}\rangle \end{array} \right\} \text{ and} \tag{3}$$

$$B = \left\{ \begin{array}{c} \langle x_1, \{\{0.2, 0.3, 0.5\}, \{0.3, 0.6, 0.9\}\}\rangle, \langle x_2, \{\{0.2, 0.3, 0.7\}, \{0.1, 0.9\}\}\rangle, \\ \langle x_3, \{\{0.6, 0.3, 0.5\}, \{0.9, 0.2, 0.3\}\}\rangle, \langle x_4, \{\{0.5\}, \{0.9\}\}\rangle \end{array} \right\}$$

be two DHFS and let $w = (0.3, 0.2, 0.1, 0.4)^T$ be the weight vector of x_i. Then, it can be easily verified that

$$\sum_{i=1}^{n}\left(\sum_{s=1}^{k_i} \frac{w_i}{k_i}\left(h_{A\sigma(s)}(x_i)h_{B\sigma(s)}(x_i)\right)\right) = 0.1289,$$

$$\left(\sum_{i=1}^{n}\sum_{s=1}^{k_i} \frac{\sqrt{w_i}}{\sqrt{k_i}} h_{A\sigma(s)}(x_i)\right) \times \left(\sum_{i=1}^{n}\sum_{s=1}^{k_i} \frac{\sqrt{w_i}}{\sqrt{k_i}} h_{B\sigma(s)}(x_i)\right) = 1.1335$$

It is obvious that

$$\sum_{i=1}^{n}\left(\sum_{s=1}^{k_i} \frac{w_i}{k_i} h_{A\sigma(s)}(x_i)h_{B\sigma(s)}(x_i)\right) \neq \left(\sum_{i=1}^{n}\sum_{s=1}^{k_i} \frac{\sqrt{w_i}}{\sqrt{k_i}} h_{A\sigma(s)}(x_i)\right) \times \left(\sum_{i=1}^{n}\sum_{s=1}^{k_i} \frac{\sqrt{w_i}}{\sqrt{k_i}} h_{B\sigma(s)}(x_i)\right).$$

Also, it can be easily verified that

$$\sum_{i=1}^{n}\left(\sum_{t=1}^{l_i} \frac{w_i}{l_i}\left(g_{A\sigma(t)}(x_i)g_{B\sigma(t)}(x_i)\right)\right) = 0.4003,$$

$$\left(\sum_{i=1}^{n}\sum_{s=1}^{l_i} \frac{\sqrt{w_i}}{\sqrt{l_i}} g_{A\sigma(t)}(x_i)\right) \times \left(\sum_{i=1}^{n}\sum_{s=1}^{l_i} \frac{\sqrt{w_i}}{\sqrt{l_i}} g_{B\sigma(t)}(x_i)\right) = 3.1862.$$

It is obvious that,

$$\sum_{i=1}^{n}\left(\sum_{t=1}^{l_i} \frac{w_i}{l_i} g_{A\sigma(t)}(x_i)g_{B\sigma(t)}(x_i)\right) \neq \left(\sum_{i=1}^{n}\sum_{t=1}^{l_i} \frac{\sqrt{w_i}}{\sqrt{l_i}} g_{A\sigma(t)}(x_i)\right) \times \left(\sum_{i=1}^{n}\sum_{t=1}^{l_i} \frac{\sqrt{w_i}}{\sqrt{l_i}} g_{B\sigma(t)}(x_i)\right).$$

Furthermore, it can be easily verified that

$$\left(\sum_{s=1}^{k_i} \frac{1}{\sqrt{k_i}} h_{A\sigma(s)}(x_1)\right)^2 = 1.02, \quad \frac{1}{k_i}\sum_{s=1}^{k_i} h^2_{A\sigma(s)}(x_1) = 0.4267,$$

$$\left(\sum_{t=1}^{l_i} \frac{1}{\sqrt{l_i}} g_{A\sigma(t)}(x_1)\right)^2 = 4.5800, \quad \frac{1}{l_i}\sum_{t=1}^{l_i} g^2_{A\sigma(t)}(x_1) = 1.6467.$$

$$\left(\sum_{s=1}^{k_1} \frac{1}{\sqrt{k_1}} h_{B\sigma(s)}(x_1)\right)^2 = 2.8232, \quad \frac{1}{k_1}\sum_{s=1}^{k_1} h^2_{B\sigma(s)}(x_1) = 0.8166.$$

$$\left(\sum_{t=1}^{l_1} \frac{1}{\sqrt{l_1}} g_{B\sigma(t)}(x_1)\right)^2 = 4.1960, \quad \frac{1}{l_1}\sum_{t=1}^{l_1} g^2_{B\sigma(t)}(x_1) = 1.4133.$$

It is obvious that

i. $\left(\sum_{s=1}^{k} \frac{1}{\sqrt{k}} h_{A\sigma(s)}(x_1)\right)^2 \neq \frac{1}{k}\sum_{s=1}^{k} h^2_{A\sigma(s)}(x_1)$

ii. $\left(\sum_{t=1}^{l} \frac{1}{\sqrt{l}} g_{A\sigma(t)}(x_1)\right)^2 \neq \frac{1}{l}\sum_{t=1}^{l} g^2_{A\sigma(t)}(x_1)$

iii. $\left(\sum_{s=1}^{k} \frac{1}{\sqrt{k}} h_{B\sigma(s)}(x_1)\right)^2 \neq \frac{1}{k}\sum_{s=1}^{k} h^2_{B\sigma(s)}(x_1)$

iv. $\left(\sum_{t=1}^{l} \frac{1}{\sqrt{l}} g_{B\sigma(t)}(x_1)\right)^2 \neq \frac{1}{l}\sum_{t=1}^{l} g^2_{B\sigma(t)}(x_1)$.

Thus, Example 1 verifies that the considered mathematical assumptions in the existing literature [29] to obtain the weighted correlation coefficient expressions (1) for DHFSs are not valid.

4. Proposed CoCf for the DHFSs

Considering the above mentioned limitation in Section 3 as a motivation, an attempt has been made to modify the existing expression (1) [29], and hence the weighted CoCf for DHFSs is proposed which is represented in expression (3).

$$\rho_{WDHFS}(A,B) = \frac{\sum_{i=1}^{n} w_i\left(\frac{1}{k_i}\sum_{s=1}^{k_i} h_{A\sigma(s)}(x_i)h_{B\sigma(s)}(x_i) + \frac{1}{l_i}\sum_{t=1}^{l_i} g_{A\sigma(t)}(x_i)g_{B\sigma(t)}(x_i)\right)}{\sum_{i=1}^{n} w_i\left(\left(\sqrt{\sum_{s=1}^{k_i}\left(\frac{h_{A\sigma(s)}}{\sqrt{k_i}}\right)^2 \sum_{s=1}^{k_i}\left(\frac{h_{B\sigma(s)}(x_i)}{\sqrt{k_i}}\right)^2}\right) + \left(\sqrt{\sum_{t=1}^{l_i}\left(\frac{g_{A\sigma(t)}(x_i)}{\sqrt{l_i}}\right)^2 \sum_{t=1}^{l_i}\left(\frac{g_{B\sigma(t)}(x_i)}{\sqrt{l_i}}\right)^2}\right)\right)} \tag{4}$$

where,

i. w_i represents the normalized weight $\left(w_i \geq 0, \text{and} \sum_{i=1}^{n} w_i = 1\right)$ of the ith element.

ii. n represents the number of elements.

iii. $h_{A\sigma(s)}(x_i)$ and $g_{A\sigma(t)}(x_i)$ are two sets of some values in $[0,1]$. Out of these two, $h_{A\sigma(s)}(x_i)$ represents the set of all the possible membership degree and $g_{A\sigma(t)}(x_i)$ represents the set of all the possible non-membership degree.

iv. k_i represents the number of values in $h_{A\sigma(s)}(x_i)$.

v. l_i represents the number of values in $g_{A\sigma(t)}(x_i)$.

5. Origin of the proposed CoCf for the DHFSs

The modified expression (3) has been obtained mathematically as follows:

$$\sum_{i=1}^{n} w_i\left(\frac{1}{k_i}\sum_{s=1}^{k_i} h_{A\sigma(s)}(x_i)h_{B\sigma(s)}(x_i) + \frac{1}{l_i}\sum_{t=1}^{l_i} g_{A\sigma(t)}(x_i)g_{B\sigma(t)}(x_i)\right)$$

$$= \sum_{i=1}^{n} w_i\left(\left(\sum_{s=1}^{k_i} \frac{h_{A\sigma(s)}(x_i)}{\sqrt{k_i}}\,\frac{h_{B\sigma(s)}(x_i)}{\sqrt{k_i}}\right) + \left(\sum_{t=1}^{l_i} \frac{g_{A\sigma(t)}(x_i)}{\sqrt{l_i}}\,\frac{g_{B\sigma(t)}(x_i)}{\sqrt{l_i}}\right)\right)$$

Assuming,

$X^{(s)} = \frac{h_{A\sigma(s)}(x_i)}{\sqrt{k_i}}$, $Y^{(s)} = \frac{h_{B\sigma(s)}(x_i)}{\sqrt{k_i}}$,

$X^{(t)} = \frac{g_{A\sigma(t)}(x_i)}{\sqrt{l_i}}$ and $Y^{(t)} = \frac{g_{B\sigma(t)}(x_i)}{\sqrt{l_i}}$

$$\sum_{i=1}^{n} w_i\left(\frac{1}{k_i}\sum_{s=1}^{k_i} h_{A\sigma(s)}(x_i)h_{B\sigma(s)}(x_i) + \frac{1}{l_i}\sum_{t=1}^{l_i} g_{A\sigma(t)}(x_i)g_{B\sigma(t)}(x_i)\right) = \sum_{i=1}^{n} w_i\left(\sum_{s=1}^{k_i} X^{(s)}\,Y^{(s)} + \sum_{t=1}^{l_i} X^{(t)}Y^{(t)}\right)$$

$$\leq \sum_{i=1}^{n} w_i\left(\sqrt{\left(\sum_{s=1}^{k_i}\left(X^{(s)}\right)^2 \times \sum_{s=1}^{k_i}\left(Y^{(s)}\right)^2\right)} + \sqrt{\sum_{t=1}^{l_i}\left(X^{(t)}\right)^2 \times \sum_{t=1}^{l_i}\left(Y^{(t)}\right)^2}\right)$$

$$\leq \sum_{i=1}^{n} w_i\left(\left(\sqrt{\sum_{s=1}^{k_i}\left(\frac{h_{A\sigma(s)}(x_i)}{\sqrt{k_i}}\right)^2 \times \sum_{s=1}^{k_i}\left(\frac{h_{B\sigma(s)}(x_i)}{\sqrt{k_i}}\right)^2}\right) + \left(\sqrt{\sum_{t=1}^{l_i}\left(\frac{g_{A\sigma(t)}(x_i)}{\sqrt{l_i}}\right)^2 \times \sum_{t=1}^{l_i}\left(\frac{g_{B\sigma(t)}(x_i)}{\sqrt{l_i}}\right)^2}\right)\right)$$

$$\Rightarrow \frac{\sum_{i=1}^{n} w_i\left(\frac{1}{k_i}\sum_{s=1}^{k_i} h_{A\sigma(s)}(x_i)h_{B\sigma(s)}(x_i) + \frac{1}{l_i}\sum_{t=1}^{l_i} g_{A\sigma(t)}(x_i)\,g_{B\sigma(t)}(x_i)\right)}{\sum_{i=1}^{n} w_i\left(\left(\sqrt{\sum_{s=1}^{k_i}\left(\frac{h_{A\sigma(s)}(x_i)}{\sqrt{k_i}}\right)^2 \times \sum_{s=1}^{k_i}\left(\frac{h_{B\sigma(s)}(x_i)}{\sqrt{k_i}}\right)^2}\right) + \left(\sqrt{\sum_{t=1}^{l_i}\left(\frac{g_{A\sigma(t)}(x_i)}{\sqrt{l_i}}\right)^2 \times \sum_{t=1}^{l_i}\left(\frac{g_{B\sigma(t)}(x_i)}{\sqrt{l_i}}\right)^2}\right)\right)} \leq 1.$$

6. Exact results of the existing real life problem

There is an investment company, which intends to invest a sum of money in the best alternative [29]. There are four available alternatives, A_1: a car company, A_2: a food company, A_3: a computer company, and A_4: an arms company. The investment company considers three attributes, C_1: the risk analysis, C_2: the growth analysis, and C_3: the environment impact analysis to consider the best alternatives. Since, there is a need to identify the best investment company among A_1, A_2, A_3 and A_4, with respect to an ideal alternative A^* on the basis of three different attributes C_1, C_2, and C_3, it is assumed that:

i. The weights assigned to the attributes C_j ($j = 1, 2$ and 3) are 0.35, 0.25 and 0.40 respectively.

ii. The DHFS $A^* = \{h^*, g^*\} = \{\{1\}, \{0\}\}$ ($j = 1, 2$ and 3) represents the ideal alternative.

iii. The $(i.j)^{th}$ element of **Table 1**, represented by a DHFS, represents the rating value of the ith alternative over the jth attribute i.e. D is a dual hesitant fuzzy decision matrix.

$$D = \begin{matrix} \\ A_1 \\ A_2 \\ A_3 \\ A_4 \end{matrix} \begin{matrix} C_1 & C_2 & C_3 \\ \left[\begin{matrix} \{\{0.5,0.4,0.3\},\{0.4,0.3\}\} \\ \{\{0.7,0.6,0.4\},\{0.3,0.2\}\} \\ \{\{0.6,0.4,0.3\},\{0.3\}\} \\ \{\{0.8,0.7,0.6\},\{0.2,0.1\}\} \end{matrix} \right. & \begin{matrix} \{\{0.6,0.4\},\{0.4,0.2\}\} \\ \{\{0.7,0.6\},\{0.3,0.2\}\} \\ \{\{0.6,0.5\},\{0.3\}\} \\ \{\{0.7,0.6\},\{0.2\}\} \end{matrix} & \left. \begin{matrix} \{\{0.3,0.2,0.1\},\{0.6,0.5\}\} \\ \{\{0.7,0.6,0.4\},\{0.2,0.1\}\} \\ \{\{0.6,0.5\},\{0.3,0.1\}\} \\ \{\{0.4,0.3\},\{0.2,0.1\}\} \end{matrix} \right] \end{matrix}.$$

Table 1.
Rating values of the alternatives over the attributes.

Existing real-life problem [29]	Existing expressions (1) [29]	Proposed expressions (3)
Best investment company among A_1, A_2, A_3, and A_4	$\rho_1((A^*, A_1)) = 0.5981$ $\rho_1((A^*, A_2)) = 0.9200$ $\rho_1((A^*, A_3)) = 0.8668$ $\rho_1((A^*, A_4)) = 0.9088$ $A_2 > A_4 > A_3 > A_1$ i.e. A_2 is the best alternative.	$\rho_1((A^*, A_1)) = 0.9670$ $\rho_1((A^*, A_2)) = 0.9822$ $\rho_1((A^*, A_3)) = 0.9852$ $\rho_1((A^*, A_4)) = 0.9935$ $A_4 > A_3 > A_2 > A_1$ i.e. A_4 is the best alternative.

Table 2.
Results of the considered real-life problem.

Then, by applying the existing expression (1) [29] the obtained preferred company is A_2 i.e. the food company is the best alternative for the investment. However it is discussed in Section 3 that the expression (1) [29] is not valid in its present form since it is scientifically incorrect. Therefore, the result of the considered real-life problem, obtained in existing literature [29], is also not exact. Thus, to obtain the exact results of the existing problem [29], the proposed CoCf represented by expression (3) is utilized and the solution is obtained successfully. Furthermore, comparison of the results of the considered real-life problem is obtained by the existing expression (1) [29] as well as by the modified expression (3), and the results are shown below in **Table 2**.

From the above obtained results as shown in **Table 2**, it is obvious that according to existing expression (1), A_2 i.e. the food company is the most preferred company to invest the money, while, according to the proposed expression (3), A_4 i.e. arms company is the most preferred company to invest the sum of the money by the investment company.

7. Advantages of the proposed measure

The proposed correlation coefficient measure is an efficient tool which has the following advantages for solving the decision-making problems under the dual hesitant fuzzy environment.

i. Dual hesitant fuzzy set is an extension of hesitant fuzzy set (HFS), and intuitionistic fuzzy set (IFS) which contains more information i.e., it has wider range of hesitancy included both in membership and non-membership of an object than the others fuzzy sets (HFSs, deals with only membership hesitant degrees and IFSs deals with both membership degree and non-membership degree).

ii. It is observed in the suggested modified approach that the correlation coefficients of HFS [22–24], IFS [9–17] are the special cases of the proposed correlation coefficients of DHFSs. Thus, it can be comprehended that the proposed correlation coefficients for DHFSs is quite efficient in solving the decision-making problems under HFS, IFS, environment, whereas the existing methods [9–17, 22–24] poses some limitations.

iii. Since DHFSs contains more information in the data in relation to the uncertainties involved in comparison to the IFS, HFS environment hence the proposed tool is efficient in giving an appropriate solution in real-life applications in decision-making problems.

8. Conclusions

This paper is an outcome of the deep analysis made in understanding the ranking measures of DHFS using CoCf. In the present paper, a deep mathematical analysis is made to study the CoCf of a DHFS and it's concluded that there exist certain limitations in the existing CoCf [29] for DHFS. These shortcomings are pointed out with a detailed mathematical derivation which suggests that there are some mathematical incorrect assumptions involved hence, it is not appropriate to apply the existing CoCf of a DHFS in its present form. This limitation encouraged to propose a valid mathematical expression for ranking DHFSs in terms of CoCfs. Therefore, a new CoCf given by expression (3) is proposed for DHFSS which is a modified form of expression (1) [29]. To validate the claim of the modified expressions of the CoCf for DHFSs the detailed mathematical derivation is stated and the results of the real-life problems considered in existing paper [29] are obtained and to validate the obtained results a systematic comparison between the results are made.

Conflict of interest

The authors declare that they do not have any financial or associative interest indicating a conflict of interest in about submitted work.

Author details

Akanksha Singh[1,2]

1 School of Mathematics, Thapar Institute of Engineering and Technology (Deemed to be University), Punjab, India

2 Chandigarh University, Mohali, Punjab, India

*Address all correspondence to: asingh3_phd16@thapar.edu1; akanksha.e10462@cumail.in

References

[1] Atanassov KT: Intuitionistic fuzzy sets. Fuzzy Sets Syst. 1986; 20:87–96.

[2] Atanassov KT, Gargov G: Interval-valued intuitionistic fuzzy sets. Fuzzy Sets Syst. 1989; 31:343–349.

[3] Yager RR: Pythagorean membership grades in multicriteria decision making. IEEE Transa. Fuzzy Syst. 2014; 22: 958–965.

[4] Torra V: Hesitant fuzzy sets. Int. J. Intell. Syst. 2010; 25:529–539.

[5] Torra V, Narukawa Y: On hesitant fuzzy sets and decision. In: The 18th IEEE International Conference on Fuzzy Systems, Jeju Island, Korea, 2009, pp. 1378–1382.

[6] Xia M, Xu Z: Hesitant fuzzy information aggregation in decision making. Int. J. Appr. Reas. 2011; 52:395–407.

[7] Zhu B, Xu Z, Xia M: Dual hesitant fuzzy sets. J. Appl. Math. 2012, Article ID 879629 (2012) 13 pages.

[8] Hung WL, Wu JW: A note on the correlation of fuzzy numbers by expected interval. Int. J. Uncertainty Fuzziness Knowl.-Based Syst. 2001; 9(4):517–523.

[9] Gerstenkorn T, Manko J: Correlation of intuitionistic fuzzy sets. Fuzzy Sets Syst. 1991; 44:39–43.

[10] Szmidt E, Kacprzyk J: Correlation of intuitionistic fuzzy sets. Lect. Notes Comput. Sci. 2010; 6178:169–177.

[11] Huang HL, Guo Y: An improved correlation coefficient of intuitionistic fuzzy sets. J Intell. Syst. 2019; 28:231–243.

[12] Szmidt E, Kacprzyk J, Bujnowski P: Pearson's coefficient between intuitionistic fuzzy sets. In: Fifteenth Int. Conf. on IFSs, Burgas, 11–12 May 2011.

[13] Ye J: Fuzzy decision-making method based on the weighted correlation coefficient under intuitionistic fuzzy environment. Eur. J. Oper. Res. 2010; 205:202–204.

[14] Hong DH, Hwang SY: Correlation of intuitionistic fuzzy sets in probability spaces. Fuzzy Sets Syst. 1995; 75:77–81.

[15] Hung WL: Using statistical viewpoint in developing correlation of intuitionistic fuzzy sets. Int. J. Uncertainty Fuzziness Knowledge-Based Syst. 2001; 9(4):509–516.

[16] Hung WL, Wu JW: Correlation of intuitionistic fuzzy sets by centroid method. Inf. Sci. 2002; 144:219–225.

[17] Zeng W, Li H: Correlation coefficient of intuitionistic fuzzy sets. J. Ind. Eng. Int. 2007; 3(5):33–40.

[18] Bustince H, Burillo P: Correlation of interval-valued intuitionistic fuzzy sets. Fuzzy Sets Syst. 1995; 74:237–244.

[19] Garg H: A novel correlation coefficients between Pythagorean fuzzy sets and its applications to decision-making processes. Int. J. Intell. Syst. 2016; 31(12):1234–1252.

[20] Zheng T, Zhang M, Li L, Wu Q, Zhou L: Correlation Coefficients of Interval-Valued Pythagorean Hesitant Fuzzy Sets and Their Applications. IEEE Access, 2020; 8:9271–9286.

[21] Garg H: Novel correlation coefficients under the intuitionistic multiplicative environment and their applications to decision-making process. J. Ind. Manag. Optim. 2018; 14 (4):1501–1519.

[22] Chen N, Xu Z, Xia M: Correlation coefficients of hesitant fuzzy sets and their applications to clustering analysis. Appl. Math. Model. 2013; 37(4):2197–2211.

[23] Lu N, Liang L: Correlation coefficients of extended hesitant fuzzy sets and their applications to decision making. Symmetry. 2017; 9(4):47.

[24] Xu Z, Xia M: On distance and correlation measures of hesitant fuzzy information. Int. J. Intell. Syst. 2011; 26 (5):410–425.

[25] Karaaslan F, Özlü S: Correlation coefficients of dual type-2 hesitant fuzzy sets and their applications in clustering analysis. Int. J of Intell. Syst. 2020; 35(7):1200–1229.

[26] Asim A, Nasar R, Rashid T: Correlation coefficient of intuitionistic hesitant fuzzy sets based on informational energy and their applications to clustering analysis. Soft Comput. 2019; 23(20):10393–10406.

[27] Wang F, LI X, Chen X: Hesitant fuzzy soft set and its applications in multicriteria decision making, J. Appl. Math. 2014; Article ID 643785, 10 pages.

[28] Wang L, Ni M, Zhu L: Correlation measures of dual hesitant fuzzy sets. J Appl Math, 2013 (2013).

[29] Ye J: Correlation coefficient of dual hesitant fuzzy sets and its application to multiple attribute decision making. Appl. Math. Model. 2014; 38:659–666.

[30] Singh A: Modified expression to evaluate the correlation coefficient of dual hesitant fuzzy sets and its application to multi-attribute decision making. Authorea Preprints. DOI: 10.22541/au.159415553.37101935

[31] Zadeh LA: Fuzzy sets. Inf. Control. 1965; 8: 38–353.

Chapter 6

Evaluating the Organizational Hierarchy Using the IFSAW and TOPSIS Techniques

Mahuya Deb

Abstract

Performance evaluations in organizations are viewed as ideal instruments for evaluating and rewarding the employee's performance. While much emphasis is laid onto the administering of the evaluation techniques, not much thought has been laid out on assessing the contributions of each hierarchical level. Moreover the manifold decision making criteria can also impact the measurement of pertinent contributions because of their ambivalent characteristics. In such a scenario, intuitionistic fuzzy multi-criteria decision making can help strategists and policy makers to arrive at more or less accurate decisions. This paper restricts itself to six decision making criteria and adopts the intuitionistic fuzzy simple additive weighting (IFSAW) method and TOPSIS method to evaluate and rank the employee cadres. The results obtained were compared and both the methods revealed that the middle management displayed impeccable performance standards over their other counterparts.

Keywords: performance evaluation, organisation, intuitionistic fuzzy, IFSAW, TOPSIS

1. Introduction

Organizational fit theories have long emphasized that appropriate selection strategies can lead to superior performance compared to firms that relatively overlook the employee selection based on fit theories [1]. The extent of fit between the individual and the organization determines the labor productivity [2–4] as well as the financial performance [5–8].

The other criterions that influence the overall organizational performance are informal learning [9], workplace competencies [10], organizational citizenship behavior [11] and the like.

While many employee focused parameters are relied on while determining the organizational performance, very few researches have essayed the contributions of each of the hierarchical cadres. Performance evaluations in organizations have traditionally focused on short-term financial and technical results. But modern organizations have not just demanded a generic short-term performance assessment, but an effective means to categorize employees as vital opportunities or threats. By using measurable performance results, with a focus on the entire organization, managers will be able to determine their progress toward longterm goals and objectives [12]. Moreover, superior performance cannot be achieved by just de-layering and de-staffing. Whilst these techniques can to a certain extent eliminate the imperfections within the system, it is the overall behaviors of the employees that need a volte-face. Explicit construal of roles of the employees and managers in particular, will ensure that the managers do not slip into the comfortable and familiar role structure of grand strategists, administrative controllers, and operational implementers. Each hierarchical level or cadre needs to exemplify its cardinal responsibilities that add distinct value to an organization [13]. Identifying, weighting and evaluating the various level of managers against various criteria can be assumed as a function of multi criteria decision making process.

While focus on HR metrics has been growing off late, there is still an element of bias and ambiguity regarding the criteria that are being used rather the greatest difficulty lies in the quantification of criteria being not clearly defined. The basis for the selection of criterions is the subjective judgements by the higher authorities in organisations. These judgements/verbal descriptions do not exhibit the characteristic of being classified into a dichotomous group and are therefore treated as linguistic variables. Also the relation between the different hierarchical levels and the criterions on the basis of which they are assessed are not known precisely. This provides a framework where a different methodology is required. Thus to understand such a structure a verbal description would suffice. A formal way of dealing with them is the linguistic approach by Zadeh [14]. Its basic feature is the use of linguistic variables which are the ones whose values are words or sentences in a language in place of numerical value and a fuzzy conditional statement for expressing the relation between linguistic variables. Here the meaning of a linguistic variable is equated with a fuzzy set while the meaning of the fuzzy conditional statement with a fuzzy relation. Since its inception about a decade ago, the theory of fuzzy sets has evolved in many directions, and is finding applications in a wide variety of fields in which the phenomena under study are too complex or too ill defined to be analyzed by conventional techniques. Fuzzy set theory (FST) [15] allows for subjective evaluation by the decision maker under conditions of uncertainty and ambiguity. It helps to express irreducible observations and measurement uncertainties which are intrinsic to the empirical data. It offers far greater resources for managing complexity and controlling computational cost and allows for conversion of linguistic variables to fuzzy numbers using membership functions. Membership functions assigns to each object a grade of membership denoted by $\mu_A(x)$ which ranges between zero and one. It maps every element of the universe of discourse X to the interval [0, 1] which is written as $\mu_A : X \to [0,1]$. Each fuzzy set is completely and uniquely defined by one particular membership function. A "direct" use of verbal descriptions of those criteria via the concepts of the fuzzy set is proposed here.

A fuzzy set is defined by

$$\overline{A} = \left\{ \left(x, \mu_{\overline{A}}(x) \right) / x \in X, \mu_{\overline{A}}(x) \in [0,1] \right\}$$

In the pair $\left(x, \mu_{\overline{A}}(x)\right)$ the first element x belong to the classical set X, the second element $\mu_{\overline{A}}(x)$belong to the interval [0, 1] which is called the membership function or grade of membership function. This membership function is represented with the help of fuzzy number. It represents the degree of compatibility or a degree of truth of x in $\overline{A}$. The idea of fuzzy numbers was given by Dubois and Prade [16].

A fuzzy subset $\overline{A}$ of the real line R with membership function $\mu_{\overline{A}}(x) : R \to [0,1]$ is called a fuzzy number if.

i. $\overline{A}$ is normal, (i.e.) there exist an element x_0 such that $\mu_{\overline{A}}(x_0) = 1$.

ii. $\overline{A}$ is fuzzy convex,

i.e. $\mu_{\overline{A}}(\lambda x_1 + (1-\lambda)x_2) \geq \min\{\mu_{\overline{A}}(x_1), \mu_{\overline{A}}(x_2)\} \quad x_1, x_2 \in R, \forall \lambda \in [0,1]$

iii. $\mu_{\overline{A}}(x)$ is upper continuous, and.

iv. supp $\overline{A}$ is bounded, where supp $\overline{A} = \{x \in R : \mu_{\overline{A}}(x) > 0\}$.

A fuzzy number $\overline{A}$ of the universe of discourse U may be characterized by a triangular distribution function parameterized by a triplet (a_1, a_2, a_3) (**Figure 1**).

Mikhailovich [17] used the fuzzy sets while solving the problem of factor causality. Dintsis [18] in his work dealt with the idea of implementing fuzzy logic for transforming descriptions of natural language to formal fuzzy and stochastic models. However, fuzzy sets lack in the idea of non -membership function. Whatever information is provided by fuzzy sets does not appear complete in context of decision making as there is no room for alternatives dissatisfying the attributes. Thus Atanassov [19] used the idea of membership value, non-membership value as well as the hesitation index to characterize an intuitionistic fuzzy set. He opined that the sum of membership value and non-membership value lies between zero and one and the hesitation index is calculated as one minus the sum of membership value and non-membership value of an element of a set. In other words some hesitation about degree of belongingness of an element of a set exists. For a fuzzy set the hesitation index is zero. The fuzzy sets along with intuitionistic fuzzy sets can depict real life application areas defined by uncertainity. Some recent applications of fuzzy systems are found in the works of [20, 21].

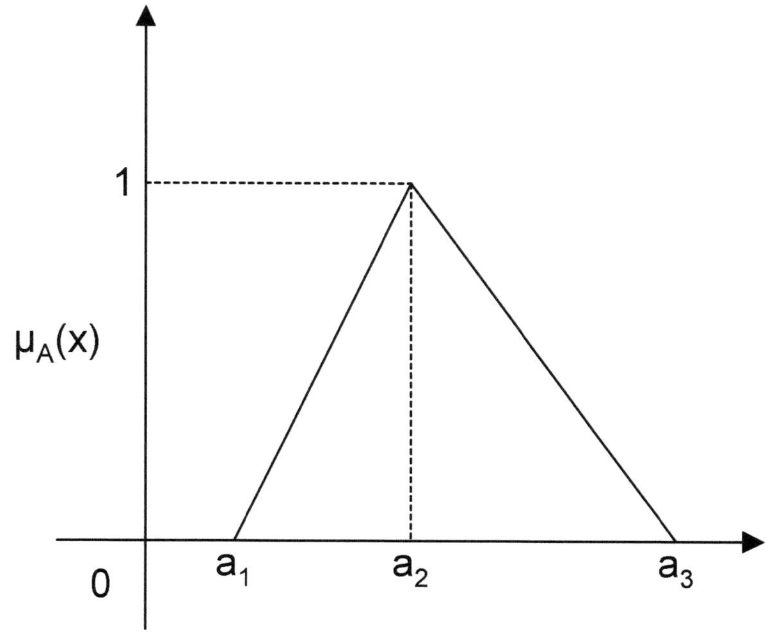

Figure 1.
Membership function of TFN.

1.1 Intuitionistic fuzzy set

Let X be a fixed set. An IFS $\tilde{A}$ in X is of the form $\tilde{A} = \{ <x, \mu_{\tilde{A}}(x), v_{\tilde{A}}(x)> : x \in X\}$, where the $\mu_{\tilde{A}}(x) : X \rightarrow [0, 1]$ and $\nu_{\tilde{A}}(x) : X \rightarrow [0, 1]$. This represents the degree of membership and of non membership respectively of the element $x \in X$ to the set $\tilde{A}$, which is a subset of the set X, for every element of $x \in X, 0 \leq \mu_{\tilde{A}}(x) + v_{\tilde{A}}(x) \leq 1$ [22].

The value of $\pi_A(X) = 1 - \mu_A(X) - v_A(X)$ represents the degree of hesitation (or uncertainty) associated with the membership of elements x ε X in IFS A. This is known as the intuitionistic fuzzy index of A with respect to element x.

1.2 Intuitionistic fuzzy number

An IFN $\tilde{A}$ is defined as follows [22]:

i. an intuitionistic fuzzy subset of the real line

ii. it is normal, i.e. there is any $x_0 \in R$ such that $\mu_{\tilde{A}}(x) = 1 \ (so \ \ v_{\tilde{A}}(x) = 0)$

iii. a convex set for the membership function $\mu_{\tilde{A}}(x)$i.e.

$$\mu_{\tilde{A}}(\lambda x_1 + (1 - \lambda)x_2) \geq \min\left(\mu_{\tilde{A}}(x_1), \mu_{\tilde{A}}(x_2)\right) \forall x_1, x_2 \in R \lambda \in [0, 1]$$

iv. a concave set for the non-membership function $v_{\tilde{A}}(x)$ i.e.

$$v_{\tilde{A}}(\lambda x_1 + (1 - \lambda)x_2) \leq \max\left(v_{\tilde{A}}(x_1), v_{\tilde{A}}(x_2)\right) \forall x_1, x_2 \in R, \lambda \in [0, 1]$$

A triangular intuitionistic fuzzy number $\tilde{A} = (a_1, a_2, a_3; a_1', a_2, a_3')$ is a subset of intuitionistic fuzzy set on the set of real number R whose membership and non membership are defined as follows:

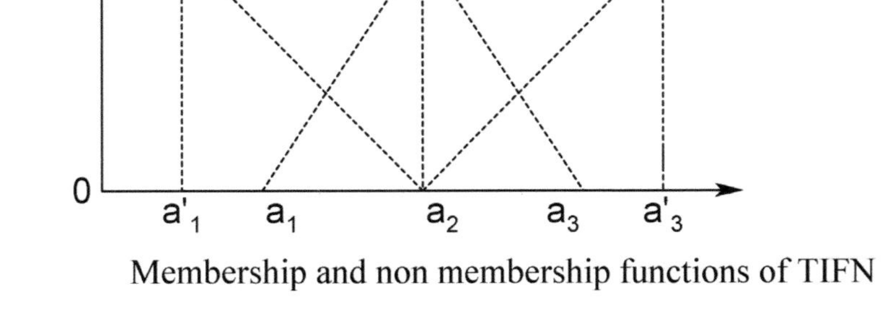

Figure 2.
A triangular intuitionistic fuzzy number.

$$\mu_{\tilde{A}}(x) = \begin{cases} \dfrac{x-a_1}{a_2-a_1}, & a_1 < x \le a_2 \\ \dfrac{a_3-x}{a_3-a_2}, & a_2 < x \le a_3 \\ 0, & otherwise \end{cases} \qquad v_{\tilde{A}}(x) = \begin{cases} \dfrac{a_2-x}{a_2-a_1'}, & a_1' < x \le a_2 \\ \dfrac{x-a_2}{a_3'-a_2}, & a_2 < x \le a_3' \\ 1, & otherwise \end{cases}$$

Intuitionistic fuzzy set is widely recognised and is being studied and applied in various fields be it in science, psychology and other growing fields like consumer behaviour, advertising and communications where decision making is crucial (**Figure 2**).

In this work two methods of intuitionistic fuzzy sets viz. SAW (simple additive weight method) and TOPSIS (Technique for Order Preference by Similarity to Ideal Solution) are used for ranking the various levels of employees in an organisation. The paper is organised as follows: Section 2 begins with the basic operations of intuitionistic fuzzy sets; Section 3 and 4 explain the intuitionistic fuzzy SAW algorithm and TOPSIS methodology which are used in the paper. Section 5 illustrates the procedure for evaluating the hierarchical level using the proposed algorithms. Section 6 is the final discussion and conclusion related to the evaluation procedure.

2. Operations on intuitionistic fuzzy sets

Let A and B are IFS s of the set X, then multiplication operator is defined as follows [19]:

$$A \otimes B = \left[\mu_A(x).\mu_B(x)_{V_A}(x) + v_B(x) - v_A(x)v_B(x), 1 - \{\mu_A(x).\mu_B(x) + (v_A(x) + v_B(x) - v_A(x)v_B(x)\}\right] \tag{1}$$

Let A = (μ, v) be an intuitionistic fuzzy number, a score function S of an intuitionistic fuzzy value can be represented as follows:

$$\mathrm{S(A)} = \mu\text{-v}, \mathrm{S(A)}\,\varepsilon[\text{-}1, 1] \tag{2}$$

If S (A_i) represents the largest among the values of {**S**(A_i)}, then the alternative A_i is the best choice.

3. Intuitionistic fuzzy simple additive weighting algorithm

This method is a simple additive weighting method developed by Hwang and Yoon [23]. According to this principle the first step ensures in obtaining a weighted sum of the performance ratings of each alternative under all attributes. Let A_1, A_2, A_3, … , A_n be n alternatives which denotes the employee cadres. Let C_1, C_2, C_3, … , C_m, be the criteria on the basis of which the evaluation is done. Further each criteria is assigned weight given by the decision makers and it is represented by a weighting vector W = {W_1, W_2, W_3, … , W_n}, where W_1, W_2, W_3, … , W_n are represented by intuitionistic fuzzy sets defined as follows:

$$\mathrm{Wj} = \left\{\mu_w(x_j), v_w(x_j), \pi_w(x_j)\right\}, \text{where } j = 1, 2, \ldots, n. \tag{3}$$

The procedure for Intuitionistic fuzzy SAW is being presented as follows:

Step 1: Construct an intuitionistic fuzzy decision matrix: $\boldsymbol{R} = (\boldsymbol{r_{ij}})_{\mathbf{m\ x\ n}}$ such that $\tilde{r}_{ij} = \left(\mu_{ij}, \nu_{ij}, \pi_{ij}\right)$

$$\tilde{R} = \begin{bmatrix} \tilde{r}_{11} & \tilde{r}_{12} & \dots. & \tilde{r}_{1n} \\ \tilde{r}_{21} & \tilde{r}_{22} & \dots & \tilde{r}_{2n} \\ \dots & .. & .. & .. \\ \tilde{r}_{m1} & \tilde{r}_{m2} & .. & \tilde{r}_{mn} \end{bmatrix}$$

(i = 1,2, … ,m; j = 1,2, … ,n),. In $\tilde{r}_{ij}$, μ_{ij} indicates the degree that the alternative A_i satisfies C_j and ν_{ij} indicates the degree that the alternative A_i does not satisfy the attribute C_j.

Step 2: This step entails performing the transformation by using Eq. (1) and obtain the total intuitionistic fuzzy scores V (A_i) for individual vendors. This is determined by the product of intuitionistic fuzzy weight vectors (W) and intuitionistic fuzzy rating matrix (R).

$$V(A_i) = R \odot W = \sum_{i=1} \left[\{\mu_{A_i}(x_j), \nu_{A_i}(x_j), \pi_{A_i}(x_j)\} \otimes \{\mu_w(x_j), \nu_w(x_j), \pi_{w_i}(x_j)\} \right] \tag{4}$$

Step 3: The third step is used for ranking the alternatives. Applying Eq. (2) a crisp score function $S(A_1)$,$S(A_2)$, … ,$S(A_n)$ is calculated for the various alternatives. The largest value of S (A_j) among $S(A_1)$,$S(A_2)$, … ,$S(A_n)$ represents the best alternative or vendor.

Step 4: This approach is compared with Jun Ye [24] on weighted correlation coefficient under intuitionistic fuzzy environment.

4. Principle of TOPSIS for decision making with intuitionistic fuzzy set

TOPSIS methodology is proposed by [25]. The fundamental principle underlying this theory is that the alternative which is chosen entails that it has the least distance from the positive ideal- solution (i.e. alternative) and its distance is the farthest from the negative ideal- solution (i.e. alternative).

Suppose there exists n decision making alternatives given by the set $A = \{A_1, A_2, \ldots, A_n\}$ from which a most preferred alternative is to be selected. These are assessed based on m attributes, both quantitative and qualitative. The set of all attributes is denoted by $X = \{x_1, x_2, \ldots, x_m\}$. The ratings of different alternatives A_j on attributes x_i are expressed with intuitionistic fuzzy sets $F_{ij} = (\mu_{ij}, \nu_{ij})$ where $\mu_{ij} \in [0,1], \nu_{ij} \in [0,1]$ and $0 \le \mu_{ij} + \nu_{ij} \le 1$. Thus, the ratings of any alternatives A_j on all m attributes x_i are expressed with intuitionistic fuzzy vector $\left(\langle \mu_{1j}, \nu_{1j} \rangle, \langle\langle \mu_{2j}, \nu_{2j} \rangle, \ldots \ldots, \langle \mu_{mj}, \nu_{mj} \rangle\right)^T$.

The intuitionistic fuzzy decision matrix is represented as $F = \left((\mu_{ij}, \nu_{ij})\right)_{mxn}$

$$= \begin{bmatrix} (\mu_{11}, \nu_{11}) & (\mu_{12}, \nu_{12}) & \dots. & ..(\mu_{1n}, \nu_{1n}).. \\ (\mu_{21}, \nu_{21}) & (\mu_{22}, \nu_{22}) & \dots\dots & (\mu_{2n}, \nu_{2n}) \\ \dots & \dots & \dots. & \dots. \\ (\mu_{m1}, \nu_{m1} & (\mu_{m2}, \nu_{m2}) & \dots & (\mu_{mn}, \nu_{mn}) \end{bmatrix} \tag{5}$$

It is assumed that the weights ω_i of the attributes $x_i \in X$ are real numbers known a priori i.e. the weight vector $\omega = (\omega_1, \omega_2, \omega_3, \ldots \omega_m)^T$ of attributes are known.

Since the weights of the attributes are not precisely defined therefore they are treated as intuitionistic fuzzy sets i.e. the weight of each factor is expressed with the intuitionistic fuzzy set $\omega_i = \{\langle x_i, \alpha_i, \beta_i \rangle\}$ where $\alpha_i \in [0, 1]$ and $\beta_i \in [0, 1]$ are respectively the degree of membership and non membership respectively of the attribute $x_i \in X$. Usually $\omega_i = \{\langle x_i, \alpha_i, \beta_i \rangle\}$ is denoted by $\omega_i = \langle \alpha_i, \beta_i \rangle$ in short. The weight of all attributes is concisely expressed in the vector format as follows:

$$\begin{aligned} \omega &= (\omega_1, \omega_2, \omega_3, \dots \omega_m)^T \\ &= (\langle \alpha_1, \beta_1 \rangle, \langle \alpha_2, \beta_2 \rangle, \dots \langle \alpha_m, \beta_m \rangle)^T \end{aligned} \tag{6}$$

4.1 Principle and process of TOPSIS

The entire methodology can be summarized as follows:

1. Identify and determine the attributes and alternatives, denoted respectively by $A = \{A_1, A_2, \dots A_n\}$ *and* $X = \{x_1, x_2, \dots, x_m\}$

2. The decision maker's opinion is obtained to get ratings of the alternatives on the attributes i.e. construct the intuitionistic fuzzy decision matrix

$$F = \left(\left\langle \mu_{ij}, \nu_{ij} \right\rangle\right)_{mxn} \tag{7}$$

3. The opinion so obtained are combined to determine the weights of the attributes expressed with intuitionistic fuzzy weight vector $\omega = (\langle \alpha_i, \beta_i \rangle)_{mx1}$

4. Next the weighted intuitionistic fuzzy decision matrix $F = \left(\left\langle \mu_{ij}, \nu_{ij} \right\rangle\right)_{mxn}$ is computed using the following formula

$$\begin{aligned} \left\langle \mu_{ij}, \nu_{ij} \right\rangle &= \omega F_{ij} \\ &= \langle \alpha_i, \beta_i \rangle \left\langle \mu_{ij}, \nu_{ij} \right\rangle \\ &= \left\langle \alpha_i \mu_{ij}, \beta_i + \nu_{ij} - \beta_i \nu_{ij} \right\rangle \end{aligned} \tag{8}$$

5. For calculating the intuitionistic fuzzy positive ideal –solution and intuitionistic fuzzy negative ideal –solution the following formulas are obtained

$$\begin{aligned} A^+ &= (\langle {\mu_1}^+, {\nu_1}^+ \rangle, \langle {\mu_2}^+, {\nu_2}^+ \rangle, \dots \langle {\mu_m}^+, {\nu_m}^+ \rangle)^T \\ A- &= (\langle {\mu_1}^-, {\nu_1}^- \rangle, \langle {\mu_2}^-, {\nu_2}^- \rangle, \dots \langle {\mu_m}^-, {\nu_m}^- \rangle)^T \end{aligned} \tag{9}$$

$$\begin{aligned} \textit{where } \mu_i^+ &= \max_{1 \le j \le n}\left\{\mu_{ij}\right\} \quad \nu_i^+ = \min_{1 \le j \le n}\left\{\nu_{ij}\right\} \\ \mu_i^- &= \min_{1 \le j \le n}\left\{\mu_{ij}\right\} \quad \nu_i^- = \max_{1 \le j \le n}\left\{\nu_{ij}\right\} \end{aligned} \tag{10}$$

6. The Euclidean distances of the various alternatives A_j(j = 1,2, ... n) from the intuitionistic fuzzy positive ideal and intuitionistic fuzzy negative ideal solution are computed using the following equations

$$D(A_j, A^+) = \sqrt{\frac{1}{2}\left(\sum_{i=1}^{m}\left[\left(\mu_{ij} - \mu_i^+\right)^2 + \left(\nu_{ij} - \nu_i^+\right)^2 + \left(\pi_{ij} - \pi_i^+\right)^2\right]\right)} \quad (11)$$

$$D(A_j, A^-) = \sqrt{\frac{1}{2}\left(\sum_{i=1}^{m}\left[\left(\mu_{ij} - \mu_i^-\right)^2 + \left(\nu_{ij} - \nu_i^-\right)^2 + \left(\pi_{ij} - \pi_i^-\right)^2\right]\right)} \quad (12)$$

7. Thereafter the relative closeness degree λ_j of the alternatives A_j $(j = 1,2 \ldots ,n)$ to the intuitionistic fuzzy positive ideal solution are obtained from

$$\lambda_j = \frac{D(A_j, A^-)}{D(A_j, A^+) + D(A_j, A^-)}, j = 1, 2, \ldots, n \quad (13)$$

8. Lastly determine the ranking order of the alternatives A_j (j = 1,2 ... n) according to the non increasing order of the relative closeness degrees λ_j and the best alternative from A.

Using the two approaches the different level of workers in the organisation are assessed. For a better understanding of the situation an example is worked out below:

5. Numerical example

The example is illustrated as below:
An organization has employed six decision making criteria in order to select the most effective hierarchical level in an organization based on the following criterions.

- Instructional effectiveness (C1)
- Decision making (C2)
- Knowledge and Proficiency (C3)
- Leadership (C4)
- Organizational Citizenship Behaviour (C5)
- Flexibility and Adaptability (C6)

The hierarchical levels of an organization were broadly restricted to four and were compared based on the six decision making criteria (as indicated in **Table 1**).

Hierarchical Levels	Criterion	C1	C2	C3	C4	C5	C6
Senior Management (HL_1)		A	B	A	A	B	A
Middle Management (HL_2)		A	A	B	A	C	B
Junior Management (HL_3)		B	A	C	B	A	C
Staff (HL_4)		B	B	C	C	A	C

A-High B-Average C-Low.

Table 1.
Comparison of Hierarchical levels.

The employees are rated (A, B, C) based on the judgement provided by experts in the organisation .

The intuitionistic fuzzy decision matrix has been constructed as below (**Table 2**):

Methods Criterion	C1	C2	C3	C4	C5	C6
Senior Management (HL_1)	(.7,.1,.2)	(.5,.3,.2)	(.8,.1,.1)	(.7,.2,.1)	(.5,.3,.2)	(.8,.1,.1)
Middle Management (HL_2)	(.7,.1,.2)	(.8,.1,.1)	(.6,.3,.1)	(.8,.1,.1)	(.3,.3,.4)	(.7,.2,.1)
Junior Management (HL_3)	(.5,.1,.4)	(.7,.1,.2)	(.3,.5,.2)	(.5,.3,.2)	(.8,.1,.1)	(.3,.3,.4)
Staff (HL_4)	(.5,.4,.1)	(.6,.3,.1)	(.3,.3,.4)	(.2,.3,.5)	(.7,.2,.1)	(.2,.3,.5)

Table 2.
Intuitionistic fuzzy decision matrix.

The weights for the criteria are as below:

	C1	C2	C3	C4	C5	C6
W_i	(.2,.4,.4)	(.2,.2,.6)	(.1,.5,.4)	(.5,.3,.2)	(.3,.4,.3)	(.2,.4,.4)

Table 3.
Weights of the criteria.

The total intuitionistic fuzzy score $V(HL_i)$ for each hierarchical level is calculated as follows:

$$V(HL_1) = [(.7, .1, .2) * (.2, .4, .4)] + [(.5, .3, .2) * (.2, .2, .6)] + [(.8, .1, .1) * (.1, .5, .4)]$$
$$+[(.7, .2, .1) * (.5, .3, .2)] + [(.5, .3, .2) * (.3, .4, .3)] + [(.8, .1, .1) * (.2, .4, .4)]$$

$$V(HL_1) = [.7 * .2, .1 + .4\text{-}.1 * .4, 1\text{-}(.7 * .2 + .1 + .4\text{-}.1 * .4)] + [.5 * .2, .3 + .2\text{-}$$
$$.3 * .2, 1\text{-}(.5 * .2 + .3 + .2\text{-}.3 * .2)] + [.8 * .1, .1 + .5\text{-}.1 * .5, 1\text{-}(.8 * .1 + .1 + .5\text{-}.1 * .5)]$$
$$+[.7 * .5, .2 + .3\text{-}.2 * .3, 1\text{-}(.7 * .5 + .2 + .3\text{-}.2 * .3)] + [.5 * .3, .3 + .4\text{-}.3 * .4, 1\text{-}(.5 * .3 + .3 + .4\text{-}.3 * .4)]$$
$$+[.8 * .2, .1 + .4\text{-}.1 * .4, 1\text{-}(.8 * .2 + .1 + .4\text{-}.1 * .4)]$$

$$V(HL_1) = [(.14, .46, .4) + (.1, .44, .46) + (.08, .55, .37) + (.35, .44, .21) + (.15, .58, .27) + (.16, .46, .38)]$$
$$V(HL_1) = [0.98, .013, .007]$$

Similarly, the intuitionistic fuzzy scores for other hierarchical levels are calculated as:

$$V(HL_2) = [0.99, .009, .001]$$
$$V(HL_3) = [0.82, .002, .178]$$
$$V(HL_4) = [0.6, .028, .372]$$

The score functions for each hierarchical level calculated using Eq. (2) stands as follows:

$$S(HL_1) = 0.98\text{-}.013 = 0.967$$
$$S(HL_2) = 0.99\text{-}.009 = 0.981$$
$$S(HL_3) = 0.82\text{-}0.002 = 0.818$$
$$S(HL_4) = 0.6\text{-}0.028 = 0.572$$

The hierarchical level with the largest score function value is HL_2 i.e. the middle management.

The ranking order is as below:

$$HL_2 > HL_1 > HL_3 > HL_4$$

The ranking order for the hierarchical levels is in agreement with Jun Ye [24] result on weighted correlation coefficient under intuitionistic fuzzy environment.

The TOPSIS methodology

	C1	C2	C3	C4	C5	C6
HL1	(0.7,0.1,0.2)	(0.5,0.3,0.2)	(0.8,0.1,0.1)	(0.7,0.2,0.1)	(0.5,0.3,0.2)	(0.8,0.1,0.1)
HL2	(0.7,0.1,0.2)	(0.8,0.1,0.1)	(0.6,0.3,0.1)	(0.8,0.1,0.1)	(0.3,0.3,0.4)	(0.7,0.2,0.1)
HL3	(0.5,0.1,0.4)	(0.7,0.1,0.2)	(0.3,0.5,0.2)	(0.5,0.3,0.2)	(0.8,0.1,0.1)	(0.3,0.3,0.4)
HL4	(0.5,0.4,0.1)	(0.6,0.3,0.1)	(0.3,0.3,0.4)	(0.2,0.3,0.4)	(0.7,0.2,0.1)	(0.2,0.3,0.5)

The weights for the criteria are as mentioned in **Table 3**.
The weighted IF decision matrix is obtained as:

	C1	C2	C3	C4	C5	C6
HL1	(0.14,0.04,0.08)	(0.10,0.06,0.12)	(0.08,0.05,0.04)	(0.35,0.06,0.02)	(0.15,0.12,0.06)	(0.16,0.04,0.04)
HL2	(0.14,0.04,0.08)	(0.16,0.02,0.06)	(0.06,0.15,0.04)	(0.45,0.03,0.02)	(0.09,0.12,0.12)	(0.14,0.2,0.16)
HL3	(0.10,0.04,0.16)	(0.14,0.02,0.12)	(0.03,0.25,0.08)	(0.25,0.09,0.04)	(0.24,0.04,0.03)	(0.06,0.12,0.16)
HL4	(0.10,0.16,0.04)	(0.12,0.06,0.06)	(0.03,0.15,0.16)	(0.10,0.09,0.10)	(0.21,0.08,0.03)	(0.04,0.12,0.20)

$$A^+ = \{(0.35, 0.04), (0.45, 0.02), (0.25, 0.02), (0.21, 0.08)\}$$

$$A^- = \{(0.08, 0.12), (0.06, 0.80), (0.03, 0.25), (0.03, 0.16)\}$$

$$D_1(1, A^+) = \frac{1}{2}\begin{bmatrix} (0.14-0.35)^2 + (0.04-0.04)^2 + (0.08-0.61)^2 + (0.14-0.45)^2 + \\ (0.04-0.45)^2 + (0.04-0.02)^2 + (0.08-0.53)^2 + (0.10-0.25)^2 + (0.04-0.02)^2 \\ +(0.16-0.73)^2 + (0.10-0.21)^2 + (0.16-0.08)^2 + (0.04-0.71)^2 \end{bmatrix}$$

$$= \frac{1}{2}\begin{bmatrix} 0.0441 + 0 + 0.2809 + 0.0961 + 0.0004 + 0.2025 + 0.0225 + 0.0004 + 0.3249 + 0.0121 \\ +0.0064 + 0.4489 \end{bmatrix}^{1/2}$$

$$= \frac{1}{2}\sqrt{1.4392}$$

$$= \frac{1}{2} \times 1.19966$$

$$= 0.59983$$

Similarly the other measures are calculated as follows:

$$D(2, A^+) = 0.6251$$

$$D(3, A^+) = 0.6462$$

$$D(4, A^+) = 0.5925$$

$$D(5, A^+) = 0.80475$$

$$D(6, A^+) = 0.67749$$

Also

$$D_1(1,A^-) = \frac{1}{2}\begin{bmatrix}(0.14-0.08)^2+(0.04-0.12)^2+(0.08-0.80)^2+(0.14-0.06)^2+\\(0.04-0.80)^2+(0.08-0.14)^2+(0.10-0.03)^2+(0.04-0.25)^2\\+(0.16-0.53)^2+(0.10-0.03)^2+(0.16-0.16)^2+(0.04-0.81)^2\end{bmatrix}$$

$$= \frac{1}{2}\begin{bmatrix}0.0036+0+0.0064+0.5184+0.0064+0.5776+0.0036+0.0049+0+0.5929\\+0.0064+0.4489\end{bmatrix}^{1/2}$$

$$= \frac{1}{2}\sqrt{1.7138}$$

$$= \frac{1}{2}\times 1.3091$$

$$= 0.6545$$

$$D(2,A^-) = 0.6900$$

$$D(3,A^-) = 0.64033$$

$$D(4,A^-) = 0.57621$$

$$D(5,A^-) = 0.619394$$

$$D(6,A^-) = 0.710$$

Now the relative closeness degree λ_j of the alternatives $A_{j\ (j\,=\,1,2\ldots,n)}$ to the intuitionistic fuzzy positive ideal solution are obtained from

$$\lambda_j = \frac{D(A_j,A^-)}{D(A_j,A^+)+D(A_j,A^-)}, j = 1,2,\ \ldots, n$$

$$\lambda_1 = \frac{0.6545}{0.6545+0.59983} = 0.52179$$

$$\lambda_2 = \frac{0.6900}{0.6900+0.6251} = 0.5246$$

$$\lambda_3 = \frac{0.64033}{0.64033+0.6462} = 0.4977$$

$$\lambda_4 = \frac{0.57621}{0.57621+0.5925} = 0.4930$$

Lastly the ranking order of the alternatives A_j (j = 1,2 ... n) according to the non increasing order of the relative closeness degrees λ_j is as follows:

HL2 > HL1 > HL4 > HL3

To obtain an overall result of the two methods for finding the effectiveness of the employees the average of the two methods is sought. This is shown in the following table as below:

Hierarchical Levels	IFSAW Method	TOPSIS Method	Average	Rating
Senior Management (HL_1)	0.967(2)	0.52179(2)	0.74439	2
Middle Management (HL_2)	0.981(1)	0.5246(1)	0.7528	1
Junior Management (HL_3)	0.818(3)	0.4977(4)	0.65785	4
Staff (HL_4)	0.572(4)	0.4930(3)	0.5325	3

6. Conclusion

In this paper, the researcher worked on a first of its kind area which explored the effectiveness of the highest and the least contributions of the organizational hierarchical levels. The usage of intuitionistic fuzzy approach in the field of HR is a completely novel way of evaluating employees based on the four hierarchical levels. The approach is novel in the sense that such classification of employees using a mathematical model has hardly been used perhaps due to the fact that the parameters defining such categories can hardly be defined in concrete mathematical forms. The results indicate that the middle management is superior in terms of their performance when compared to their counterparts. The proposed method can effectively provide significant implications to policy makers, strategists and human resource professionals which help them to effectively conduct appraisals, take staffing decisions, and allocate work responsibilities and the like when the relevant information is not available or imprecise. It can also provide the decision maker the freedom to minimize the worse or maximize the better case. The method so discussed can be used for performance evaluation of individual employees as well when the attributes measuring their performance are loosely defined i.e. defined in ambiguous terms. Above all the use of intuitionistic fuzzy set in evaluating employees at various organisational levels involves computational complexity as two types of uncertainties are used. But computational complexity is no hindrance in the route to efficient results.

Author details

Mahuya Deb
Department of Commerce, Gauhati University, India

*Address all correspondence to: mahuya8@gmail.com

References

[1] Gupta AK, Govindarajan V. Resource sharing among SBUs: Strategic antecedents and administrative implications. Academy of Management journal. 1986;**29**(4):695-714

[2] Griffith, R., & Simpson, H. (2004). Characteristics of foreign-owned firms in British manufacturing. In Seeking a Premier Economy: The Economic Effects of British Economic Reforms, 1980–2000 (pp. 147–180). University of Chicago Press.

[3] Griliches Z, Regev H. Firm productivity in Israeli industry 1979–1988. Journal of econometrics. 1995; **65**(1):175-203

[4] Oulton N. Competition and the dispersion of labor productivity amongst UK companies. Oxford Economic Papers. 1998;**50**(1):23-38

[5] Brown S, Mc Hardy J, McNabb R, Taylor K. Workplace performance, worker commitment. and loyalty. Journal of Economics & Management Strategy. 2011;**20**(3):925-955

[6] Machin SJ, Stewart MB. Unions and the financial performance of British private sector establishments. Journal of Applied Econometrics. 1990;**5**(4):327-350

[7] McNabb R, Whitfield K. The impact of financial participation and employee involvement on financial performance. Scottish Journal of Political Economy. 1998;**45**(2):171-187

[8] Munday M, Peel MJ, Taylor K. The Performance of the Foreign-Owned Sector of UK Manufacturing: Some Evidence and Implications for UK Inward Investment Policy. Fiscal Studies. 2003;**24**(4):501-521

[9] Van der Klink, M., Boon, J. & Schlusmans, K. (2012). All by myself. Research into employees' informal learning experiences. International Journal of Human Resources Development and Management, 12 (1/2), 77–91.

[10] Vazirani N. Review Paper: Competencies and Competency Model– A Brief overview of its Development and Application. SIES Journal of Management. 2010;7(1):121-131

[11] Yaghoubi NM, Salehi M, Moloudi J. Improving Service Quality by Using Organizational Citizenship Behavior: Iranian Evidence. Iranian Journal of Management Studies. 2011;**4**(4):79-97

[12] Meybodi MZ. Alignment of Strategic Benchmarking Performance Measures: A Lean Manufacturing Perspective. Advances in Competitiveness Research. 2013;**21**(1–2):14

[13] Bartlett CA, Ghoshal S. The myth of the generic manager: new personal competencies for new management roles. California management review. 1997;**40**(1):92-116

[14] Zadeh LA. Outline of a new approach to the analysis of complex system and decision processes ,IEEE, Trans.Syst.. Lan and Cyber. C-3. 1973:28

[15] Zadeh LA. Fuzzy sets. Information and control. 1965;**8**(3):338-353

[16] Dubois D, Prade H. *Operation on fuzzy numbers*. International Journal of Systems Science. 1978;**9**(6):613-626

[17] Mikhailovich,D.M. The Fuzzy Logic Methodology for Evaluating the Causality of Factors in Organization Managemen. In: Fuzzy Logic ,DOI: 10.5772 /IntechOpen,77460,ISBN: 978–1–78984-232-6. 2019

[18] Dintsis, Daniel. Implementing Complex Fuzzy Analysis for Business

Planning Systems, Modern Fuzzy Control Systems and Its Applications. In: doi: 10.5772/67974. 2017

[19] Atanassov KT. Intuitionistic fuzzy sets. Fuzzy sets and Systems. 1986; **20**(1):87-96

[20] Volosencu C, editor. Fuzzy Logic. London, UK: Intech Open Ltd.; 2020 978-1-78984-231-9

[21] Voloşencu, C., Properties of Fuzzy Systems, WSEAS Transactions on Systems, Issue 2, Volume 8, Feruary, 2009, ISSN: 1109-2777, pg. 210–228

[22] Mahapatra GS, Roy TK. *Reliability evaluation using triangular Intuitionistic fuzzy numbers arithmetic operations*, International Journal of Computer and Information. Engineering. 2009;**3**(2): 350-357

[23] Hwang CL, Yoon K. Multiple Attribute Decision Making: Methods and Applications, A State of the Art Survey. New York, NY: Springer-Verlag; 1981

[24] Ye J. Fuzzy decision-making method based on the weighted correlation coefficient under intuitionistic fuzzy environment. European Journal of Operational Research. 2010;**205**(1):202-204

[25] Li F.D., Decision and Game Theory in Management with Intuitionistic Fuzzy Sets, Studies in Fuzziness and Soft Computing , Vol. 308, Springer Heildelberg New York Dordrecht London

Section 3

Clustering

Chapter 7

Data Clustering for Fuzzyfier Value Derivation

JaeHyuk Cho

Abstract

The fuzzifier value m is improving significant factor for achieving the accuracy of data. Therefore, in this chapter, various clustering method is introduced with the definition of important values for clustering. To adaptively calculate the appropriate purge value of the gap type −2 fuzzy c-means, two fuzzy values m_1 and m_2 are provided by extracting information from individual data points using a histogram scheme. Most of the clustering in this chapter automatically obtains determination of m1 and m_2 values that depended on existent repeated experiments. Also, in order to increase efficiency on deriving valid fuzzifier value, we introduce the Interval type-2 possibilistic fuzzy C-means (IT2PFCM), as one of advanced fuzzy clustering method to classify a fixed pattern. In Efficient IT2PFCM method, proper fuzzifier values for each data is obtained from an algorithm including histogram analysis and Gaussian Curve Fitting method. Using the extracted information form fuzzifier values, two modified fuzzifier value m_1 and m_2 are determined. These updated fuzzifier values are used to calculated the new membership values. Determining these updated values improve not only the clustering accuracy rate of the measured sensor data, but also can be used without additional procedure such as data labeling. It is also efficient at monitoring numerous sensors, managing and verifying sensor data obtained in real time such as smart cities.

Keywords: fuzzifier value determining, sensor data clustering, fuzzy C-means, histogram approach, interval type-2 PFCM

1. Introduction

In the majority of cases, fuzzy clustering algorithms have been verified to be a better method than hard clustering in dealing with discrimination of similar structures [1], dataset in dimensional spaces [2], and is more useful for unlabeled data with outliers [3]. Fuzzy C-means proved to offer better solutions in machine learning, and image processing than hard clustering such as Ward's clustering and the k mean algorithm [4–9]. Generally, fuzzy c-mean has 66% accuracy while Gustafson-Kessel scored 70% [10]. Fuzzy c-mean is one of the most largely applied and modified techniques in pattern recognition applications [11] even though the sensitivity of fuzzy C-means is counted as a weak point of outcome to the prototypes and also the optimizing process [12–14].

Classification algorithms are generally subject to various sources of uncertainty that should be appropriately managed. Fuzzy clustering can be used with datasets where the variables have a high level of overlap. Therefore, membership functions are represented as a fuzzy set which can be either Type-I, Type-II or Intuitionistic.

Data are generated by a possible distribution or collected from various resources; Since Euclidean distance leads to clustering outcomes of spherical shapes, which is suitable for most cases, it is a top choice for many applications, it is the measurement used in most clustering algorithms to decide new centers [15].

2. Basic notions

- Degree of membership: The degree of likelihood of one dataset belonging to several centers. The sum of membership degrees is equivalent to 1.

- Data: Data can be categories, compounded or numbers. Data in matrix form contains themes and features of various units. For instance, value and time.

- Clusters: Cluster is a group of data points or datasets that share similarities. Distance or distance norm is a mathematic interpretation of likeness. The point of the model clustering algorithms is the data structure.

- Fuzzifier value: The fuzzifier value is essential to find the clustering membership function when the density or volume of a given cluster is dissimilar to those of another cluster. It is assumed that all of the relative distances to the cluster center are equally 0.5, which implies that the fuzzifier value m is 1 and take account of a decision boundary. With these explained conditions, the fuzzy area does not exist.

Figure 1(a) the case where a small m value is set in two clusters with different volumes. Because the section with a fuzzy membership value extends to a bulky C_2 cluster, applying it to the C_1 cluster allot a lot of relatively unnecessary patterns. **Figure 1(b)** large m value is set. It seems to have good performance since similar membership values are assigned, but the center value of the C_1 cluster tends to move to the C_2 cluster, **Figure 1(c)** Fuzzy area in accordance with Interval type-2 m value. Instead of the fuzzy area according to the value of m_1 and m_2 using the characteristics of the Interval type-2 membership set, uncertainty can be reduced and a proper fuzzy area for the cluster volume can be formed.

As presented above, deciding the lowest and highest boundary range values of the fuzzier value extracted from particular data has been suggested by some methods. The following is about PFCM membership function for deciding the fuzzifier value's range. The membership function at k-th data point for cluster i is presented in Eq. (1). d_{ik}/d_{ij} signifies Euclidean distance value between cluster and data point.

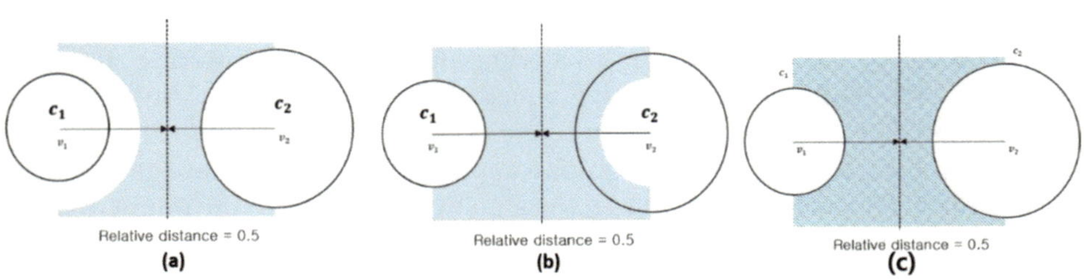

Figure 1.
Fuzzy area between clusters according to m. (a) the case where a small m value, (b) large m value is set, (c) instance of appropriate fuzzy area using Interval type-2.

$$u_{ik} = \frac{1}{\sum_{j=1}^{c} \left(d_{ik}/d_{ij}\right)^{2/(m-1)}} \tag{1}$$

The neighbor membership values are computed, employing the membership value presented in Eq. (1) in order to decide the fuzzifier value's range. Summarization with an expression including fuzzifier value indicates Eq. (2). It obtains the lower and upper boundary values of the fuzzy constant which includes the number of clusters as C and the fuzzifier value as m.

$$1 + \frac{C-1}{C} \cdot \frac{2}{\delta} \cdot |\Delta| \le m \le \frac{2 \log d}{\log\left(\frac{\delta}{1-\delta} \cdot \frac{1}{c-1}\right)} + 1 \text{ where } \Delta = \frac{d_i - d_i^*}{d_i^*} \text{ and } \delta \text{ is threshold} \tag{2}$$

3. Conventional fuzzy clustering algorithm

3.1 Fuzzy C- means (FCM)

FCM includes the concept of a fuzzifier m being used to determine the membership value of data X_k in a specific cluster with cluster prototype. Specifically, the equation of FCM is consist of the cluster center vi and the membership value of data X_k, representing $k = 1, 2...n$ and $i = 1, 2...c$, where n indicates the number of patterns and c indicates the number of clusters. FCM requests the knowledge of the initial number of desired clusters. The membership value is by the relative distance between the pattern X_k and the cluster center V_i. However, one of the main weaknesses by using FCM is its noise sensitivity as well as its limited memberships. The weighting exponent m; is referred to the being effective on the clustering performance of FCM algorithm [16].

3.2 PCM

In order to solve problems of FCM method, PCM uses a parameter given by value estimated from the dataset itself. PCM applies the possibilistic approach which obviously means that the membership value of a point in a class represents the typicality of the point in the class. It also means the possibility of data X_k in the class with cluster prototype V_i where k = 1, 2...n and i = 1, 2...c. Then, the noise points are comparatively less typical, using typicality in PCM algorithm. Furthermore, noise sensitivity is significantly reduced [17, 18]. However, the PCM algorithm also has the problem that the clustering outcome is sensitively reacted according to the initial parameter value [19].

3.3 PFCM

The PFCM algorithm is a mixture of PCM algorithm and FCM algorithm [20]. Although the representative value limit (or constraint = 1) was mitigated, the heat constraints on the membership value were preserved, so the PFCM algorithm generated both membership and possibility, and solved the noise sensitivity problem as seen in the FCM [21]. The PFCM is based on the fuzzy value m, which determines the membership value, and the PFCM also uses constants to define the relative importance of fuzzy membership and typicality values in the objective function. The PFCM utilizes more parameters to determine the optimal solution for clustering, which increases the degree of freedom and thus controls better results than the above-mentioned study. However, when considering fuzzy sets and other parameters in certain algorithms, we face the potential for fuzzy of these parameters. In this paper, we describe the fuzziness of the fuzzy value m and the possible value of the bandwidth parameter and generate FOU of uncertainty for both considering the fuzzy m interval, i.e. the m_1 and m_2 intervals and the fuzzy interval. Existing studies have been implemented to measure the optimal range along the upper and lower bounds of fuzzy values through multiple iterations [22]. This study is ongoing, but the same fuzzy constant range cannot be applied to all data [23].

3.4 Type-1 fuzzy set (T1FS)

Type 1 fuzzy logic was first introduced by Jade (1965). Fuzzy logic systems are based on Type 1 fuzzy sets (T1FS), and have demonstrated their capabilities in many applications, especially for control of complex nonlinear systems that are difficult to model analytically [24, 25]. Since the Type 1 fuzzy logic system (T1FS) uses a clear and accurate type 1 fuzzy set, T1FS can be used to model user behavior under certain conditions. Type 1 fuzzy sets deal with uncertainty using precise membership functions that users think capture uncertainty [26–30]. When the Type 1 membership function is selected, all uncertainties disappear because the Type 1 membership function is completely accurate. The Type 2 fuzzy set concept was presented by Jade as an extension of the general fuzzy set concept., i.e. a type 1 fuzzy set [31]. All fuzzy sets are identified as membership functions. In a type 1 fuzzy set, each element is identified as a two-dimensional membership function. The membership rating for Type 1 fuzzy sets is [0, 1], which is an accurate number. The comparison of membership function and uncertainty extracted from the result of the conventional fuzzy clustering algorithm is shown as below [32].

FCM	$J_{FCM}(V,U,X) = \sum_{i=1}^{c}\sum_{k=1}^{n} u_{ik}^{m}\|x_k - v_i\|\ 1<m<\infty$
PCM	$J_{PCM}(V,U,X) = \sum_{i=1}^{c}\sum_{k=1}^{n} u_{ik}^{m}d_{ik}^{2} + \sum_{i=1}^{c}\eta_i \sum_{k=1}^{n}(1-u_{ik})^{m}$ $\eta : scale, typicality\ \eta = \frac{\sum_{k=1}^{n} u_{ik}^{m}\|x_k - v_i\|^2}{\sum_{k=1}^{n} u_{ik}^{m}}$
FPCM	$J_{FPCM}(U,T,V) = \sum_{i=1}^{c}\sum_{k=1}^{n}(u_{ik}^{m} + t_{ik}^{\eta})\|x_k - v_i\|^2$
PFCM	$J_{PFCM}(U,T,V) = \sum_{i=1}^{c}\sum_{k=1}^{n}(au_{ik}^{m} + bt_{ik}^{\eta})\|x_k - v_i\|^2 + \sum_{i=1}^{c}\delta_i \sum_{k=1}^{n}(1+\tau_{ik})\eta$
T1FC	$J_{T1FC}(X,U,C) = \sum_{i=1}^{c}\sum_{k=1}^{n} u_j(x_i)^m d_{ij}^2$

4. Advanced fuzzy clustering algorithm

Fuzzy c-means (FCM) is an unsupervised form of a clustering algorithm where unlabeled data $X = \{x_1, x_2..., x_N\}$ is grouped together in accordance with their fuzzy membership values [33, 34]. Since, data analysis and computer vision problems, analyzing and dealing the uncertainties are a very important issue, FCM is being widely used in these fields. Several methods of other IT2 approach for pattern recognition algorithms have been successfully reported [35–41]. Type-1 fuzzy sets cannot deal uncertainties therefore; type-2 fuzzy sets were defined to represent the uncertainties associated with type-1 fuzzy sets. As shown in **Figure 2**, the type-reduction process in IT2 FSs requires a relatively large amount of computation as type-2 fuzzy

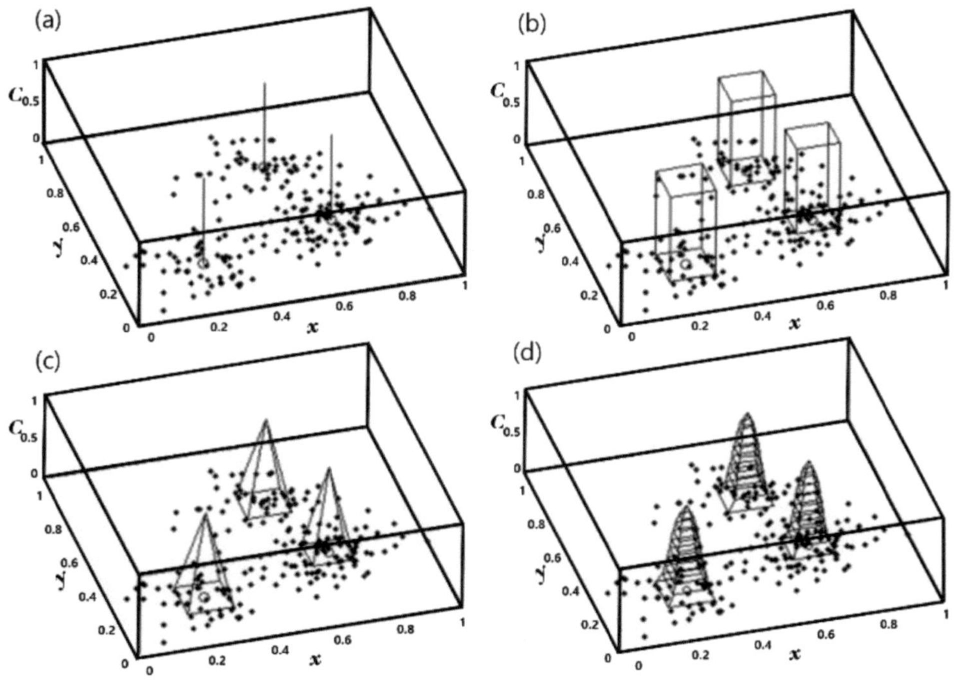

Figure 2.
(a) Cluster position uncertainty for T1FCM, (b) 1 T2 FCM, (c) QT2 FCM, (d) GT2 FCM algorithms.

methods increase the computational complexity due to the numerous combinations of embedded T2 FSs. Methods for reducing the computational complexity have been proposed, such as, the increase in computational complexity of T2 FSs may be less costly for improved performance by applying satisfactory results using T1 FSs. In [42], it was suggested that two Fuzzifier m values is used and the centroid type reduction algorithm for center update is incorporated for interval type-2 (IT2) fuzzy approach to FCM clustering. The IT2 FCM was suggested to clear up the complication with FCM for clusters with different number of volumes and patterns. Moreover, it was suggested that miscellaneous uncertainties were linked with clustering algorithms such as FCM and PCM [43]. Motivation of the success IT2 FSs has made on T1 FSs algorithms.

4.1 Type-2 fuzzy set (T2 FS)

Due to their potential to model various uncertainties, Type-2 fuzzy sets (T2 FSs) have primarily received interest of increased research [44]. Type-2 fuzzy sets are characterized by a three-dimensional fuzzy membership function. The [0, 1] fuzzy set is the membership grade for each element of a type-2 fuzzy set. The extra third dimension provides extra degrees of freedom to get more information about the expressed term. Type-2 fuzzy sets are valuable in situations where it is difficult to resolve the exact membership function of the fuzzy set. This helps to incorporate uncertainty [45].

The computational complexity of the Type-2 fuzzy set is higher than that of the Type 1 fuzzy set. However, the results gained by the Type-2 fuzzy set are much better than those gained by the Type 1 fuzzy set. Therefore, if type-2 fuzzy sets can significantly improve performance (depending on the application), the increased computational complexity of the type-2 fuzzy sets can be an affordable price to pay [46].

4.2 Type-2 FCM (T2-FCM)

Type-2 FCM (T2-FCM), whose type-2 membership is promptly generated by extending a scalar membership degree to a T1-FS. When limiting the secondary fuzzy set to have a triangular membership function, T2-FCM extends the scalar membership u_{ij} to a triangular secondary membership function [47, 48].

4.3 General type-2 FCM

The GT2 FCM algorithm accepts a linguistic description of the fuzzifier value expressed as a set of T1 fuzzy- upper and lower value [49]. The linguistic fuzzifier value is denoted as a T1 fuzzy set of m. **Figure 3** is shown as two examples of encoding the linguistic nation of the appropriate Fuzzifier value for the GT2 FCM algorithm using three linguistic terms.

4.4 Interval type 2 fuzzy sets (IT2 FSs)

In order to model uncertainty associated to a type-1 fuzzy set with an interval type 2 fuzzy set, a membership interval with all secondary grades of the primary memberships equaling to one can represent the primary membership $J_{x'}$ of a sample point x' [18, 50].

Figure 3(a) represents an instance of an interval type 2 fuzzy set where the gray shaded region indicates FOU. In the figure, the membership value for a sample x' is represented by the interval between upper $\overline{\mu}_{\tilde{A}}(x')$, and lower $\underline{\mu}_{\tilde{A}}(x')$ membership. Therefore, each x' has a primary membership interval as

$$J_{x'} = \left[\underline{\mu}_{\tilde{A}}(x'), \overline{\mu}_{\tilde{A}}(x')\right] \tag{3}$$

In the **Figure 3(b)** shown as the vertical slice x′, where the secondary grade for the primary membership of each x' equals one, in accordance with the property of interval type-2 fuzzy sets. This interval is defined as the FOU. An interval type 2 fuzzy set A can be expressed as

$$\tilde{A} = \{((x,u), \mu_{\tilde{A}}(x,u)) | \forall x \in A, \forall u \in J_x \subseteq [0,1], ((x,u), \mu_{\tilde{A}}(x,u)) = 1\} \tag{4}$$

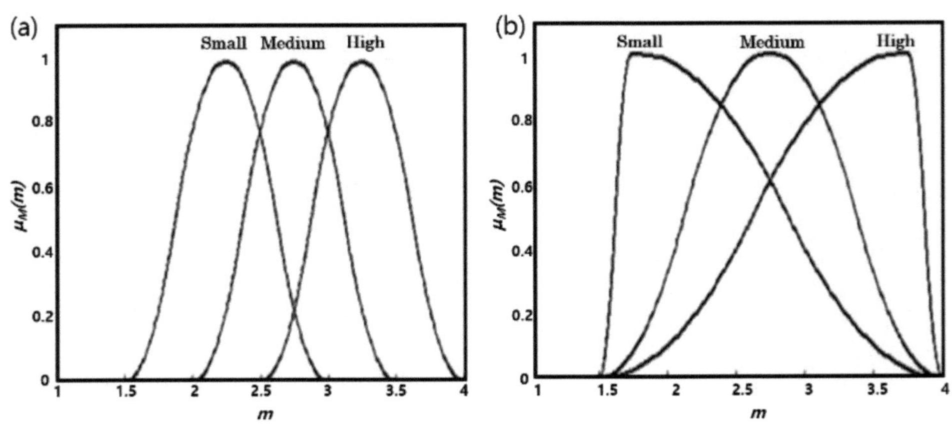

Figure 3.
Two possible linguistic representation of the Fuzzifier M using T1 fuzzy sets. (a) membership value for a sample x′ (b) vertical slice x′.

4.5 Interval type-2 FCM (IT2-FCM)

In fuzzy clustering algorithms such as FCM, the fuzzy fire value m plays a significant [50] role in determining clustering uncertainty. However, it is generally difficult to properly determine the value of m. IT2-FCM regards fuzzy fire values as intervals $[m_1, m_2]$ and settles two optimization matters [51].

First, an interval type 2 FCM is used to obtain a rough estimate of which data points belong to which cluster.

In Eq. (3) is minimized with respect to u_{ij} to provide upper and lower membership values.

$$\bar{u}_j(x_i) = \begin{cases} \dfrac{1}{\sum_{k=1}^{c}\left(d_{ij}/d_{ik}\right)^{2/(m_1-1)}}, \text{if } 1/\sum_{k=1}^{c}\left(d_{ij}/d_{ik}\right) < \dfrac{1}{c} \\ \dfrac{1}{\sum_{k=1}^{c}\left(d_{ij}/d_{ik}\right)^{2/(m_2-1)}}, \text{otherwise} \end{cases} \tag{5}$$

$$\underline{u}_j(x_i) = \begin{cases} \dfrac{1}{\sum_{k=1}^{c}\left(d_{ij}/d_{ik}\right)^{2/(m_1-1)}}, \text{if } 1/\sum_{k=1}^{c}\left(d_{ij}/d_{ik}\right) \geq \dfrac{1}{c} \\ \dfrac{1}{\sum_{k=1}^{c}\left(d_{ij}/d_{ik}\right)^{2/(m_2-1)}}, \text{otherwise} \end{cases} \tag{6}$$

After this cluster prototypes are calculated, then type reduction and then classification is done. Qiu et al. (2014) proposed this complete method of interval type-2 FCM for finding the clusters in each class of the histogram in individual dimensions is acquired with these labeled clusters. This histogram is smoothed by the mean of moving window (using a triangular window in my case). The curve fitting of this smoothed histogram gets the membership function. Histograms with values greater than the membership value are assigned as histograms for higher membership, and histograms for values less than membership value are saved as histograms for lower membership. Curve fitting is carried out severally in the top and bottom histograms to supply the top and bottom member values [52]. This membership value is suggested to estimate the values of fuzzifiers m_1 and m_2. Fixed-point iteration is a method of expressing the transcendental equation $f(x) = 0$ in the form of $x = g(x)$ and then solving this expression iteratively for x in iterative relationship.

$$x_{i+1} = g(x_i), I = 0, 1, 2, \ldots \tag{7}$$

where x_0 being some initial guess. Rewriting the equation to express Eq. (5) and (6) in the form of (7) and dropping the upper and lower term,

$$u_j = \frac{1}{\sum_{k=1}^{c}\left(d_{ij}/d_{ik}\right)^{2/(m-1)}} \tag{8}$$

$$\Rightarrow \frac{1}{u_j} = \sum_{k=1}^{c}\left(d_{ij}/d_{ik}\right)^{2/(m-1)}$$

log on both sides, Eq. (8) can be rewritten as

$$log\left(\frac{1}{u_j}\right) = log\left(\sum_{k=1}^{c}\left(d_{ij}/d_{ik}\right)^{2/(m-1)}\right) \tag{9}$$

$$\because log\,(a+c) = log\,a + log\left(1 + \frac{c}{a}\right)$$

Extending this logarithmic identity to the sum of N elements,

$$\Rightarrow log\left(a_0+\sum_{k=1}^{N}a_k\right)=log\,a_0+log\left(1+\sum_{k=1}^{N}\frac{a_k}{a_0}\right) \tag{10}$$

$$\log\left(\frac{1}{u_j}\right)=\frac{2}{(m-1)}\log\left(\frac{d_{ij}}{d_{1j}}\right)+\log\left(1+\sum_{k=2}^{c}\left(\frac{d_{ij}}{d_{ik}}\right)^{2/\left(m_{i_{old}}-1\right)}\right) \tag{11}$$

Rearranging Eq. (11) and expressing it in terms of m, gives us Eq. (12).

$$\gamma=\frac{\log\left(\frac{1}{u_j}\right)-\log\left(1+\sum_{k=2}^{c}\left(\frac{d_{ij}}{d_{ik}}\right)^{2/\left(m_{i_{old}}-1\right)}\right)}{\log\left(\frac{d_{ij}}{d_{1j}}\right)} \tag{12}$$

$$m_{jnew}=1+\frac{2}{\gamma}$$

So, Eq. (13) gives m_{1jnew} and m_{2jnew}, where $m_{1j\ new}\geq m_{2j\ new}$. Eq. (12) is used to calculate fuzzifier values of each data. In some cases, the value of fuzzifier of particular data shows relatively large variation. Here, upper (m_{upper}) and a lower (m_{lower}) fuzzifier is necessary, using Eq. (2). If the curtain data point has a fuzzy fire value below the lower bound, the fuzzy fire value is set to the m_{lower} bound, and if it exceeds the upper bound, the fuzzy fire value is set to the m_{upper} bound. In the end, a mean of these fuzzifiers is taken to get the last fuzzifier values m_1 and m_2.

4.6 Multiple kernels PFCM algorithm

Typically, the kernel method uses a spatial conversion function to convert input data from input property space to kernel property space [53]. This is to change the kernel property space to a kernel property space so that it is easy to distinguish between overlapping data and having a nonlinear boundary surface in the input property space. If the data in the input space is $X_i, i=1, \ldots, N$, the data converted to the kernel property space through the function is represented by $\Phi(X_j), j=1\ldots N$. Alike as general PFCM, in the case of Kernels-PFCM, the goal is to minimize the following objective function.

$$J^{\phi}=\sum_{k=1}^{n}\sum_{i=1}^{c}\left(au_{ik}^{m}+bt_{ik}^{\eta}\right)\times d_{ij}^{2}+\sum_{i=1}^{c}\gamma\sum_{k=1}^{n}(1-t_{ik})^{\eta} \tag{13}$$

In the input space for kernel K, the pattern x_i and the distance d_{ij} in the kernel attribute space of cluster prototype v_j are expressed as Eq. (14) by the kernel function.

$$\begin{aligned} d_{ij}&=\left\|\Phi(x_j)-\Phi(v_j)\right\|^2\\ &=\Phi(x_j)\Phi(x_j)+\Phi(v_j)\Phi(v_j)-2\Phi(x_j)\Phi(v_j)\\ &=K(x_j,x_j)+K(v_j,v_j)-2k(x_j,v_j)\end{aligned} \tag{14}$$

Commonly, the new Gaussian multi-kernel $\tilde{k}$ using a Gaussian kernel assumes a multi-kernel with the number of kernels S, and the formula is as follows [54].

$$\tilde{k}^{(j)} = (x_j, v_j) = \sum_{l=1}^{s} \frac{w_{il}}{\sigma_l} \frac{\exp\left(-\frac{\|x_j - v_j\|^2}{2\sigma_l^2}\right)}{\sum_{t=1}^{s} \frac{w}{\sigma_t}} \tag{15}$$

From [55] way, using e FCM-MK, normalized kernel is defined to recognize weights by cluster prototypes, resolution and membership values. Using this optimization way, following PFCM objective equation should be minimized. By minimizing the objective function, cluster prototype v_i, resolution-specific weight w_{il}, and membership value u_{ij} are defined.

$$J_{m,\eta}(U, T, V; X) = 2\sum_{k=1}^{n}\sum_{i=1}^{c}(au_{ik}^{m} + bt_{ik}^{\eta} \times$$
$$\left(1 - \sum_{l=1}^{s} \frac{w_{il}}{\sigma^2} \exp\left(-\frac{\|x_j - v_i\|^2}{2\sigma_l^2}\right) \times \frac{1}{\sum_{t=1}^{s} \frac{w}{\sigma_t}}\right) \tag{16}$$
$$+ \sum_{i=1}^{c} \gamma_i \sum_{k=1}^{n} ((1 - t_{ik})^{\eta})$$

Here, ρ is a gradient descent way to learn rate parameter. Finally, using type reduction and hard partitioning, clustering is performed as described in the Interval Type-2 PFCM [56].

4.7 Interval type-2 fuzzy c-regression clustering

Let the regression function be represented by Eq. (17)

$$y_i = f^z(x_i, \alpha_j) = a_1^z x_{1i} + a_2^z x_{2i} + \cdots + a_M^z x_{Mi} + b_0^z \tag{17}$$

where, $x_i = [x_{1i}, x_{2i}, \ldots, x_{Mi}]$ represents points of data, the number of data indicates $_i$ = 1,.. .,n, the number of clusters (or rules) indicates j = 1,.. ., c, the number of variables in each regression indicates q = 1,.. .,M and the number of regression functions indicates z = 1,.. ., r. By a_j, regression coefficients are denoted. We use weighted least square method (WLS) for calculating regression coefficients a_j, In this way, membership grades of partition matrix P are worked for weights. In Eq. (18), Xi is a data point matrix with inputs, y is a data point matrix with outputs.

$$x_i = \begin{bmatrix} x_{1,i} \\ x_{2,i} \\ \vdots \\ x_{M,i} \end{bmatrix}^T, y = \begin{bmatrix} y_{1,} \\ y_2 \\ \vdots \\ y_M \end{bmatrix}^T, P_j = \begin{bmatrix} u_j(x_1) & 0 & \ldots & 0 \\ 0 & u_j(x_1) & \ldots & 0 \\ \vdots & \vdots & \ddots & \vdots \\ 0 & 0 & \ldots & u_j(x_1) \end{bmatrix} \tag{18}$$
$$\alpha_j = \left[X^t P_j X\right]^{-1} X^T P_j y$$

The partition matrix P is acquired through Gaussian mixture distribution which is the first stage for computing regression coefficients. We consider two fuzzifiers or weighting exponent m_1 and m_2 for indicating the problem into IT2F. However, there is a difference that this model is FCM although our model is FCRM. These two fuzzy fires divide the objective function into two separate functions. The aim is to minimize the total error from Eq. (19) shows these two objective functions. It should be mentioned that the following proof is an extended and modified version of'type-1, which has been presented in [57].

$$\begin{cases} J_{m_1}(U,v) = \sum_{i=1}^{n}\sum_{j=1}^{C} u_j(x_i)^{m_1} E_{ji}(\alpha_j) \\ \\ J_{m_2}(U,v) = \sum_{i=1}^{n}\sum_{j=1}^{C} u_j(x_i)^{m_2} E_{ji}(\alpha_j) \end{cases} \tag{19}$$

Where type-1 FCRM, E_{ji} is the total error, which indicates the distance between actual output and estimated regression equation, and it is presented by Eq. (20).

$$E_{ji}(\alpha_j) = \left(y_i - f_j(x_i,\alpha_j)\right)^2 \tag{20}$$

Eq. (21) represents the Lagrangian of the objective functions of IT2 FCRM model. We expend the type-1 NFCRM algorithm to interval type-2 NFCRM.

$$\begin{cases} L_1(\lambda_1,u_j) = \sum_{i=1}^{n}\sum_{j=1}^{C} u_j(x_i)^{m_1} E_{ji}(\alpha_j) - \lambda_1\left(\sum_{j=1}^{c} u_j - 1\right) \\ \\ L_2(\lambda_2,u_j) = \sum_{i=1}^{n}\sum_{j=1}^{C} u_j(x_i)^{m_2} E_{ji}(\alpha_j) - \lambda_2\left(\sum_{j=1}^{c} u_j - 1\right) \end{cases} \tag{21}$$

The partial derivatives with respect to u_j of Eq. (21) are set to 0 in Eq. (22) and (23) for minimizing the objective function.

$$\begin{Bmatrix} \frac{\partial L_1}{\partial u_1(x_i)} = m_1 u_1(x_i)^{m_{1-1}} E_{1i}(\alpha_1) - \lambda_1 = 0 \\ \vdots \\ \frac{\partial L_1}{\partial u_C(x_i)} = m_1 u_C(x_i)^{m_{1-1}} E_{Ci}(\alpha_C) - \lambda_1 = 0 \end{Bmatrix} \tag{22}$$

$$\begin{Bmatrix} \frac{\partial L_2}{\partial u_1(x_i)} = m_2 u_1(x_i)^{m_{2-1}} E_{1i}(\alpha_1) - \lambda_2 = 0 \\ \vdots \\ \frac{\partial L_2}{\partial u_C(x_i)} = m_2 u_C(x_i)^{m_{1-1}} E_{Ci}(\alpha_C) - \lambda_2 = 0 \end{Bmatrix} \tag{23}$$

Next, the partial derivatives with respect to k_1 and k_2 are performed.

$$\frac{\partial L_1}{\partial \lambda_1} = -\left(\sum_{j=1}^{c} u_j(x_i) - 1\right) = 0 \tag{24}$$

To adapt KPCM to IT2 KPCM, three steps are included. In other words, we update the prototype location via initialization, two different fuzzy devices, high and low membership or typicality value calculation, format reduction, and de-fuzzing for data patterns. In the way we propose, by using IT2FS, our point lies in the development of a prototype update process that can solve the cluster matching problem caused by KPCM. Cluster matching usually results in a set of patterns containing clusters that are relatively close to each other. This allows by definition a type 1 fuzzy set to obtain a type reduction via an embedded fuzzy set, but a type-reduced fuzzy set can be obtained by a combination of central intervals estimated from the embedded fuzzy set. This approach is a standard method for obtaining reduced fuzzy set types from IT2FS. However, this approach avoids due to its huge computational requirements, which include a number of embedded fuzzy sets. Therefore, we consider the KM algorithm as an alternative type reduction method. Since KM is an iterative algorithm which estimates both ends of an interval, calculating the left (right) interval v_L (v_R) can be found without using all of the embedded fuzzy sets.

Form KERNELS SFCM ALGORITHM in **Figure 4**,

The kernel distance,

$$\|\Phi(x_k) - v_i\|^2 \tag{25}$$

can be derived using the kernel way as

$$\|\Phi(x_k) - v_i\|^2 = K(x_k, x_k) - 2\frac{\sum_{j=1}^{N} u_{ij}^m K(x_k, x_j)}{\sum_{j=1}^{N} u_{ij}^m} + \frac{\sum_{j=1}^{N}\sum_{l=1}^{N} u_{ij}^m u_{il}^m K(x_j, x_l)}{\left(\sum_{j=1}^{N} u_{ij}^m\right)^2} \tag{26}$$

The inverse mapping of prototypes is also needed to approximate the prototypes expressions v_i in the feature space. The objective equation can be written as

$$V(\hat{v}_i, v_i) = \sum_{i=1}^{C} \|\Phi(\hat{v}_i) - v_i\| = \sum_{i=1}^{C} \left(\Phi(\hat{v}_i)^T \Phi(\hat{v}_i) - 2\Phi(\hat{v}_i)v_i + v_i^T v_i \right. \tag{27}$$

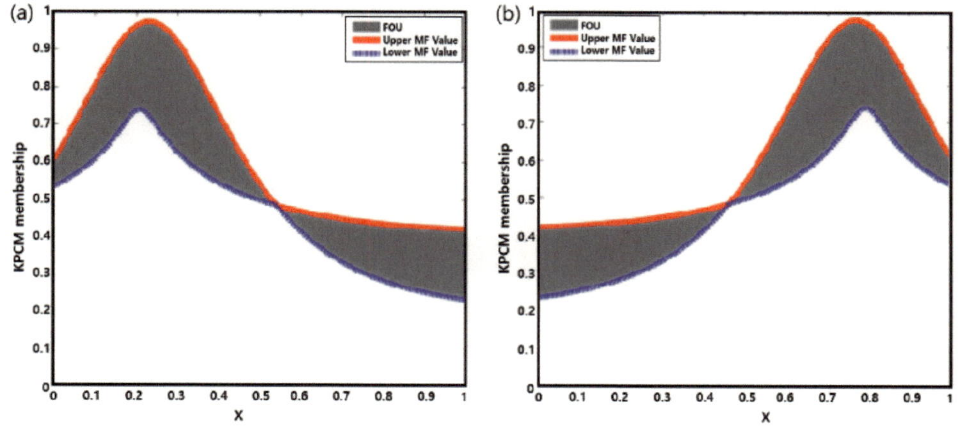

Figure 4.
FOU representation for our proposed IT2 KPCM approach with $m_1 = 2$, $m_2 = 5$ *and variance* = 0.5; *(a) FOU of cluster 1 (b) FOU of cluster 2 [58].*

While, the final location for $\hat{v}_i$ in the KPCM algorithm becomes,

$$\hat{v}_i = \frac{\sum_{k=1}^{N} u_{ik}^m K(x_k, \hat{v}_i) x_k}{\sum_{k=1}^{N} u_{ik}^m K(x_k, \hat{v}_i)} \tag{28}$$

The left (right) interval of the centroids can be found by employing the KM algorithm on the ascending order of a pattern set and its associated interval memberships. The result of the KM algorithm can be expressed as,

$$v_i = 1.0/[v_L v_R] \tag{29}$$

While the procedure to calculate the left value of interval set v_L and right value v_R, defuzzification is used next to calculate the crisp centers and is defined as the midpoint between v_L, v_R. We can now compute the defuzzified output that is a crisp value of the prototypes by using the expression.

$$v_i = \frac{\sum_{v \in J_{Y_i}} (u(v))v}{\sum_{v \in J_{Y_i}} (u(v))} = \frac{v_L + v_R}{2} \tag{30}$$

Hard partitioning is used to classify test patterns using the resulting prototype of the procedure above. Euclidian distance is now used to hard partition patterns because the prototype is in feature space. The pattern is assigned to a cluster prototype with a minimum Euclidean distance. Experimental results presented in the following sections will demonstrate the validity of the proposed IT2 approach to KPCM clustering.

4.8 Interval type-2 possibilistic fuzzy C-means (IT2PFCM)

In order to solve the uncertainty existing in the fuzzifier value m in the general PFCM algorithm, Multiple Kernels PFCM algorithm should be extended to the Interval Type-2 fuzzy set. If there are N data, W set of resolution-specific weight, U partition matric, C clusters, V set of cluster prototype and S kernels, the cluster prototype can be obtained from minimizing the Gaussian kernel objective function as follows.

$$w_{il}^{(new)} = w_{il}^{(old)} - \rho \frac{\partial J}{\partial w_{il}} \tag{31}$$

$$d_{ij}^2 = \left(2 - 2 \sum_{i=1}^{S} \frac{w_{il}}{\sigma_l} \frac{\exp\left(- \frac{\|x_j - v_j\|^2}{2\sigma_l^2} \right)}{\sum_{t=1}^{S} \frac{w}{\sigma_t}} \right) \tag{32}$$

Where,

$$v_i = \left(2 - 2 \sum_{i=1}^{S} \frac{w_{il}}{\sigma_l} \frac{\exp\left(- \frac{\|x_j - v_j\|^2}{2\sigma_l^2} \right)}{\sum_{t=1}^{S} \frac{w}{\sigma_t}} \right) \tag{33}$$

The cluster prototype is calculated to optimize the objective function for the center v_i of each cluster [23].

Where,

$$\overline{K}^{(i)}\left(x_j, v_i\right) = \left(\sum_{i=1}^{S} \frac{w_{jl}}{\sigma_l^3} \frac{\exp\left(\left\|x_j - v\right\|^2\right)}{\sum_{t=1}^{S} \frac{w}{\sigma_t}}\right) \tag{34}$$

optimized membership value- the smallest membership value and the largest membership value for each pattern using the Interval Type-2 fuzzy set- is used for calculating the crisp value v_i. In order to compute v_R and v_L, determination of the upper or lower bound of fuzzifier is essential. It is organized as follows by given Eq. (38) [59] .

$$J(U, V, W) = 2\sum_{i=1}^{C}\sum_{j=1}^{N} u_{ij}^m d_{ij}^2 \tag{35}$$

Using the final v_R and v_L, the crisp center value is obtained from defuzzification as follows.

$$\begin{aligned}&\text{For } v_R,\\ &\text{if } (v(i<k)) \text{ then } u_{ij} = \overline{u}_{ij}\\ &\text{else } u_{ij} = \underline{u}_{ij}\end{aligned} \tag{36}$$

Using the cluster Prototype v_i, obtained through the optimization function and the membership value u_{ij}, the resolution-specific weight value w_{il} is re-obtain as follows.

$$\frac{\partial J}{\partial w_{il}} = -2\sum_{i=1}^{N} \frac{u_{ij}^m}{\sum_{t=1}^{S} \frac{w_t}{\sigma_t}} \left(K\left(x_j, v_i - \overline{K}^{(i)}\left(x_i, v_j\right)\right)\right) \tag{37}$$

Where

$$v_{iR} = \frac{\sum_{j=1}^{N} u_{ij}^m \overline{K}^{(i)}\left(x_{\,j}, v_i\right) x_j}{\sum_{j=1}^{N} u_{ij}^m \overline{K}^{(i)}\left(x_j, v_i\right)} \tag{38}$$

To define the Interval Type-2 fuzzy set and calculate uncertainty for membership, the input data, the primary fuzzy set, is needed to assign into the Interval Type-2 fuzzy set. Eventually, the upper and lower membership function are created from the primary membership functions.

After calculating the upper and lower membership for each cluster, we need to update the new center values. The membership is obtained from the Type-2 fuzzy set, however, the center value is a crisp value, the value cannot be calculated from the above method. Therefore, in order to compute the center value, type reduction is performed by the Type-1 fuzzy set. In addition, defuzzification is accomplished to change the value of Type-1 to a crisp value.

5. Heuristic method: histogram analysis

The goal of heuristic method is to extract information from data, and then adaptively calculates the fuzzifier value. In this approach, some heuristic type- 1 membership function is used appropriately for given dataset. The parameters are defined as the upper and lower membership is decided according to following rules. First, given that the membership values are determined, the IT2 PFCM algorithm

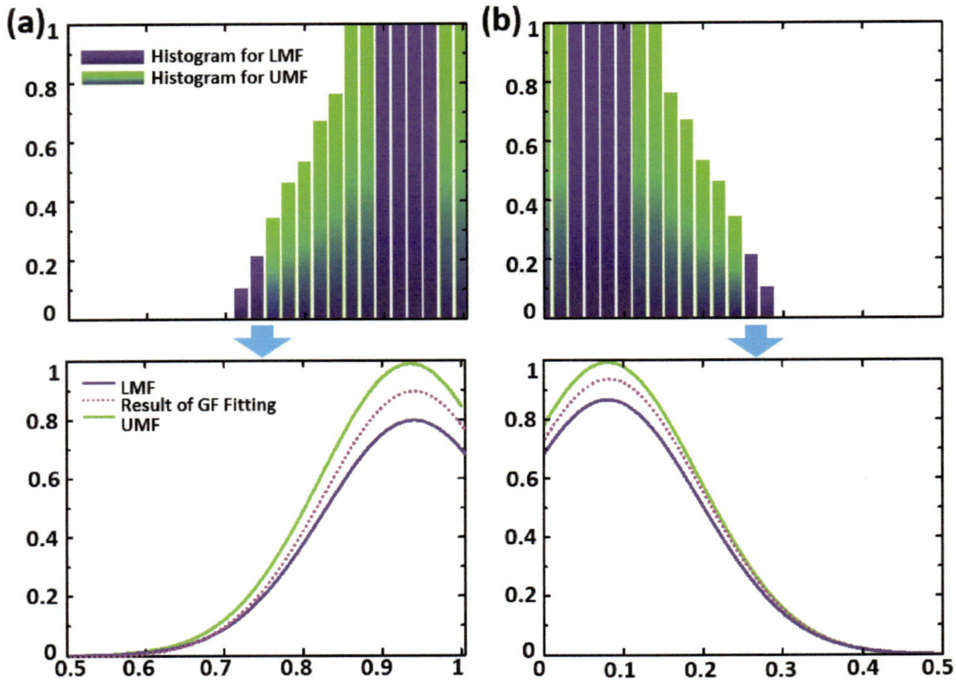

Figure 5.
FOU obtained for individual class and dimension updated fuzzifier value m1 and m2 are obtained (a) class 1 dimension 1, and (b) class 2 dimension 1.

calculates roughly in which cluster the data belongs to and then secure a histogram based on the classified clusters. The histogram from IT2 PFCM tends to be gentler and smoother through the membership function by curve fitting of the same histogram. Curve fitting is enforced separately on upper and lower histograms to obtain upper and lower membership values. In order to reach to the IT2 FS, determination of FOU is necessary, which is generally the set of membership values of the T2 FS. Given that, the greater values of the histogram than the membership value are allocated as the highest membership histogram while the opposite case is calculated. **Figure 5** shows histograms and FOU determined by classification and dimensional calculation. To find X, satisfying $f(X) = 0$, it can be expressed as $X = g(X)$ using fixed-point iteration, where X is,

$$X_{i+1} = g(X), \quad i = 0, 1, \ldots, N \tag{39}$$

Eq. (7) and (8) of the membership function u_i can be shown in the form of Eq. (38) as follows.

$$u_i = \frac{1}{\sum \left(\frac{d_{ik}}{d_{ij}}\right)^{\frac{2}{m-1}}} \tag{40}$$

Where fuzzifier value m is a value that determines the degree of final clustering fuzzifier as the value of the fuzzy parameter. This value of m_1 and m_2 is then applied into the algorithm for calculate updated clusters and this routine is repeated repeatedly. The detailed algorithm is as follows:

1. Set the initial fuzzifier value of m_1 and m_2.

2. Apply m_1 and m_2 to interval type-2 FCM and obtain the membership of data.

3. Generate a histogram of each cluster from the membership.

4. Curve fit the histogram to get primary memberships.

5. Create histogram of upper and lower membership.

6. Use curve fitting over upper and lower histograms to calculate upper and lower memberships.

7. Normalize the memberships according to upper membership.

8. Fuzzifier m_{1i} and m_{2i} are obtained using Eq. (13).

9. Average m_{1i} and m_{2i} and update m_1 and m_2 from the average.

10. The algorithm is iteratively performed using updated m_1 and m_2.

The Upper Membership Function (UMF) Histogram and Lower Membership Function (LMF) Histogram are drawn in **Figure 5**. A new membership function obtained from the Gaussian Curve Fitting (GF-F) method as.

From simply log process on both sides in Eq. (39), Eq. (40) can be expressed as follows:

$$\log\left(\frac{1}{u_1}\right) = \frac{2}{m-1}\log\left(\frac{d_{ki}}{d_{1i}}\right) + \log\left(1 + \sum_{j=2}^{c}\left(\frac{d_{ki}}{d_{ji}}\right)^{\frac{2}{m_{dd}-1}}\right). \tag{41}$$

Rearranging Eq. (40) and calculate it in terms of m, gives us Eq. (41), (42).

$$\gamma = \frac{\log\left(\frac{1}{u_j}\right) - \log\left(1 + \sum_{k=2}^{C}\left(\frac{d_{ij}}{d_{ik}}\right)^{2/m_{old}-1}\right)}{\log\left(\frac{d_{ij}}{d_{ik}}\right)} \tag{42}$$

$$m_{jnew} = 1 + \frac{2}{\gamma} \tag{43}$$

As in the above process, the membership value $u_i \in \{u_i(X_k)\}$ and mj_{new} is used as a function to get the u_i. Where Eq. (9) is applied to each clustered data and updated, m_{1inew} and m_{2inew} values is easily calculated, averaging the fuzzifier value by Eq. (42), the new fuzzifier value m_1 and m_2 are finally calculated as follow

$$m_1 = \left(\sum_{i=1}^{N} m_{1i}\right)/N, m_2 = \left(\sum_{i=1}^{N} m_{2i}\right)/N \tag{44}$$

6. Comparing performances algorithms

Algorithms can be compared in previous experiences using the following criteria:

Root Mean Squared Error (RMSE): The evaluation metric used by all algorithms of clustering is RMSE. RMSE is calculated by the root of the averaging all squared errors between the original data (X) and the corresponding predicted values data ($\overline{X}$).

$$RMSE = \sqrt{\frac{\sum_{k=1}^{n}\sum_{i=1}^{c}(\boldsymbol{x}_{ik} - \overline{\boldsymbol{x}}_{ik}}{\boldsymbol{n}}} \tag{45}$$

where n is the total number of patterns in a given data set and c is the number of clusters; x_{ik} *and* $\overline{x}_{ik}$ the actual and predicted rating values data respectively.

Accuracy is one metric for evaluating classification models. Informally, accuracy is the fraction of predictions the model got right. Formally, accuracy has the following definition:

$$Accuracy = \frac{number\ of\ correct\ samples}{total\ number\ of\ samples} * 100 \tag{46}$$

Acknowledgements

This work was supported by Institute of Korea Evaluation Institute of Industrial Technology (KEIT, Next Generation Artificial Intelligence Semiconductor R&D Program) grant funded by the Korea government (Ministry of Trade, Industry & Energy, MOTIE) (Project No. 20010098, Development of Mixed Signal SoC with complex sensor for Smart Home Appliances).

Author details

JaeHyuk Cho
Department of Electronic Engineering, Soongsil University, Seoul, Korea

*Address all correspondence to: chojh@ssu.ac.kr; wogur900@gmail.com

References

[1] M.J. Bayley, V.J. Gillet, P. Willett, J. Bradshaw and D.V.S. Green, "Computational Analysis of Molecular Diversity for Drug Discovery", Proceeding of the 3rd Annual Conference on Research in Computational Molecular Biology, pp. 321–330, 1999.

[2] J. M. Barnard and G.M. Downs, "Clustering of Chemical Structures on the Basis of Two-Dimensional Similarity Measures", Journal of Chemical Information and Computer Science, vol. 32, pp. 644–649, 1992.

[3] M. Feher and J.M. Schmidt, "Fuzzy Clustering as a Means of Selecting Representative Conformers and Molecular Alignment", Journal of Chemical Information and Computer Science, vol. 43, pp. 810–818, 2003.

[4] R. Guthke, W. Schmidt-Heck, D. Hahn and M. Pfaff, "Gene Expression data Mining for Functional Genomics using Fuzzy Technology" in Advances in Computational Intelligence and Learning Methods and Applications, Kluwer, pp. 475–487, 2002.

[5] S.L. Rodgers, J.D. Holliday and P. Willet, "Clustering Files of Chemical Structures Using the Fuzzy k-Means Clustering Method", Journal of Chemical Information and Computer Science, vol. 44, pp. 894–902, 2004.

[6] Rohit Rastogi et al., "GA-Based Clustering of Mixed Data Type of Attributes (Numeric Categorical Ordinal Binary and Ratio-Scaled)", BVICAM's International Journal of Information Technology, vol. 7, no. 2, 2015.

[7] M. Ramze Rezaee, B.p.f. Lelieveldt and J.h.c. Reiber, "A New Cluster Validity Index for the Fuzzy C-mean", Pattern Recognition Letters, vol. 19, no. 3–4, pp. 237–46, 1998.

[8] J. C. Dunn, "A fuzzy relative of the ISODATA process and its use in detecting compact well-separated clusters", J. Cybern., vol. 3, no. 3, pp. 32–57, 1973.

[9] L. J. Hubert and P. Arabie, "Comparing Partitions", J. Classification, vol. 2, pp. 193–218, 1985.

[10] J. C. Bezdek, "Mathematical models for systematics and taxonomy", Proceedings of the 8th International Conference on Numerical Taxonomy, 1975.

[11] Veit Schwämmle and Ole N. Jensen, "A Simple and Fast Method to Determine the Parameters for Fuzzy c means Cluster Validation", arXiv preprint arXiv:1004.1307, 2010.

[12] J.C. Bezdek, "Cluster Validity with Fuzzy Sets", J. Cybernet., pp. 58–73, 1974.

[13] L. Zadeh, "Fuzzy sets", Inform. Control, vol. 8, pp. 338–353, 1965.

[14] U. Kaymak and M. Setnes, "Extended Fuzzy Clustering Algorithm", ERIM Report Series Research in Management, pp. 1–23, 2000.

[15] J. Dunn, "A Fuzzy Relative of the ISODATA Process and Its Use in Detecting Compact Well-Separated Cluster", Journal of Cybernetics, vol. 3, no. 3, pp. 32–57, 1973.

[16] Şahinli, F., 1999. Kümeleme analizine fuzzy set teorisi yaklaşımı. Yüksek Lisans Tezi, Gazi Üniversitesi Fen Bilimleri Enstitüsü, Ankara, 119.

[17] J.Bezdek,Pattern Recognition with Fuzzy Objective Function Algorithms. New York: Plenum, 1981.

[18] J.ValentedeOliveiraandW.Pedrycz (Eds.),AdvancesinFuzzyClustering and its Applications. Hoboken, NJ: Wiley, 2007.

[19] F. Ch-H. Rhee, "Uncertain fuzzy clustering: Insights and recommendations," IEEE Comput. Intell. Mag., vol. 2, no. 1, pp. 44–56, Feb. 2007.

[20] Pal, N. R., Pal, K., Keller, J. M. and Bezdek, J. C., 2005. A possibilistic fuzzy c-means clustering algorithm. IEEE Transactions on Fuzzy Systems, 13, 517–530.

[21] R. Hathaway, J. C. Bezdek, and W. Pedrycz, "A parametric model for fusing heterogeneous fuzzy data," IEEE Trans. Fuzzy Syst., vol. 4, no. 3, pp. 270–281, Aug. 1996.

[22] W. Pedrycz, J. C. Bezdek, R. J. Hathaway, and G. W. Rogers, "Two nonparametric models for fusing heterogeneous fuzzy data," IEEE Trans. Fuzzy Syst., vol. 6, no. 3, pp. 411–425, Aug. 1998.

[23] W. Pedrycz, "Shadowed sets: Representing and processing fuzzy sets," IEEE Trans. Syst., Man, Cybern. B, Cybern., vol. 28, no. 1, pp. 103–109, Feb. 1998.

[24] Zadeh L.A. (1973), Outline of a New Approach to the Analysis of Complex Systems and Decision Processes, IEEE Transactions on Systems, Man, and Cybernetics, 28–44, 1973.

[25] KING, Peter J.; MAMDANI, Ehmat H. The application of fuzzy control systems to industrial processes. Automatica, 1977, 13.3: 235–242.

[26] ZADEH, Lotfi A. From imprecise to granular probabilities. Fuzzy Sets and Systems, 2005, 154.3: 370–374.

[27] KLIR, G. J.; FOLGER, T. A. Fuzzy sets, uncertainty, and information. Hall1988, 1988.

[28] KARNIK, Nilesh N.; MENDEL, Jerry M. Centroid of a type-2 fuzzy set. information SCiences, 2001, 132.1–4: 195–220.

[29] WU, Hongwei; MENDEL, Jerry M. Uncertainty bounds and their use in the design of interval type-2 fuzzy logic systems. IEEE Transactions on fuzzy systems, 2002, 10.5: 622–639.

[30] CASTILLO, Oscar; MELIN, Patricia. Intelligent systems with interval type-2 fuzzy logic. International Journal of Innovative Computing, Information and Control, 2008, 4.4: 771–783.

[31] ZADEH, Lotfi A. The concept of a linguistic variable and its application to approximate reasoning—I. Information sciences, 1975, 8.3: 199–249.

[32] LINDA, Ondrej; MANIC, Milos. General type-2 fuzzy c-means algorithm for uncertain fuzzy clustering. IEEE Transactions on Fuzzy Systems, 2012, 20.5: 883–897.

[33] Bezdek, J. C., Ehrlich, R., & Full, W. (1984). FCM: The fuzzy c-means clustering algorithm. Computers & Geosciences, 10 (2–3), 191–203. http://doi.org/10.1016/ 0 098–30 04(84)90 020–7.

[34] Cannon, R. L., Dave, J. V., & Bezdek, J. C. (1986). Efficient implementation of the fuzzy c-means clustering algorithms. IEEE Transactions on Pattern Analysis and Machine Intelligence, PAMI-8 (2), 248–255. http://doi.org/10.1109/ TPAMI.1986. 4767778.

[35] C. Hwang and F. Rhee, "Uncertain fuzzy clustering: interval type-2 fuzzy approach to C-means," IEEE Transactions on Fuzzy Systems, vol. 15, pp. 107–120, 2007. (Pubitemid 46444307)

[36] F. Rhee, "Uncertain fuzzy clustering: insights and recommendations," IEEE Computational Intelligence Magazine, vol. 2, pp. 44–56, 2007. (Pubitemid 46718757)

[37] D. Neog, M. Raza and F. Rhee., "An interval type 2 fuzzy approach to multilevel image segmentation," in proc. 20th IEEE Int. Conf. Fuzzy Syst. (FUZZ-IEEE2011), pp. 1164–1170, Taipei, Taiwan, June 27–30, 2011.

[38] C. Hwang and F. Rhee, "An interval type-2 fuzzy C spherical shells algorithm," in proc. IEEE Int. Conf. Fuzzy Syst. (FUZZ-IEEE2002), vol. 2 pp. 1117–1122, 2004. (Pubitemid 40028054)

[39] F. Rhee and H. Cheul, "A type-2 fuzzy C-means clustering algorithm," in Proc. 2001 Joint Conf. IFSA/NAFIPS, pp. 1926–1929, July 2001.

[40] F. Rhee and C. Hwang, "An interval type-2 fuzzy K-nearest neighbor," in Proc. 2003 Int. conf. Fuzzy Syst., vol. 2, pp. 802–807, May 2003.

[41] F. Rhee and C. Hwang, "An interval type-2 fuzzy perceptron," in Proc.2002 Int. Conf. Fuzzy Syst., vol. 2, pp. 1331–1335, May 2001. (Pubitemid 35466316)

[42] J. Min, E. Shim, and F. Rhee, "An interval type-2 fuzzy PCM algorithm for pattern recognition," in proc. IEEE Int. Conf. Fuzzy Syst. (FUZZ-IEEE2009), pp. 480–483.

[43] Y. Long, L. Yue, and X. Shixiong, "Robust interval type-2 possibilistic C-means clustering and its application for fuzzy modeling," in Sixth Int. Conf. on Fuzzy Systems and Knowledge Discovery, pp. 360–365, 2009.

[44] W. Pedrycz, "Interpretation of clusters in the framework of shadowed sets," Pattern Recog. Lett., vol. 26, no. 15, pp. 2439–2449, Nov. 2005.

[45] Mendel, J. M. (2001). Uncertain rule-based fuzzy logic systems: Intro- duction and new directions. Prentice Hall PTR Retrieved from https: // www.pearsonhighered.com/ program/ Mendel-Uncertain- Rule- Based- Fuzzy- Logic-Systems- Introduction- and- New-Directions/PGM139804.html.

[46] Mendel, J. M. (2004). Computing Derivatives in interval type-2 fuzzy logic systems. IEEE Transactions on Fuzzy Systems, 12 (1), 84–98. http://doi.org/10.1109/TFUZZ. 2003.822681.

[47] O. Linda and M. Manic, "Interval type-2 fuzzy voter design for fault tolerant systems," Inf. Sci., vol. 181, no. 14, pp. 2933–2950, Jul. 2011.

[48] O. Linda and M. Manic, "Interval type-2 fuzzy voter design for fault tolerant systems," Inf. Sci., vol. 181, no. 14, pp. 2933–2950, Jul. 2011.

[49] General Type-2 Fuzzy C-Means Algorithm for Uncertain Fuzzy Clustering

[50] F. Rhee and C. Hwang, "A type-2 fuzzy C-means clustering algorithm," in Proc. Joint Conf. Int. Fuzzy Syst. Assoc./ North Am. Fuzzy Inf. Process. Soc., Jul., 2001, pp. 1926–1919.

[51] L. A. Zadeh, "The concept of a linguistic variable and its approximate reasoning-II," Inf. Sci., vol. 8, pp. 301–357, 1975.

[52] RUBIO, Elid; CASTILLO, Oscar. Interval type-2 fuzzy clustering for membership function generation. In: 2013 IEEE Workshop on Hybrid Intelligent Models and Applications (HIMA). IEEE, 2013. p. 13–18.

[53] N. Karnik and J. M. Mendel, "Centroid of a type-2 fuzzy set," Inf. Sci., vol. 132, pp. 195–220, 2001.

[54] O. Linda and M. Manic, "Uncertainty-robust design of interval type-2 fuzzy logic controller for delta parallel robot," IEEE Trans. Ind. Inf., vol. 7, no. 4, pp. 661–671, Nov. 2011.

[55] N. Karnik and J. M. Mendel, "Centroid of a type-2 fuzzy set," Inf. Sci., vol. 132, pp. 195–220, 2001.

[56] H. B. Mitchell, "Pattern recognition using type-II fuzzy sets," Inf. Sci., vol. 170, no. 2–4, pp. 409–418, Feb. 2005.

[57] Zhang, D., Zhang, D., & Chen, S. (2003). Kernel-based fuzzy and possibilistic c-means clustering. In International conference on artificial neural networks (ICANN03) (pp. 122–125). Retrieved from http://citeseerx.ist.psu.edu/viewdoc/summary? doi= 10.1.1.491.540.

[58] J. M. Mendel, "Advances in type-2 fuzzy sets and systems," Inf. Sci., vol. 177, pp. 84–110, 2007.

[59] J. M. Mendel, R. John, and F. Liu, "Interval type-2 fuzzy logic systems made simple," IEEE Trans. Fuzzy Syst., vol. 14, no. 6, pp. 808–821, Dec. 2006.

Section 4

Adaptive Neural Fuzzy Inference Systems

Chapter 8

Compensatory Adaptive Neural Fuzzy Inference System

Rabah Mellah, Hocine Khati, Hand Talem and Said Guermah

Abstract

The traditional approach to fuzzy design is based on knowledge acquired by expert operators formulated into rules. However, operators may not be able to translate their knowledge and experience into a fuzzy logic controller. In addition, most adaptive fuzzy controllers present difficulties in determining appropriate fuzzy rules and appropriate membership functions. This chapter presents adaptive neural-fuzzy controller equipped with compensatory fuzzy control in order to adjust membership functions, and as well to optimize the adaptive reasoning by using a compensatory learning algorithm. An analysis of stability and transparency based on a passivity framework is carried out. The resulting controllers are implemented on a two degree of freedom robotic system. The simulation results obtained show a fairly high accuracy in terms of position and velocity tracking, what highlights the effectiveness of the proposed controllers.

Keywords: control, fuzzy logic, neural-fuzzy, compensatory fuzzy, Kalman filter, manipulator robot

1. Introduction

The advantage of fuzzy control is that the fuzzy system can model any continuous (sufficiently smooth) nonlinear function in a compact set and the modeling error decreases [1]. Fuzzy logic resembles human analysis in its use of inaccurate information to create decisions. Many such problems can be formulated as the minimization of functional defined over a class of admissible domains [2]. However, the difficulty in deploying fuzzy clustering strategies along with the high calculating cost and without update the parameters were their disadvantage [3].

On the other hand, a major concern of researchers is turned towards the combination of fuzzy logic and neural network. In this combination, a fuzzy reasoning is followed within multilayered hierarchical neural network. The parameters are represented by connection weights or involved in unit functions. They are learned the actual data [4]. In the near past, ANFIS (Adaptive Neural Fuzzy Inference System) models have become very popular for two reasons: the first reason is that in calibrating of non-linear relationships they offer more advantages over conventional modeling techniques, namely the ability to handle large amounts of noisy data from dynamic and non-linear systems, particularly when the underlying physical relationships are not fully understood. The second reason is that they facilitate the solving of linear systems which include the interpolation modeling such as time series [5].

The reason why authors used ANFIS is that it not only includes the characteristics of both methods but also eliminates some disadvantages in case of their lonely use [6]. Unfortunately, conventional neural fuzzy systems can only optimize the fuzzy membership functions under specially defined fuzzy operators which are unchangeable forever, which makes it use the local optimization technique rather than the global optimization technique [7–8]. Thus an adaptive neural fuzzy controller with compensatory fuzzy is most suitable in an environment where system dynamics change dramatically, become highly nonlinear, and in principle not fully known.

On the ground of these observations, several optimal and systematic methods have been developed for the design of neural fuzzy controllers with compensatory fuzzy. Among these methods, we have retained the compensatory adaptive neural fuzzy inference system approach which consists in adjusting not only fuzzy membership functions but also dynamically optimize the adaptive fuzzy reasoning. Besides that, ANFIS is a class of adaptive networks that are functionally equivalent to a first order Takagi-Sugeno fuzzy model.

Recently compensatory adaptive neural fuzzy inference system control has gained more attention from the control Community in general, as adaptive fuzzy systems are of crucial importance in several areas. The compensatory adaptive neural fuzzy inference system is preferred to deal with nonlinearities and complexity by working on data characterized by incompleteness and inaccuracy. Therefore, it offers powerful skills, such as adaptive adjustment, parallelism, tolerance error and generalization for the neural fuzzy controller. Thus the optimal methods are used to adjust and optimize the parameters of neural fuzzy controllers through an optimization algorithm in order to improve the control performance [9].

In this chapter, we will present and analyze in section 2 the structure of adaptive neural fuzzy inference system (ANFIS), based on concepts such as fuzzy logic, optimization techniques. This approach is carried out in order to remove a control constraint relating to the need to have a model as faithful as possible, knowing that the modeling errors and the imperfections of the models, contribute to significantly degrade the performance of the conventional control laws [10].

Section 3 presents the mathematical formalism appropriate to the compensatory neural fuzzy inference system controller proposed. The effectiveness of the proposed control is highlighted by some simulation results in Section 4. Finally this paper is concluded with a summary and an outlook to future research directions in Section 5.

2. Presentation of adaptive neural fuzzy inference system (ANFIS)

Jang was the first to present ANFIS as a multi-layer adaptive network-based fuzzy inference system [11]. One can compare this method to a fuzzy inference system besides that it uses back- propagation in minimizing the errors. The operation of a FIS is similar to that of both fuzzy logic (FL) and artificial neural networks (ANN). In both (ANN) and (FL), the input passes through the input layer (via the input membership function) and the output appears in output layer (via the output membership function). This type of advanced fuzzy logic uses neural networks. Hence, a learning algorithm can be used to change the parameters until an optimal solution is found. It follows that ANFIS uses either back-propagation or a combination of least squares estimation and back-propagation to estimate the membership function parameters [12]. Neural-Fuzzy system has newly known more attraction in research communities than other types of fuzzy expert systems. The reason of it combines the advantages of learning ability of neural network and reasoning ability of fuzzy logic to successfully solve many non-linear and complex real-world problems [13].

The regulator ANFIS is computationally very efficient, as it favors mathematical analysis, and works well with linear, adaptive, and optimization techniques. The fuzzy reasoning is performed with operators min and prod [14].

The conclusions of fuzzy rules are numeric values calculated from the inputs, so the final value is obtained by performing a weighted average of the conclusions [15, 16].

To simplify understanding and without loss of generalities, let us consider a fuzzy regulator with two inputs e_1 and e_2 and one output u. The entry x_1 is associated with two fuzzy sets A_1 and A_2. As for the two fuzzy sets associated with the second entry x_2 are B_1 and B_2. The output u is modeled by a fuzzy Sugeno-type system, composed of the following four rules [17]:

$$\text{Rule 1}: if\ x_1\ is\ A_1\ and\ x_2\ is\ B_1\ then\ u_1 = f_1(x_1, x_2) = a_1x_1 + b_1x_2 + c_1 \tag{1}$$

$$\text{Rule 2}: if\ x_1\ is\ A_1\ and\ x_2\ is\ B_2\ then\ u_2 = f_2(x_1, x_2) = a_2x_1 + b_2x_2 + c_2 \tag{2}$$

$$\text{Rule 3}: if\ x_1\ is\ A_2\ and\ x_2\ is\ B_1\ then\ u_3 = f_3(x_1, x_2) = a_3x_1 + b_3x_2 + c_3 \tag{3}$$

$$\text{Rule 4}: if\ x_1\ is\ A_2\ and\ x_2\ is\ B_2\ then\ u_4 = f_4(x_1, x_2) = a_4x_1 + b_4x_2 + c_4 \tag{4}$$

Let us denote $O_{k,i}$ the node in the *ith* position of the *kth* layer. The node functions in the same layer are of the same function family as defined below.

The input layer is denoted Layer 1 and any node i in this layer is a square node with a node function that describes the membership function. Hence $O_{1,i}$ is the membership function of A_i, and it specifies the degree to which a given variable x satisfies its quantifier A_i. We select the membership function in such a way the maximum of which is equal to unity and the minimum equal to zero.

The structure of the regulator ANFIS is given by the following figure (**Figure 1**):

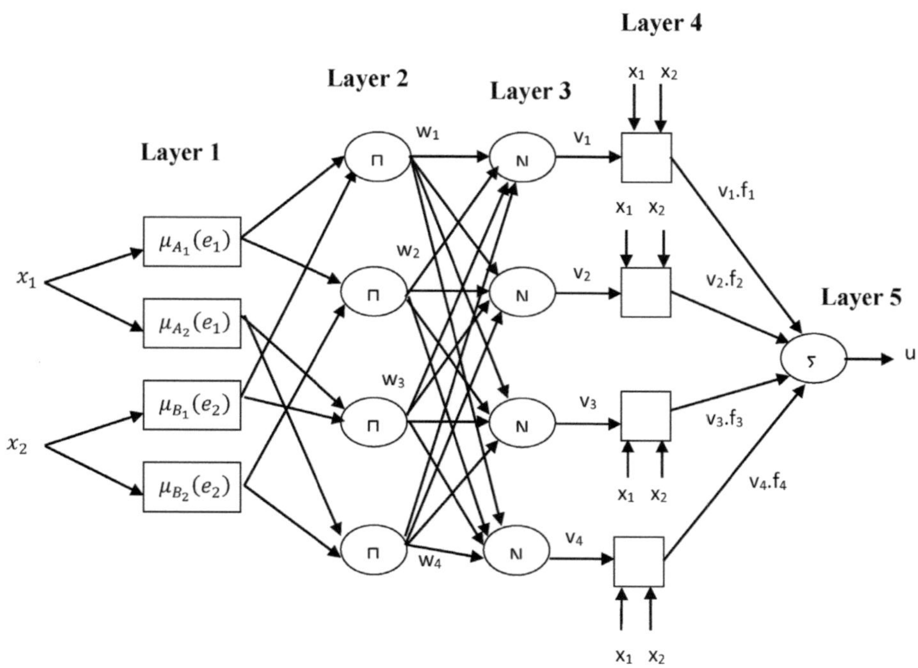

Figure 1.
Structure of the regulator ANFIS.

Through this structure we can see five layers described as follows:

Layer 1: The function of node at this layer is identical to the membership function in the fuzzification process:

Layer 2: generate the degree of activation of a rule.

$$\begin{cases} O_{2,1} = O_{1,1}.O_{1,3} = w_1 \\ O_{2,2} = O_{1,1}.O_{1,4} = w_2 \\ O_{2,3} = O_{1,2}.O_{1,3} = w_3 \\ O_{2,4} = O_{1,2}.O_{1,4} = w_4 \end{cases} \tag{5}$$

Layer 3: Each node of this layer is a circular node denoted by N. The output node represents the normalized activation degree according to the *i*th rule.

$$O_{3,i} = \frac{w_i}{\sum_{k=1}^{4} w_k} \text{ for } i = 1, 4 \tag{6}$$

Layer 4: Each node of this layer is a square node with a function described as follows:

$$O_{4,i} = O_{3,i} f_i = v_i(a_i e + b_i \Delta e + c_i) \tag{7}$$

where v_i is the output of the node I of layer 3 and $\{a_i, b_i, c_i\}$ is the set of update parameters.

Layer 5: In this layer, there is only one node that determines the overall output by using the following expression:

$$O_{5,i} = \sum_i O_{4,i} f_i \text{ for } i = 1..4 \tag{8}$$

Considering x_1 and x_2 are the position error e and its derivative Δe: $[x_1, x_2] = [e, \Delta e]$. We associate two fuzzy sets for each of the inputs x_1 and x_2 namely *N (Negative)* and *P (Positive)*. μ_N and μ_P represent the degrees of membership appropriate to variables x_i with respect to the fuzzy subsets A_i and B_i, defined by the following membership functions (**Figure 2**) [17]:

For $i = 1, 2$

$$\mu_N(x_i) = \begin{cases} 1, & \text{if } x_i < -1 \\ -0.5x_i + 0.5, & \text{if } -1 < x_i < 1 \\ 0, & \text{if } x_i > 1 \end{cases} \tag{9}$$

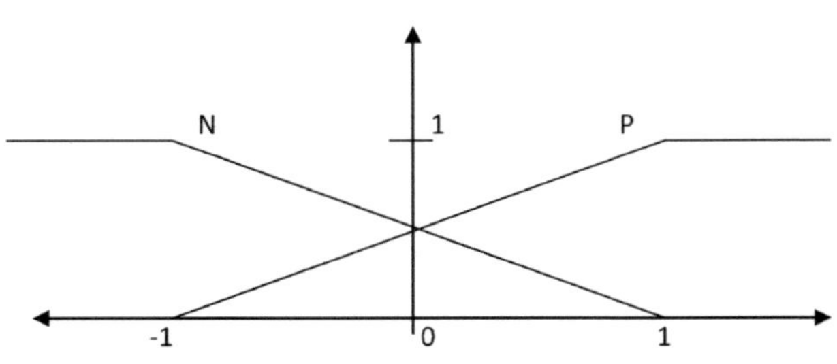

Figure 2.
Membership functions.

$$\mu_P(x_i) = \begin{cases} 0, & if\ x_i < -1 \\ 0.5x_i + 0.5, & if\ -1 < x_i < 1 \\ 1, & if\ x_i > 1 \end{cases} \tag{10}$$

2.1 Learning algorithm

The learning process consists of identifying the consequence parameters a_i, b_i and c_i for $i = 1,2, \ldots .4$. Thus, let us assume y_d and y are respectively the desired and actual outputs of system. In this work, we consider that the consequence parameters are adjusted by the minimization of the following objective function:

$$\mathrm{e(k)} = \frac{1}{2}(\mathrm{e})^2 \tag{11}$$

where $e = y_d - y$.

In addition, let be Φ_i the vector of parameters to be adjusted. Our objective is to find the parameters a_i, b_i and c_i of the vector Φ_i using the gradient descent method combined with the approach of extended Kalman filter. This is equivalent to writing:

$$\Phi_i(k+1) = \Phi_i(k) - \alpha(k)\frac{\partial J}{\partial \Phi_i} \tag{12}$$

We have:

$$\frac{\partial J}{\partial \Phi_i} = -e\frac{\partial y}{\partial \Phi_i} = -e\frac{\partial y}{\partial u}\frac{\partial u}{\partial \Phi_i} \tag{13}$$

From Eqs. (12) and (13), it follows:

$$\Phi_i(k+1) = \Phi_i(k) + \alpha(k)\frac{\partial y}{\partial u}\frac{\partial u}{\partial \Phi_i}e \tag{14}$$

In our case, $\frac{\partial y}{\partial u}$ cannot be evaluated, but can be estimated using the extended Kalman filter equations. Consequently, Eq. (14) can be written as:

$$\Phi_i(k+1) = \Phi_i(k) + K'\Psi_i e \tag{15}$$

where

$$K' = \alpha(k)\frac{\partial y}{\partial u} \tag{16}$$

$$\Psi_i = \frac{\partial u}{\partial \Phi_i} = \begin{bmatrix} \frac{\partial u}{\partial a_i} \\ \frac{\partial u}{\partial b_i} \\ \frac{\partial u}{\partial c_i} \end{bmatrix} \tag{17}$$

The Eq. (15) can be identified to extended Kalman filter equation:

$$\Phi_i(k+1) = \Phi_i(k) + K(k)e \tag{18}$$

Where $K(k)$ is the Kalman gain defined as follows:

$$K(k) = \frac{P(k)H^T(k)}{H(k)P(k)H^T(k) + R(k)} \tag{19}$$

Where $H(k)$ is the Jacobian matrix (observation matrix of the system); $P(k)$ is the covariance estimation matrix of the error and is the covariance matrix of the process noise.

Taking $H(k) = (\Psi_i)^T, P(k) = \lambda_1$ and $R(k) = \lambda_2$, the gain $K(k)$ can be written:

$$K(k) = \frac{\lambda_1}{(\Psi_i)^T \lambda_1 (\Psi_i) + \lambda_2}(\Psi_i) = \frac{\lambda_1}{\lambda_1 (\Psi_i)^T (\Psi_i) + \lambda_2}(\Psi_i) \tag{20}$$

Hence Eq. (18) reduces:

$$\Phi_i(k+1) = \Phi_i(k) + \frac{\lambda_1}{\lambda_1 (\Psi_i)^T (\Psi_i) + \lambda_2}(\Psi_i)e \tag{21}$$

By identification between Eqs. (15) and (21), we have:

$$K' = \frac{\lambda_1}{\lambda_1 (\Psi_i)^T (\Psi_i) + \lambda_2} \tag{22}$$

Finally, the vector of consequence parameters Φ_i can be adjusted by the following relation:

$$\Phi_i(k+1) = \Phi_i(k) + K'(\Psi_i)e \tag{23}$$

2.2 Stability analysis of the control system

From Eq. (23), we can consider for a very short time T_e, this relation:

$$\dot{\Phi}_i = \frac{\Phi_i(k+1) - \Phi_i(k)}{T_e} = \frac{K'(\Psi_i)e}{T_e} = K_1(\Psi_i)e \tag{24}$$

Where $K_1 = \frac{K'}{T_e}$

Hence:

$$\dot{\Phi}_i = K_1(\Psi_i)e_u \tag{25}$$

Where $e_u = K_1 e$ is the error between the controller's desired output u_d and actual output u.

Let be $\tilde{\Phi}_i = \Phi_{id} - \Phi_i$, where Φ_i is the vector of the consequence parameters and Φ_{id} the vector of the desired consequence parameters.

$$\dot{\tilde{\Phi}}_i = \dot{\Phi}_{id} - \dot{\Phi}_i = -(\Psi_i)e_u \tag{26}$$

For linear variation, the error e_u is defined by [18]:

$$e_u = u_d - u = \sum_{i=1}^{4}\left((\Psi_i)^T \Phi_{id} - (\Psi_i)^T \Phi_i\right) = \sum_{i=1}^{4}\left((\Psi_i)^T (\Phi_{id} - \Phi_i)\right) = \sum_{i=1}^{4}\left((\Psi_i)^T \tilde{\Phi}_i\right) \tag{27}$$

Consider the following Lyapunov function [19–21]:

$$V = \frac{1}{2}\sum_{i=1}^{4}\left(\left(\tilde{\Phi}_i\right)^T\left(\tilde{\Phi}_i\right)\right) \tag{28}$$

Differentiating V with respect to time yields, we obtain [17]:

$$\dot{V} = \sum_{i=1}^{4}\left(\left(\dot{\tilde{\Phi}}_i\right)^T\left(\tilde{\Phi}_i\right)\right) \tag{29}$$

From Eqs. (26), (27) and (29), we obtain:

$$\dot{V} = \sum_{i=1}^{4}\left((-(\Psi_i)e_u)^T\left(\tilde{\Phi}_i\right)\right) = -(e_u)^T\sum_{i=1}^{4}\left((\Psi_i)^T\left(\tilde{\Phi}_i\right)\right) = -(e_u)^T(e_u) \tag{30}$$

Consequently, from Eq. (30), we find that $\dot{V} \leq 0$, so we conclude that the system is asymptotically stable in the sense of Lyapunov according to the LaSalle theorem.

3. Compensatory adaptive neural fuzzy inference system (CANFIS)

The other class of inference systems that can deal this type of analytic information in conclusion of rules inference was proposed by Sugeno and his staff.

As for our contribution here, it consists in adding a compensatory fuzzy part to adjust consequence parameters and as well to dynamically optimize the adaptive fuzzy reasoning. In addition to this, ANFIS represents a class of adaptive networks that are functionally equivalent to a first order Takagi-Sugeno fuzzy model. As performed above, by taking a center-average deffuzzifier mapping, the crisp value of the output u is given as:

$$u = \frac{\sum_{i=1}^{4}(a_i e + b_i \Delta e + c_i)w_i}{\sum_{i=1}^{4} w_i} \tag{31}$$

We consider the pessimistic and optimistic operation given respectively as follows:

$$z_i = w_i \tag{32}$$

$$m_i = [w_i]^{\frac{1}{2}} \tag{33}$$

By using these two operations, our contribution is to add the compensatory form formulated as [7]:

$$C_i\left(z_i, q_i, \gamma_i\right) = (z_i)^{1-\gamma_i}(m_i)^{\gamma_i} \tag{34}$$

Where $\gamma_i \epsilon$ $[0 \;\; 1]$ is compensatory degree. Finally, the crisp value of the compensatory neural-fuzzy inference is derived as [13, 14]:

$$u = \frac{\sum_{i=1}^{4}(a_i e + b_i \Delta e + c_i)[w_i]^{1-\frac{\gamma_i}{2}}}{\sum_{i=1}^{4}[w_i]^{1-\frac{\gamma_i}{2}}} \tag{35}$$

For simplicity, we define:

$$\alpha_i = 1 - \frac{\gamma_i}{2} \tag{36}$$

Then we have:

$$u = \frac{\sum_{i=1}^{4}(a_i e + b_i \Delta e + c_i)[w_i]^{\alpha_i}}{\sum_{i=1}^{4}[w_i]^{\alpha_i}} \tag{37}$$

The structure of the CANFIS controller with compensatory fuzzy for two input and one output, is shown by **Figure 3** [9].

3.1 Learning algorithm

Consider as for ANFIS two dimensional data vectors, $x = [e, \Delta e]$ and one dimensional output data vector u_2. In order to limit the computation time, we have optimally adjusted the consequence parameters and compensatory degree by minimizing the following objective function:

$$J = \frac{1}{2}(y_d - y)^2 \tag{38}$$

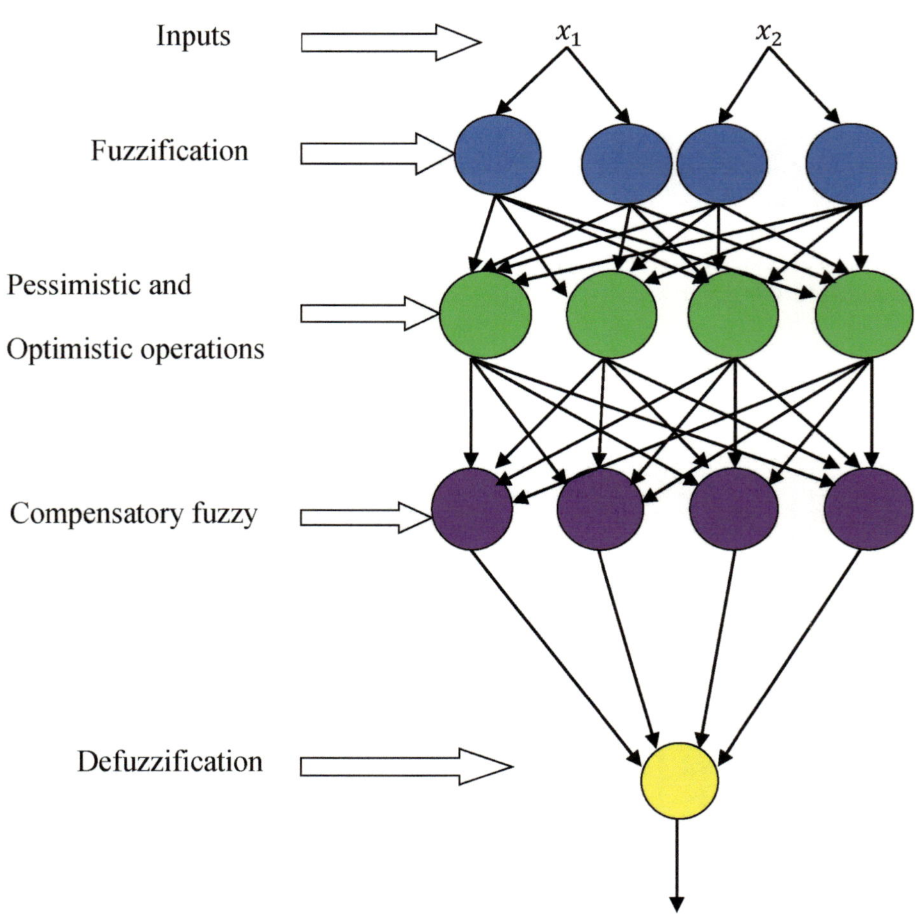

Figure 3.
Structure of CANFIS controller.

Where y_d and y are respectively desired and actual values of output system. Now let Φ_{2i} for i = 1….4, be the vector of update parameters. We aim to determine vector Φ_{2i} through the extended Kalman filter which consists in linearizing the output around the control input at each sampling period. This is equivalent to writing [16, 19, 20]:

$$\frac{\partial J}{\partial \Phi_{2i}} = \frac{\partial J}{\partial u_2} = -\left(y_d - y\right)\frac{\partial y}{\partial u_2}\frac{\partial u_2}{\partial \Phi_{2i}} = -K'.\Psi_{2i}e \tag{39}$$

In which

$$\Psi_{2i} = \frac{\partial u_2}{\partial \Phi_{2i}} \tag{40}$$

$$K_2 = \frac{\lambda_1}{\lambda_1 \Psi_{2i}^T \Psi_{2i} + \lambda_2} \tag{41}$$

Where λ_1 and λ_2 are adaptation gains for varying the convergence rate. Further, to eliminate the constraint $\gamma_i \in [0, 1]$, we redefine γ_i as follows [7]:

$$\frac{\left(p_i\right)^2}{\left(p_i\right)^2 + \left(r_i\right)^2} \tag{42}$$

Where p_i and r_i are update parameters such that $\gamma_i \in [0, 1]$. Consequently, the vector of update parameters is given as $(\Phi_{2i})^T = \left[a_i, b_i, c_i, p_i, r_i\right]$ for CANFIS. According to the definition, we have [17]:

$$\frac{\partial u_2}{\partial a_i} = \frac{e\left[w_i\right]^{\alpha_i}}{\sum_{i=1}^{4}\left[w_i\right]^{\alpha_i}} \tag{43}$$

$$\frac{\partial u_2}{\partial b_i} = \frac{\Delta e\left[w_i\right]^{\alpha_i}}{\sum_{i=1}^{4}\left[w_i\right]^{\alpha_i}} \tag{44}$$

$$\frac{\partial u_2}{\partial c_i} = \frac{\left[w_i\right]^{\alpha_i}}{\sum_{i=1}^{4}\left[w_i\right]^{\alpha_i}} \tag{45}$$

$$\frac{\partial u_2}{\partial \gamma_i} = -\frac{1}{2}\left[\sum_{i=1}^{4}(a_i e + b_i \Delta e + c_i)\right]\frac{z_i\, ln\,(w_i)}{\sum_{i=1}^{4} z_i} \tag{46}$$

$$\frac{\partial u_2}{\partial p_i} = -\left\{\frac{2p_i\left(q_i\right)^2}{\left(p_i\right)^2 + \left(q_i\right)^2}\right\}\frac{\partial u_2}{\partial \gamma_i} \tag{47}$$

$$\frac{\partial u_2}{\partial r_i} = \left\{\frac{2q_i\left(p_i\right)^2}{\left(p_i\right)^2 + \left(r_i\right)^2}\right\}\frac{\partial u_2}{\partial \gamma_i} \tag{48}$$

Finally, the vector of parameters Φ_{2i} is adjusted using the following equation:

$$\Phi_{2i}(k+1) = \Phi_{2i}(k) + K'\Psi_{2i}e \tag{49}$$

Where $(\Psi_{2i})^T = \left[\frac{\partial u_2}{\partial a_i}, \frac{\partial u_2}{\partial b_i}, \frac{\partial u_2}{\partial c_i}, \frac{\partial u_2}{\partial p_i}, \frac{\partial u_2}{\partial r_i}\right]$.

3.2 Stability analysis of the control system

From Eq. (48), we can consider for a very short time T_e, this relation:

$$\dot{\Phi}_{2i} = \frac{\Phi_{2i}(k+1) - \Phi_{2i}(k)}{T_e} = \frac{K'(\Psi_{2i})e}{T_e} = K_1(\Psi_{2i})e \tag{50}$$

Where $K_1 = \frac{K'}{T_e}$

Hence:

$$\dot{\Phi}_{2i} = K_1(\Psi_{2i})e_u \tag{51}$$

Where $e_u = K_1 e$ is the error between the controller's desired output u_d and actual output u.

Let be $\tilde{\Phi}_{2i} = \Phi_{2id} - \Phi_{2i}$, where Φ_{2i} is the vector of the consequence parameters and Φ_{2id} the vector of the desired consequence parameters.

$$\dot{\tilde{\Phi}}_{2i} = \dot{\Phi}_{2id} - \dot{\Phi}_{2i} = -(\Psi_{2i})e_u \tag{52}$$

For linear variation, the error e_u is defined by:

$$e_u = u_d - u = \sum_{i=1}^{4}\left((\Psi_{2i})^T\Phi_{2id} - (\Psi_{2i})^T\Phi_{2i}\right)$$

$$= \sum_{i=1}^{4}\left((\Psi_{2i})^T(\Phi_{2id} - \Phi_{2i})\right) = \sum_{i=1}^{4}\left((\Psi_{2i})^T\tilde{\Phi}_{2i}\right) \tag{53}$$

Consider the following Lyapunov function:

$$V = \frac{1}{2}\sum_{i=1}^{4}\left((\tilde{\Phi}_{2i})^T(\tilde{\Phi}_{2i})\right) \tag{54}$$

Differentiating V with respect to time yields, we obtain:

$$\dot{V} = \sum_{i=1}^{4}\left(\left(\dot{\tilde{\Phi}}_{2i}\right)^T(\tilde{\Phi}_{2i})\right) \tag{55}$$

From Eqs. (51), (52) and (54), we obtain:

$$\dot{V} = \sum_{i=1}^{4}\left((-(\Psi_{2i})e_u)^T(\tilde{\Phi}_{2i})\right) = -(e_u)^T\sum_{i=1}^{4}\left((\Psi_{2i})^T(\tilde{\Phi}_{2i})\right) = -(e_u)^T(e_u) \tag{56}$$

Consequently, from Eq. (55), we find that $\dot{V} \leq 0$, so we conclude that the system is asymptotically stable in the sense of Lyapunov according to the LaSalle theorem.

4. Simulation results and interpretation

We applied in simulation the neural-fuzzy command equipped with a compensator explained above, to the two-joint robot in a performance environment described by the following joint trajectories:

$$q_{ir} = \frac{\pi}{6}(1 - \cos(6t)) \tag{57}$$

With $i = 1 \ldots .2$.

The compact form of the dynamic model relating to the two-joint robot is given as follows:

$$\tau = \underbrace{\begin{bmatrix} M_{11}\, M_{12} \\ M_{21}\, M_{22} \end{bmatrix}}_{M(q)} \ddot{q} + \underbrace{\begin{bmatrix} C_{11}(q,\dot{q})\; C_{12}(q,\dot{q}) \\ C_{21}(q,\dot{q})\; C_{22}(q,\dot{q}) \end{bmatrix}}_{C(q,\dot{q})} \dot{q} + \underbrace{\begin{bmatrix} G_1(q) \\ G_2(q) \end{bmatrix}}_{G(q)} \tag{58}$$

Where.

$q = [q_1; q_2]$: Vector of joint position variables.
$\dot{q} = [\dot{q}_1; \dot{q}_2]$: Vector of joint velocity variables.
$\ddot{q} = [\ddot{q}_1; \ddot{q}_2]$: Vector of joint acceleration variables.
$\tau = [\tau_1; \tau_2]$: Vector of torques applied to joint.
$M(q)$: Inertial matrix.
$C(q, \dot{q})$: Matrix of terms centripetal and coriolis.
$G(q)$: Vector of gravitational effects

$$M_{11}(q) = m_1 l_{c1}^2 + m_2 l_1^2 + m_2 l_{c2}^2 + m_2 l_1 l_{c2} \cos(q_2) + I_1 + I_2$$
$$M_{12}(q) = M_{21}(q) = m_2 l_{c2}^2 + m_2 l_1 l_{c2} \cos(q_2) + I_2$$
$$M_{22}(q) = m_2 l_{c2}^2 + I_2$$
$$C_{11}(q,\dot{q}) = -m_2 l_1 l_{c2} \sin(q_2) \dot{q}_2$$
$$C_{12}(q,\dot{q}) = -m_2 l_1 l_{c2} \sin(q_2) [\dot{q}_1 + \dot{q}_2]$$
$$C_{21}(q,\dot{q}) = m_2 l_1 l_{c2} \sin(q_2)$$
$$C_{22}(q,\dot{q}) = 0$$
$$G_1(q) = [m_1 l_{c1} + m_2 l_1] g \sin(q_1) + m_2 g l_{c2} \sin(q_1 + q_2)$$
$$G_2(q) = m_2 g l_{c2} \sin(q_1 + q_2)$$

The parameters relating to the dynamic model of this robot are given in the following table (**Table 1**):

Parameter	Value
Mass of the link 1 (m_1)	6.5225 kg
Mass of the link 2 (m_2)	2.0458 kg
Link length 1 (l_1)	0.26 m
Link length 2 (l_2)	0.26 m
Gravity (g)	9.81 m/s^2
Distance to the center of mass of link 1 (l_{c1})	0.0983 m
Distance to the center of mass of link 2 (l_{c2})	0.0229 m
Moment of inertia of the center of mass m_1	0.1213 $kg.m^2$
Moment of inertia of the center of mass m_2	0.0116 $kg.m^2$

Table 1.
Robot parameters.

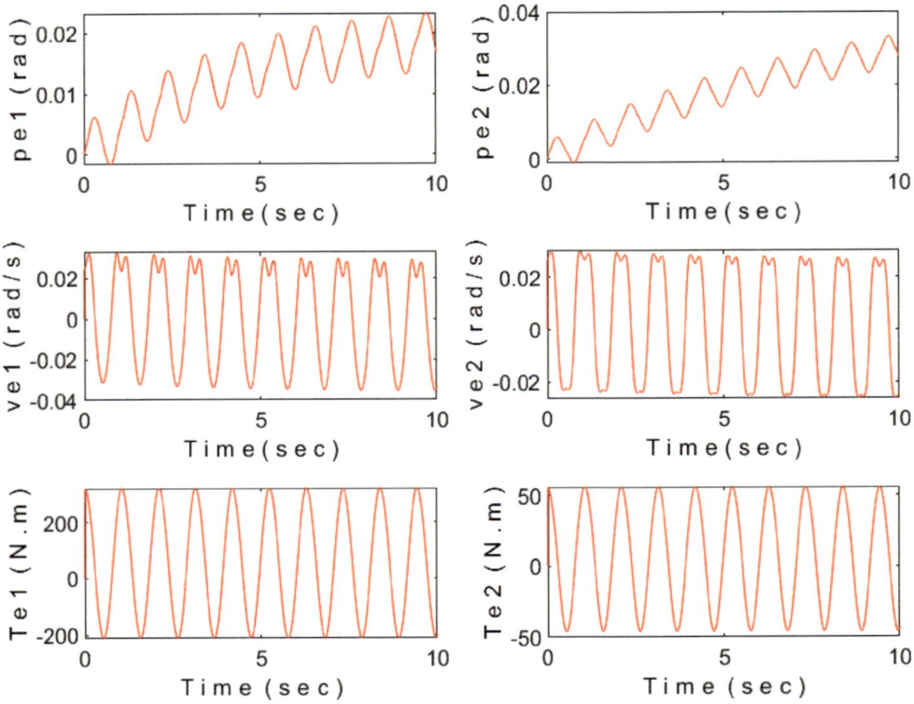

Figure 4.
Motion errors tracking and torques behavior with neural-fuzzy without disturbances.

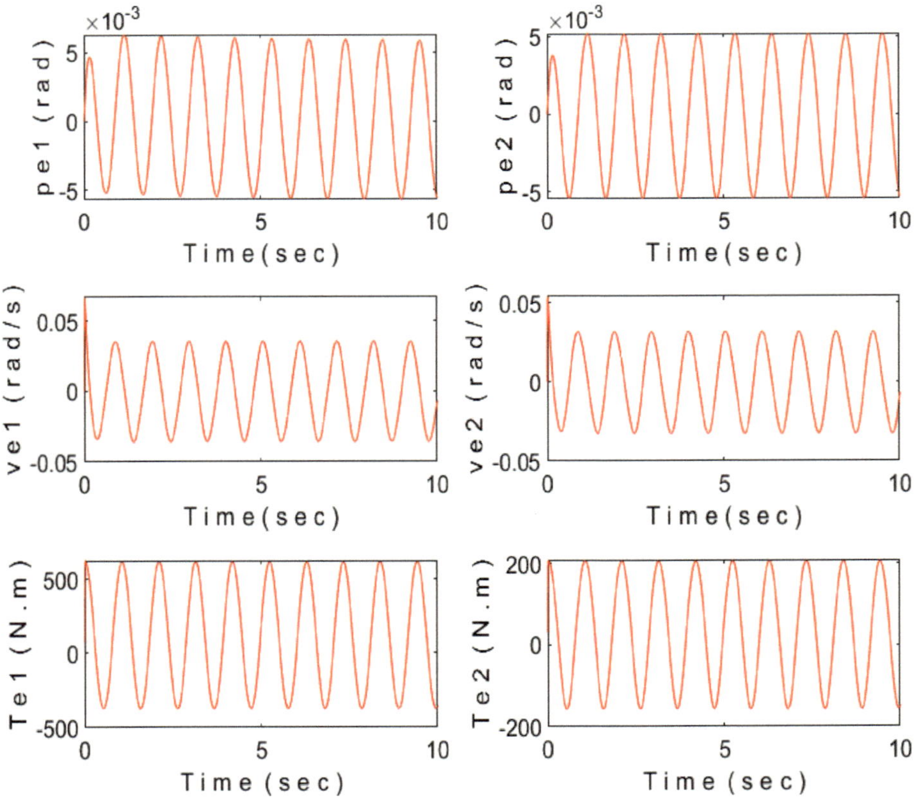

Figure 5.
Motion errors tracking and torques behavior with compensatory neural-fuzzy with without disturbances.

Figures 4 and **5** show the time evolution position and velocity errors and the torque applied to each joint of manipulator Robot with neuro-fuzzy controller and neuro-fuzzy controller respectively. Through these graphics, we can see that, the Neural –fuzzy controllers and compensatory Neural-fuzzy controllers provide a good tracking performance.

On the one hand, we can observe that the tracking errors are limited by low values, and the dynamics of the errors in position vary little compared to that of the errors in speed. This is physically explained by the fact that the position depends only on the environment while the velocity in addition to the environment depends on the Jacobean matrix. On the other hand, the command paths are smooth, which facilitates their implementation. This is achieved through the appropriate choice of parameters of the control structures.

In order to test the capacity of adaptation and robustness of the proposed approach, we have added in our simulation at time $t = 5s$ the combined friction and external torque disturbance for each joint, given as follows.

$$\tau_{fi} = 38.3\dot{q}_i + 18.9\cos(q_i) \tag{59}$$

The results obtained are illustrated by **Figures 6** and **7**, where we note that the tracking errors show peaks especially at the moment of the introduction of the disturbances, which are rejected quickly, by the Neural – fuzzy controllers and Compensatory Neural-fuzzy structure of the regulator, which allows to conclude that the tracking performance is very little affected by these disturbances. This is due to the low sensitivity to disturbance of the input data of the proposed control strategy.

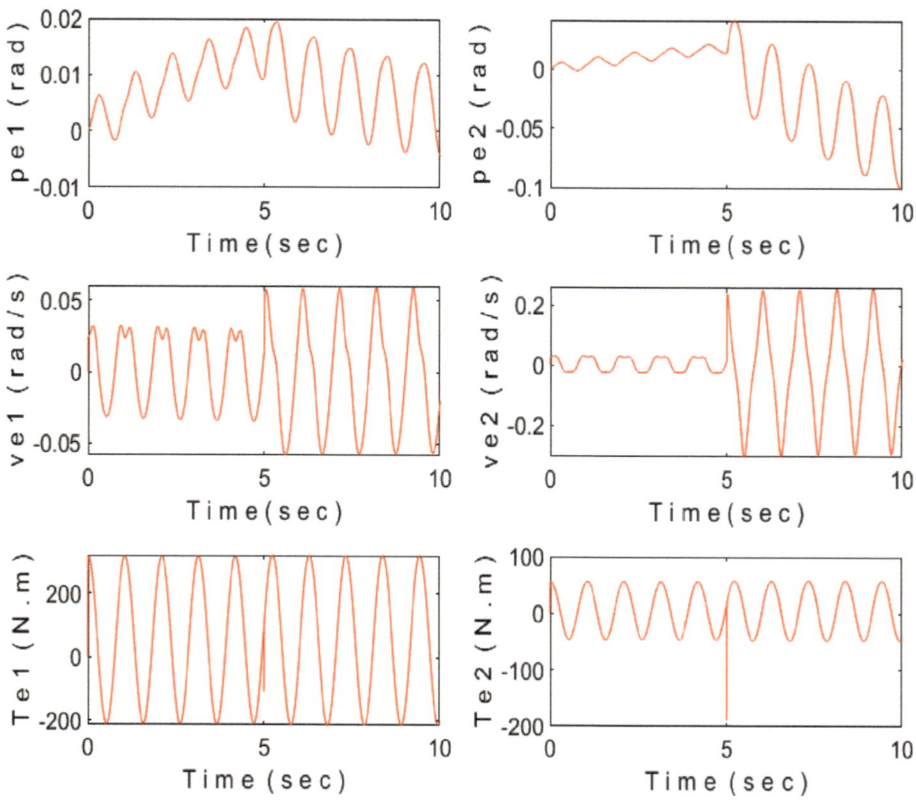

Figure 6.
Motion errors tracking and torques behavior with neural –fuzzy with disturbances.

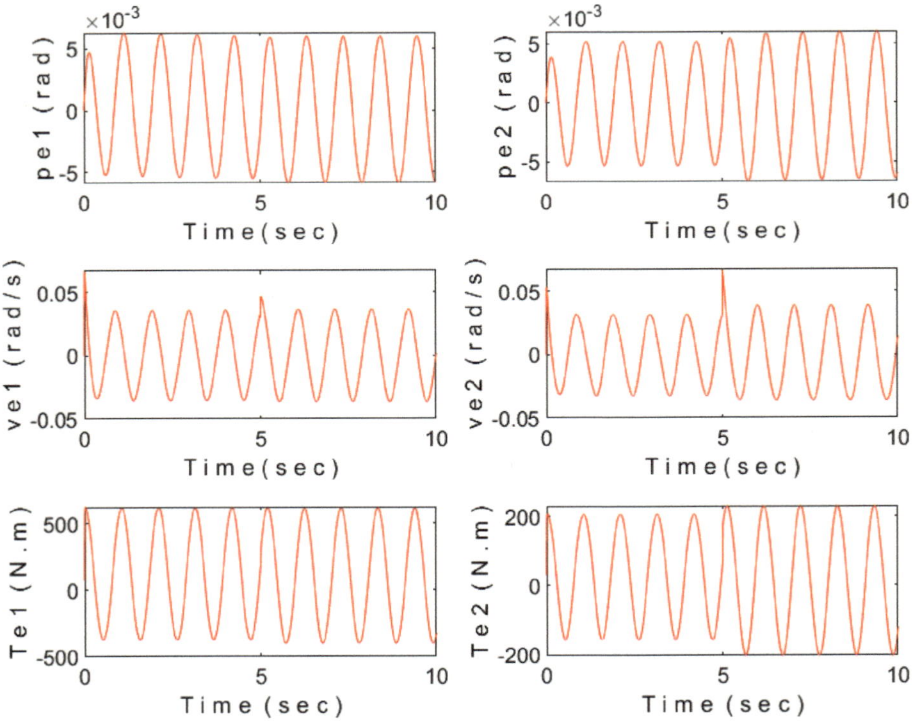

Figure 7.
Motion errors tracking and torques behavior with compensatory neural –fuzzy with disturbances.

5. Conclusion

In this chapter, we have proposed and presented a control strategy based on a neuro-fuzzy inference system regulator with a fuzzy compensator to control the manipulator robot with two joints.

It is important to note that the compensatory neural fuzzy inference system is more powerful than fuzzy systems or neural networks, since it can incorporate these advantages:

- Adaptive fuzzy reasoning method using fuzzy compensator can make the fuzzy system adaptive and efficient more.

- The neuro-fuzzy compensation system tolerates errors, because it is effective regardless of the choice of initial fuzzy rules (good or bad) for learning.

- The speed of convergence of the learning algorithm is faster than that of the back propagation algorithm.

- The learning algorithm not only adjusts membership functions, but also optimizes the dynamics of fuzzy reasoning by adjusting the degree of compensation. As a result, fuzzy neuro systems with a fuzzy compensator are more efficient than conventional neuro-fuzzy systems.

This control strategy has the advantage of requiring only measurements of output variables. The simulations indicate that a complete stabilization of the system is indeed observed.

In conclusion, we wanted, at the end of this chapter, to try to identify a broad-spectrum methodology that opens the field to a possible standardization of the design of control laws based on a neuro-fuzzy structure equipped with a compensator, in order to controlling a poorly understood and imprecise dynamic system. It would be interesting in the context of experimental tests to judge the performance of the methods proposed on real systems.

Author details

Rabah Mellah*, Hocine Khati, Hand Talem and Said Guermah
Design and Control Laboratory Automated Systems, Faculty of Electrical Engineering and Computer Science, UMMTO, Tizi-Ouzou, Algeria

*Address all correspondence to: mellah.rab@gmail.com

References

[1] JYH-SHING R., CHUEN-TSAI S., "*Neuro-Fuzzy Modeling and Control*" Procedings of the IEEE, Vol. 83, N^0 3, March 1995 pp. 378–404.

[2] S. Ghosh, M. Majumder. "Prediction of Wave Energy Potential in India: a Fuzzy–ANN Approach." Intech-2019, DOI: http://dx.doi.org/10.52772/intechopen.84676.

[3] S.D. Nguyen. "ANFIS: Establishing and Applying to Managing Online Damage." Intech-2019, DOI: http://dx.doi.org/10.5772/intechopen.83453.

[4] B. Ergün, C. Sahin and U. Kaplan. "ANFIS Definition of Focal Length for Zoom Lens Via Fuzzy Logic Functions." Intechopen-2017, DOI: http://dx.doi.org/10.5772/67823

[5] L. Moradi and R. Ehtehamrasi. "ANFIS to Quantify Maintenance Cost of IT Services in Telecommunication Company." Intechopen-2020, DOI: http://dx.doi.org/10.5772/intechopen.82827

[6] HAO Y. "*Constructing Nonlinear Variable Gain Controllers via the Takagi - Sugeno Fuzzy Control*." IEEE Transactions on Fuzzy Systems, Vol. 6, N°. 2, May 1998, pp. 226–234.

[7] YAN-QING., ABRAHIM K., "*Compensatory Neurofuzzy Systems with Fast Learning Algorithms*." IEEE Transactions On Neural Networks, Vol. 9, N^0 1, January 1998 pp. 83–105.

[8] BYUNG K., WOON C. H., "*Adapive Control of Robot Manipulator Using Fuzzy Compensator*." IEEE Transactions on Fuzzy Systems, Vol. 8, N°. 2, April 2000.

[9] R. Mellah, S. Guermah, and R. Toumi. "Adaptive control of bilateral teleoperation system with compensatory neural fuzzy controllers." International Journal of Control, Automation and Systems, vol. 15, no. 4, pp. 1949–1959, 2017.

[10] R. Mellah and R. Toumi, "Compensatory neuro-fuzzy control of bilateral teleoperation system," *Proc. of 20th Internationale Conference on Methods and Models in Automation an Robotic*, 24–27 August 2015, Miedzyzdroje, Poland, pp. 382–387.

[11] J. R. Jang. "ANFIS: adaptive-network based fuzzy inference system." IEEE Transactions on Systems, Man and Cybernetics N° 23, vol. 13 pp 665–685, 1993.

[12] S. R. Jang and C. T. Sun. "Neuro-fuzzy and Soft Computing: A Computational Approach to Learning and Machine Intelligence." Prentice Hall, Englewood Cliffs, NJ, 1997.

[13] M. N. M. Salleh, N. Talpur and K. H. Talpur. "A Modified Neuro-Fuzzy System Using Metaheuristic Approach for Data classification." Intechopen-2018, DOI: http://dx.doi.org/10.5772/intechopen.75575

[14] KARAKASOGLU A., SUNDARESHAN M., "A Recurrent Network-based Adaptive Variable structure Model- following Control of Robotic Manipulators." Automatica, Vol. 31, N0 10, pp. 1495–1507, 1995.

[15] YIH-GUANG L., WEI-YEN W., TSU-TIAN L., *"Robust Adaptive Fuzzy-Neural Controllers for Uncertain Nonlinear Systems"* IEEE Transactions on Robotics and Automation, Vol. 15, N°. 5 Oct 1999.

[16] JEAN B. M., "Robust Neuro-Fuzzy Sensor-Based Motion Control Among Dynamic Obstacles for Robot Manipulators." IEEE Transactions on Fuzzy Systems, Vol. 11, NO. 2, April 2003, pp. 249–261.

[17] MENG J., "*Robust Adaptive Control of Robot Manipulators Using Generalized Fuzzy Neural Networks*." IEEE on Industrial Electronics, Vol. 50, N°. 3, June 2003.

[18] R. Mellah and R. Toumi, "Control bilateral teleoperation by compensatory ANFIS," Advanced Mechatronics Solutions, vol. 393, pp. 167–172, 2016.

[19] LIMIN P., PENG-YUNG W., "*Neural-Fuzzy Control System for Robotic Manipulators*." IEEE Control Systems Magazine, February 2002. pp. 53–63

[20] H Khati, R Mellah, H Talem. "Ne uro-fuzzy Control of a Position-Position Teleoperation System Using FPGA." 24th International Conference on Methods and Models in Automation and Robotics, 26–29 August 2019, Miedzyzdroje, Poland, pp. 64–69.

[21] H Khati, H Talem, R Mellah, A Bilek. "Neuro-fuzzy control of bilateral teleoperation system using FPGA." Iranian Journal of Fuzzy Systems vol. 16, N°. 6, pp. 17–32, 2019.

Section 5

Process Monitoring

Chapter 9

Fuzzy Logic Expert System for Health Condition Assessment of Power Transformers

Teruvai Manoj and Chilaka Ranga

Abstract

In the present chapter, a new fuzzy logic (FL) model is proposed to evaluate the overall health index (OHI) of power transformers. The most significant attributes such as dissolved gases, acidity, 2-furfuraldehyde, water content, breakdown voltage and dissipation factor that influence the health condition of transformers solid and liquid insulations are considered. These attributes are further divided into three different sets. Based on these sets, three different sub fuzzy models i.e. F_1, F_2 and F_3 are designed in order to reduce the possible combinations of fuzzy rules. It results in reducing the complexity issues of the proposed OHI model. In addition, consideration of all significant testing parameters makes the model more reliable and accurate. Further, the proposed fuzzy model helps in initiating appropriate and early action on faulty conditions of the transformers. Conventional fuzzy logic models generally utilize large number of inputs and more number of rules in a single fuzzy model. It makes the models complex and inaccurate. Such shortcomings of existing conventional models are successfully overcame by the present proposed model. Furthermore, the results obtained from the proposed model are compared with the results obtained from expert model proposed by Abu-Elanien et al. This comparison ensures the reliability of the proposed method. Also, it is envisioned that the proposed model can be easily implemented by both the experienced and the inexperienced utility managers.

Keywords: Transformer, insulation, condition monitoring, fuzzy logic, membership function

1. Introduction

Power transformers are vital components of power system. The total service life of power transformers is majorly depends on the life spans of their liquid and solid insulations [1]. The deterioration of the insulation is caused due to the various electrical and thermal stresses present inside the transformers. These stresses accumulate several dissolved gases within the transformer oil, and produce partial discharge, overheating and arcing [2, 3]. Generally, the dissolved gases namely, hydrogen (H_2), acetylene (C_2H_2), ethane (C_2H_6), ethylene (C_2H_4), methane (CH_4), carbon monoxide (CO) and carbon dioxide (CO_2) are induced in the transformer oil. The severity of these gas concentrations identifies the type of faults present in the transformer [4, 5]. Moreover, these gas concentrations are employed to examine the overall health condition (HC) of the transformers. Generally, it is termed as health index (HI). Further, there are some significant attributes that influences the health of transformer solid and liquid insulations. Attributes such as dissipation factor, water content, acidity, breakdown voltage and dissolved gases decide the lifetime of oil insulation [5]. Similarly, solid insulation's life depends on 2-furfuraldehyde. During the operation of transformers, amount of these attributes increase, and lead to excessive deterioration of transformers. Further, it may cause failure of transformers resulting in a huge revenue loss to the customers. To avoid such problems, continuous condition monitoring of power transformers is essential. It is only possible if all the significant tests are performed frequently to obtain attribute values in regular time intervals [6].

Over the past few decades, various fuzzy logic (FL) models are developed by diagnostic experts to evaluate the health condition of transformers [5–7]. These models incorporated the various diagnostic attributes such as furan content, degree of polymerization (DP), dissipation factor (DF), acidity, water content (WC), breakdown voltage (BDV) and total dissolved combustible gases (TDCG) concentration. Using the dissolved gas concentrations, the thermal and electrical criticalities of oil and paper insulations are determined [8]. The recent fuzzy logic models reported in the literature have their own strengths in determining the health index, however, none of them has fully utilized all the significant attributes of the transformers, thereby remain with some backlogs. Hence, these FL models constrained with some limitations [9, 10].

In the present chapter, a new fuzzy logic (FL) model is proposed to determine the overall health index (OHI) of transformer. To validate the efficacy of this model, the test samples and results have been collected from Himachal Pradesh State Electricity Board (HPSEB). For an easy understanding, the present chapter is divided into different sections as follows. The brief explanation of transformer attributes is given in Section 2. Section 3 details about the present proposed fuzzy logic model. Finally, results of the model are discussed in the Section 4. And the complete chapter is concluded in Section 5.

2. Transformer diagnostic attributes

The major deterioration of transformer insulation is caused by attributes such as breakdown voltage, dissipation factor, water content, acidity, dissolved combustible gases and 2-furfuraldehyde. The brief introduction of these attributes is given in this section.

Breakdown voltage (BDV) is defined as voltage at which breakdown occurs between the two electrodes while oil is exposed to an electric field under critical conditions [11]. For insulation system of a transformer, electric strength is the basic parameter which indicates the presence of contaminants like perceptible sludge, moisture and sediment [12]. Dissipation factor (DF) is defined as the sine of loss angle. Also, it is important parameter to test the quality of insulation [13]. Some harmful contaminants such as oxidation products, water and de-polymerization of paper insulation are induced due to high value of DF [14].

The presence of dissolved water in the oil is termed as water content and it is expressed in parts per million (ppm) by weight [15]. The existence of moisture content in the oil is detrimental since it adversely affects the electrical characteristics of oil. Also, excess amount of moisture accelerates deterioration of insulating materials [16]. The measurement of free organic and inorganic acids accumulate in the transformer oil is defined as acidity, and is measured in milligrams of potassium hydroxide. It is required to neutralize the total free acid in one gram of oil [17].

When the influence of abnormal thermal and electrical stresses on transformer oil is not very high, the gases generated as a consequence of decomposition of insulating oil will get enough time to dissolve in the oil. In dissolved gas analysis, the percentage of gas concentrations present in oil is determined and analyzed [18]. These dissolved gas percentages helps in finding out the internal condition of transformer [19]. Solid dielectrics present in the two essential parts of transformer i.e. core and winding which is made of cellulose. Cellulose consists of long chain of molecule structure [20]. During the operation of transformer, these long chains are generally broken into several numbers of minute particles, as per the aging. These furan compounds belong to the fur-furaldehyde group. 2-Furfural is the most predominant among all furfurals compounds. The condition assessment and life estimation of paper insulation is done by using the rate of rise of furfural products with respect to time in oil. Damage in few grams of paper in oil is detected even for a large sized transformer. Therefore, fur-furaldehyde analysis is very sensitive [21]. When the transformer oil is soaked into solid dielectric, furan particles along with gases CO_2 and CO dissolved in the oil due to heat.

3. Proposed fuzzy logic model

Fuzzy logic is a very helpful tool in obtaining accurate output, and easy in implementation [22]. Also, it facilitates more effective and reasonable decision making for transformers in order to ensure maintainability and reliability [23, 24]. The purpose of the proposed FL model is to integrate various diagnostic test results with the experience of transformer diagnosis experts [23]. Different stages of the FL models are discussed in the following sub-sections. The curve which converts the precise (crisp) inputs in to imprecise fuzzy sets with degrees of membership function (DOM) in the range 0 and 1. Generally, membership functions (MFs) have different shapes namely, triangular, sigmoidal or trapezoidal, Gaussian and Gauss2. Trapezoidal MF is widely used MF because of its simplicity [20, 21]. It is shown in **Figure 1**, and given by Eq. (1). The trapezoidal MF consists of a truncated triangular curve and a flat top.

$$Membership\,function = \max\left\{\min\left(\frac{x-a}{c_1-a}, 1, \frac{b-x}{b-c_2}\right), 0\right\} \tag{1}$$

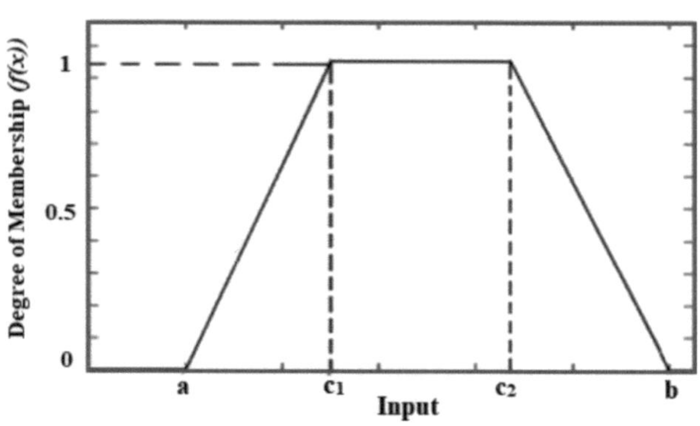

Figure 1.
The trapezoid shaped membership function.

Where, input variable is denoted as 'x', lower and upper limits are denoted as 'a' and 'b', respectively. Similarly, 'c_1' and 'c_2' are the centers of the trapezoidal MF [21]. If the input value lies in between the range of'c_1'and'c_2'of the MFs, then the corresponding MF attains the maximum DOM of unity [25]. Whereas, input values lie between the range of 'a'and'c_1'and between the range of'c_1'and'b', will have DOMs less than unity. Likewise, all the crisp input values (precise) are converted into fuzzy (imprecise) values in fuzzification stage. Where the fuzzy values range lie between 0 and 1.

In the present proposed model, fuzzy logic (FL) is used to determine the overall health index (OHI) of power transformers. The six parameters (Section 2) are considered as inputs in the present proposed model to determine the overall health index (OHI) of transformers. To make the model simple, three sub-fuzzy models viz. F_1, F_2 and F_3 are designed separately. Two parameters namely, water content and acidity are assigned as inputs for F_1, whereas BDV and DF are for F_2. Similarly, DCG and 2-FAL are considered as inputs for F_3. Furthermore, the outputs obtained from these three sub-models are considered as the inputs to a single fuzzy model called F_4. The final output obtained from the model F_4 is OHI of transformers. All the inputs of F_1, F_2 and F_3 used trapezoidal shaped MFs and their limits are assigned in accordance to [1]. These limits for MFs of water content input in F_1 are shown in **Figure 2**. Likewise, MFs are designed with trapezoidal shape for remaining sub models. The values of the six significant attributes have been listed in **Table 1**.

However, input 2-FAL in F_3 consists of 5 MFs as per [1]. These MFs are Very bad, Bad, High-moderate, Low-moderate and Good. The lower and upper limits of these MFs, and their centers are [0 0 0.2 0.2], [0.2 0.2 1 1.5], [1 1.5 3 3.5], [3 3.5 6 7.5] and [6 7.5 10 10] respectively. In case of F_4, the output MFs used in each of the three sub-models was used as input MFs. The corresponding input MFs of water content are same as described in **Figure 2**.

The block diagram consisting of the three sub fuzzy models and a main fuzzy model for transformer sample 12 is depicted in **Figure 3**. The lower and the upper limits along with the two centers of input MFs are specified in **Table 2**. Similarly, the output for each of the four models F_1, F_2, F_3 and F_4 was divided in to four MFs as specified in **Figure 4**.

In fuzzification stage, the input values are converted in to fuzzy values by using the Eq. (1) [26]. Consider transformer sample 12, where the value of acidity is 0.23 mgKOH/g. It lies in the range of Bad (**Table 2**). Therefore, the limits i.e. a, c_1, c_2 and b

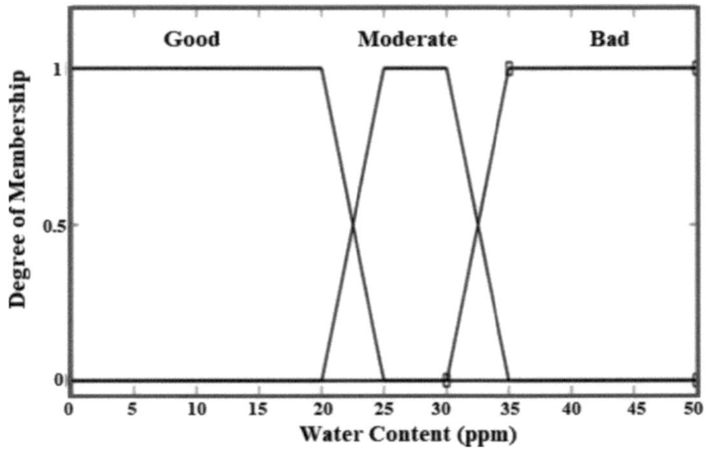

Figure 2.
Membership functions for water content input in F_1.

Sample Number	Acidity	DF	2-FAL	BDV	WC	DCG
1	0.07	0.15	0.52	73	15.4	38
2	0.04	0.18	0.31	64.5	19	8
3	0.02	0.06	1.24	27.9	27.9	501
4	0.14	0.19	7.45	36.5	14	51
5	0.03	0.08	0.85	29	21.3	489
6	0.07	0.66	15.5	29.7	31	32
7	0.04	0.15	0.22	53	13.6	77
8	0.09	0.36	0.21	39.5	27	194
9	0.09	0.89	0.61	56	26.1	292
10	0.06	0.21	0.57	37.2	26.3	25
11	0.07	0.13	5.35	31.4	25.8	321
12	0.23	0.43	5.54	47.7	21.8	215
13	0.13	0.19	9.34	26.6	15.3	76
14	0.06	0.25	0.13	61.5	19.4	61
15	0.17	0.26	0.74	70.5	16.1	147
16	0.11	0.22	0.34	43.8	23.5	33
17	0.08	0.22	0.65	67.2	13	28
18	0.41	0.27	6.62	55.2	17	51
19	0.02	0.12	0.01	73	8	127
20	0.17	0.22	8.56	22.7	15.2	38

Table 1.
Significant attribute values of the 20 test case transformers.

of acidity membership are 0.15, 0.2, 0.3 and 0.3. Substitute the values in Eq. (1) gives the fuzzified value as 1. Similarly, the imprecise values are calculated for all the samples and summarized in **Table 3**.

After fuzzification stage, the inputs are mapped with output by specially designed rules in the fuzzy inference stage. In the present work, a widely used Mamdani maximum–minimum fuzzy inference method is used [21, 25]. Using the fuzzified set of inputs and the designed fuzzy rules, the output is determined in this method. Further, the method truncates the output MF at its minimum DOM value. Initially the inputs are fuzzified using Eq. (1). Further, the truncated output from each of the three models are obtained based on the specially designed expert fuzzy rules and fuzzified inputs. In the present work, the fuzzy rules possible between the inputs of F_1 are designed consisting two inputs each with three MFs generate a total of nine combinations. Similar combinations are also obtained for F_2.

The rule base designed for sub fuzzy model, F_1 is given below:

Rule 1: If Water content is Good and Acidity is Good then output is Excellent.
Rule 3: If Water content is Good and Acidity is Bad then output is Poor.
Rule 6: If Water content is Moderate and Acidity is Bad then output is Worst.
Rule 9: If Water content is Bad and Acidity is Bad then output is Worst.

In case of F_3, the three input MFs in DCG, and five MFs in 2-FAL make a total of fifteen fuzzy rules.

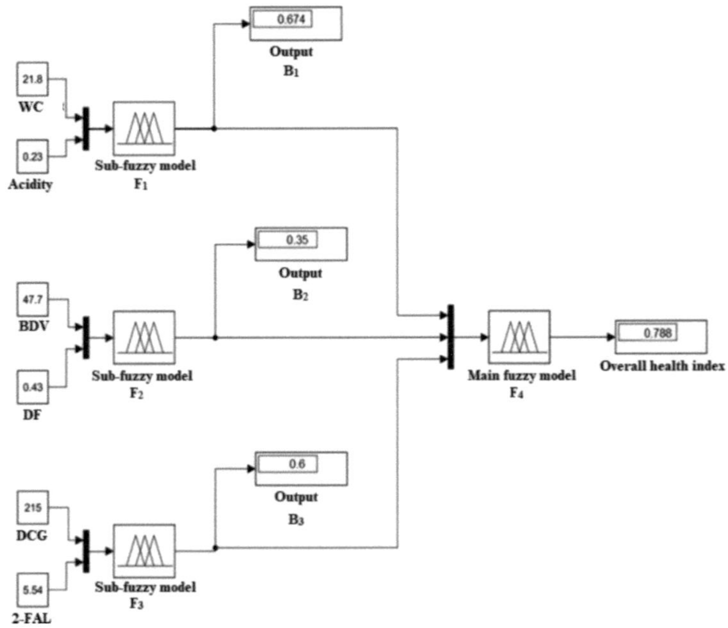

Figure 3.
Block diagram of the proposed fuzzy logic model.

Input MF ranges	Good				Moderate				Bad			
	a	c_1	c_2	b	a	c_1	c_2	b	a	c_1	c_2	b
Acidity	0	0	0.03	0.05	0.03	0.05	0.15	0.2	0.15	0.2	0.3	0.3
DF	0	0	0.05	0.1	0.05	0.1	0.8	1	0.8	1	1.5	1.5
BDV	52	53	75	75	23	24	53	54	0	0	23	24
DCG	0	0	300	400	300	400	1100	1400	1100	1400	2000	2000

Table 2.
Lower and upper limits for MFs of all inputs.

The rule base designed for sub fuzzy model F_3 is given as below:

Rule 1: If DCG is Good and 2-FAL is Very bad then Output is Excellent.
Rule 3: If DCG is Good and 2-FAL is Low-Moderate then Output is Poor.
Rule 6: If DCG is Moderate and 2-FAL is Bad then Output is Worst.
Rule 9: If DCG is Good and 2-FAL is High-Moderate then Output is Good.
Rule 12: If DCG is Moderate and 2-FAL is High-Moderate then Output is Good.
Rule 15: If DCG is Bad and 2-FAL is Good then Output is Worst.

Similarly, the possible combinations (sixty four rules) of input MFs in case of F_4 were generated. The final rule base for main fuzzy model (F_4) with all three inputs (B_1, B_2, B_3) is given below:

Rule 1: If B_1 is Excellent and B_2 is Excellent and B_3 is Excellent then Output is Excellent.
Rule 8: If B_1 is Good and B_2 is Excellent and B_3 is Worst then Output is Poor.
Rule 16: If B_1 is Good and B_2 is Poor and B_3 is Worst then Output is Poor.
Rule 24: If B_1 is Worst and B_2 is Excellent and B_3 is Worst then Output is Worst.

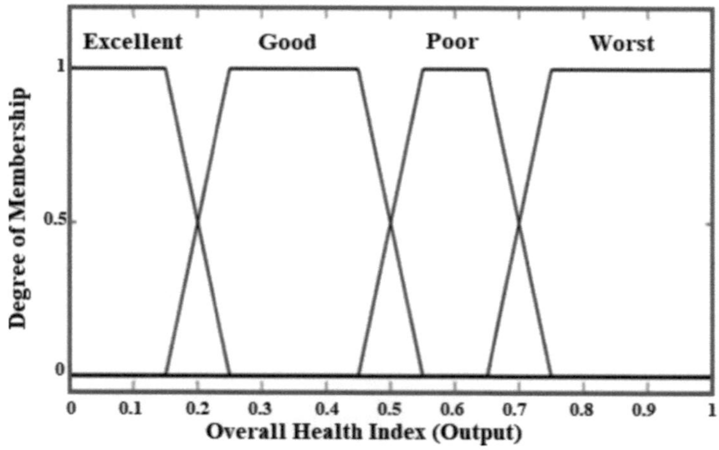

Figure 4.
Membership functions of output F_4.

Sample Number	Acidity	DF	2-FAL	BDV	WC	DCG
1	1	1	1	1	1	1
2	0.5	1	1	1	1	1
3	1	0.8	0.52	1	1	1
4	1	1	0.03	1	1	1
5	1	1	1	1	0.74	1
6	1	1	1	1	0.8	1
7	0.5	1	1	1	1	1
8	1	1	1	1	1	1
9	1	0.45	1	1	1	1
10	1	1	1	1	1	1
11	1	1	1	1	1	0.79
12	1	1	1	1	0.64	1
13	1	1	1	1	1	1
14	1	1	1	1	1	1
15	0.4	1	1	1	1	1
16	1	1	1	1	0.7	1
17	1	1	1	1	1	1
18	1	1	0.59	1	1	1
19	1	1	1	1	1	1
20	0.4	1	1	1	1	1

Table 3.
Fuzzified values (imprecise) of the attributes obtained in fuzzification stage.

Rule 32: If B_1 is Worst and B_2 is Poor and B_3 is Worst then Output is Worst.
Rule 40: If B_1 is Poor and B_2 is Excellent and B_3 is Worst then Output is Worst.
Rule 48: If B_1 is Poor and B_2 is Poor and B_3 is Poor then Output is Worst.
Rule 56: If B_1 is Excellent and B_2 is Good and B_3 is Worst then Output is Poor.
Rule 64: If B_1 is Excellent and B_2 is Worst and B_3 is Excellent then Output is Good.

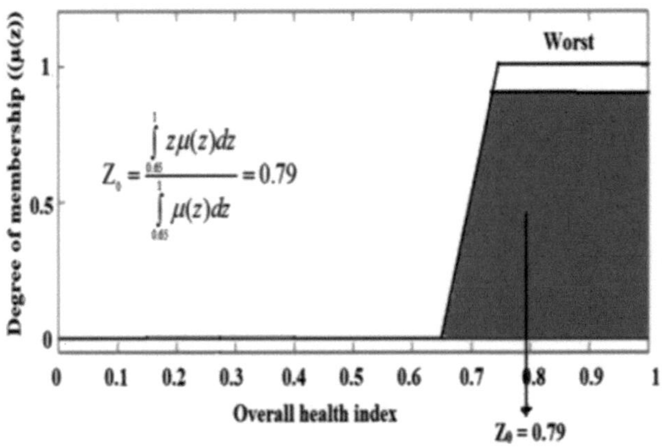

Figure 5.
Center of gravity method applied on transformer sample 12.

The rules are framed according to their severity level deteriorating transformer insulation. Since DCG and 2-FAL are very harmful attributes, the highest priority in determining OHI is given to B_3 (sub fuzzy model F_3). B_1 (sub fuzzy model F_1) has been considered as key factor next to B_3. And, least priority has been given to B_2 (sub fuzzy model F_2) among all the three inputs.

Defuzzification is the last stage of this method where a precise quantitative value from the truncated output MF is determined [27]. Center of gravity is the most popular and efficient defuzzification method [20]. This method is used in the present work. It determines the center of gravity or the centroid (Z_0) of the area bounded by the truncated output MFs [20, 21]. It is obtained by

$$Z_0 = \frac{\int z.\mu(z)dz}{\int \mu(z)dz} \tag{2}$$

Where the output variable is denoted by 'z' and 'μ(z)' is the DOM of the truncated output MF. The crisp output is obtained by using Eq. (2). The centroid representation of the output value (sample 12) is depicted in **Figure 5**. Similarly, the outputs for remaining samples are obtained using the above equation.

4. Results and discussion

For an easy understanding of proposed method, consider sample 12 and its diagnostic values are detailed in **Table 2**. The data related to all diagnostic attributes has been collected from Himachal Pradesh State Electricity Board (HPSEB). In sample 12, 0.23 mgKOH/g of acidity, 0.43 of DF, 5.54 ppm of 2-FAL, 21.8 ppm of water content, 47.7 KV of BDV and 215 ppm of DCG were initially fuzzified in the fuzzification stage. Further, these values are converted into outputs depending upon rule base given in Section 4.

The outputs of sub fuzzy models F_1, F_2 and F_3 are represented by B_1, B_2 and B_3, respectively. After the defuzzification stage, the outputs obtained from F_1, F_2 and F_3 are 0.674, 0.35 and 0.6 using Eq. (2). These three outputs are utilized and converted to inputs for F_4 (i.e. B_4 or RHI). From the F_4 model, the final output for sample 12 is 0.788. Likewise, OHI for all the remaining transformer samples are determined and summarized in **Table 4** (column 2). Also the health indices for each of these transformers were determined in accordance to [1], and are given in the same table (column 4).

Sample Number	HI obtained using Present Proposed Method	HC of Transformers using Proposed Method	HI obtained using Method in [1]	HC of Transformers using the Method in [1]
1	0.25	E	0.3	G
2	0.24	E	0.22	VG
3	0.25	E	0.53	M
4	0.60	P	0.93	VB
5	0.35	G	0.36	G
6	0.84	W	0.94	VB
7	0.24	E	0.3	G
8	0.35	G	0.3	G
9	0.41	G	0.3	G
10	0.27	E	0.3	G
11	0.85	W	0.78	B
12	0.79	W	0.78	B
13	0.62	P	0.94	VB
14	0.11	E	0.3	G
15	0.24	E	0.3	G
16	0.26	E	0.3	G
17	0.25	E	0.3	G
18	0.85	W	0.83	VB
19	0.10	E	0.2	VG
20	0.78	W	0.94	VB

Table 4.
Overall health indices obtained for 20 test case transformers.

Where in **Table 4**, E-Excellent, G-Good, P-Poor, W-worst, VG-Very good, M-Moderate, B-Bad and VB-Very bad.

Four output MFs have been designed in the proposed model, whereas five MFs were considered in Ref. model. The Excellent health condition of proposed model has been compared to the Very good and Good health conditions produced by the reference model [1]. Similarly, Good health condition of proposed model is compared with Moderate condition of reference model. And, comparison has been done Bad with Poor and Very bad with Worst. The overall comparison of all 20 transformer health index by the proposed model and model proposed in [1] is given in **Table 5**.

From **Table 5**, a curious difference has been found out while comparing the test results of proposed model with reference model [1]. It is noted that, out of total 20 test case transformers 11 test results of proposed model are matched with results obtained in [1]. From the comparison, it is observed that the proposed method has better results. To support the statement, consider test sample 12, the HC obtained using model proposed in [1] is Bad. But, the quantities of most influential parameters WC and DCG are 21 and 215 ppm, respectively. These quantities indicate that the transformer insulation is in critical condition and replacement is required. From the test results from proposed model, the HC of sample 12 is worst. It is most suitable condition for the health of transformer. It is proved that the results acquired from the proposed model provide accurate health condition. Also, all the fuzzy models in the proposed model are designed by analyzing the impact of significant diagnostic attributes on transformer insulation. These modifications make the proposed model efficient.

Proposed method/ Method in [1]	Worst	Poor	Good	Excellent	Total
Very Bad	3	2			5
Bad	2				2
Moderate				1	1
Very Good/Good			3	9	12
Total number of transformers					**20**

Table 5.
Comparison of the results obtained from both the methods.

5. Conclusion

In the present chapter, a novel fuzzy logic model has been proposed to find the overall health index of oil-immersed transformers. Parameters that influence the health condition of transformer insulation such as acidity, BDV, DF, DCG, water content and 2-FAL are used to test the HC of transformer. Three sub fuzzy models are created namely, F_1 with water content and acidity as inputs, F_2 with BDV and DF as inputs, F_3 with DCG and 2-FAL as inputs. Further, the individual outputs of three fuzzy models are taken as inputs for the final fuzzy model F_4. All the inputs of sub fuzzy models are designed with three MFs except for 2-FAL which has five. Also, the rule base is formed with nine, nine, and fifteen rules for F_1, F_2 and F_3 sub-fuzzy models, respectively. And, sixty-four rules designed for main fuzzy model F_4. The comparison has been done between the proposed model and fuzzy model designed in [1]. The fuzzy model designed in [1] consists of six inputs and thirty expert rules only. After comparing the two models, it is observed that the results of proposed model are more accurate. In addition, a complete rule base fulfilling all probable situations in determining HI is incorporated in the present model. Hence this model is most efficient, reliable and easily implemented by utilities and industries in order to obtain the health indices of their transformers which is a significant advantage.

Author details

Teruvai Manoj and Chilaka Ranga*
National Institute of Technology, Srinagar, Jammu and Kashmir, India

*Address all correspondence to: chilakaranga@nitsri.ac.in

References

[1] Abu-Elanien B, Salama MMA, Ibrahim M. Calculation of a health index for oil immersed transformers rated under 69kv using fuzzy logic. IEEE Transactions on Power Delivery. 2012;27:2029-2036.DOI: 10.1109/TPWRD.2012.2205165

[2] Jahromi R, Piercy S, Cress J, Wang Fan F. An approach to power transformer asset management using health index. IEEE Electrical Insulation Magazine. 2009;25,20-34. DOI: 10.1109/MEI.2009.4802595

[3] Zeinoddini-Meymand H, Behrooz V. Health index calculation for power transformers using technical and economical parameters. IET Science, Measurement and Tchnology. 2016;10:823 – 830. DOI: 10.1049/iet-smt.2016.0184

[4] Dehghani Ashkezari A, Ma H, Saha TK. Application of fuzzy support vector machine for determining the health index of the insulation system of in-service power transformers. IEEE Transactions on Dielectrics and Electrical Insulation. 2013;20:965 – 973. DOI: 10.1109/TDEI.2013.6518966

[5] Qiu J, Wang H, Lin D, He B, Zhao W, Xu W. nonparametric regression-based failure rate model for electric power equipment using lifecycle data. IEEE Transactions on Smart Grid. 2010;6:955 – 964. DOI: 10.1109/TSG.2015.2388784

[6] Dehghani Ashkezari A, Ma H, Saha T, Cui Y. Investigation of feature selection techniques for improving efficiency of power transformer condition assessment. IEEE Transactions on Dielectrics and Electrical Insulation. 2012;21:836 – 844. DOI: 10.1109/TDEI.2013.004090

[7] Ortiz F, Fernandez I, Ortiz A, Renedo C, Delgado F, Fernandez H. Health indexes for power transformers: A case study. IEEE Electrial Insulation Magazine. 2012;32:7-17. DOI: 10.1109/MEI.2016.7552372

[8] Shaban KB, H. El-Hag A, Benhmed A. prediction of transformer furan levels. IEEE Transactions on Power Delivery. 2016;31(4):1778-1779. DOI: 10.1109/TPWRD.2016.2521320

[9] Pompili M, Scatiggio F. Classification in iso-attention classes of HV transformer fleets. IEEE Transactions on Dielectrics and Electrical Insulation. 2015;22:2676-2683. DOI: 10.1109/TDEI.2015.005252

[10] Wang M, Vandermaar AJ, Srivastava KD. Review of condition assessment of power transformers in service. IEEE Electrical Insulation Magazine. 2002;18(6):12-25. DOI: 10.1109/MEI.2002.1161455

[11] Abu-Elanienand EB, Salama MMA. Asset management techniques for transformers. Electric Power Systems Research. 2010;80(4):456-464. DOI: 10.1016/j.epsr.2009.10.008

[12] Hughes D. Condition based risk management (CBRM)- enabling asset condition information to be central to corporate decision making. CIRED 2005 - 18th International Conference and Exhibition on Electricity Distribution; 6-9 June 2005; Turin, Italy; 2005. DOI: 10.1049/cp:20050884

[13] Singh J, Sood YR, Jarial RK. Condition monitoring of power transformers—Bibliography survey. IEEE Electrical Insulation Magazine. 2008;24(3):11-25. DOI: 10.1109/MEI.2008.4591431

[14] Hughes DT. The use of health indices to determine end of life and estimation remnant life for distribution assets. 17th International Conference on Electricity Distribution; 12-15 May 2003; Barcelona, Spain; 2003.

[15] Hjartarson T, Otal S. Predicting future asset condition based on current health index and maintenance level. ESMO 2006 - 2006 IEEE 11th International Conference on Transmission & Distribution Construction, Operation and Live-Line Maintenance; 15-19 October 2006; Albuquerque; 2006.

[16] Mei D, Min H. A fuzzy information optimization processing technique for monitoring the transformer in neural-network on-line. IEEE International Conference on Dielectric Liquids, ICDL 2005; Coimbra, Portugal; 2005. p. 273-281.

[17] Piercy R, Cress S, Service J, Fan W. An approach to power transformer asset management using health index. IEEE Electrical Insulation Magazine. 2009;25:20-34. DOI: 10.1109/MEI.2009.4802595

[18] BHEL Transformers, 2nd ed. 2003.

[19] http://www.electrical4u.com/dga-or-dissolved-gas-analysis-of-transformer-oil-furfural-or-furfuraldehyde-analysis.

[20] Emsley M, Xiao. X, Heywood RJ, Ali H. Degradation of cellulosic insulation in power transformers. Part 3: Effects of oxygen and water on ageing in oil. IEE Proceedings - Science, Measurement and Technology. 2000;147:115-119. DOI: 10.1049/ip-smt:20000021

[21] Dominelli N. Equipment Health Rating of Power Transformers, Conference Record of the 2004 IEEE International Symposium on Electrical Insulation; 19-22 September 2004; Indianapolis, IN, USA; 2004. p. 163-168.

[22] Arshad M, Islam SM, Khaliq M. Fuzzy logic approach in power transformers management and decision making. IEEE Transactions on Dielectrics and Electrical Insulation. 2014;21:2343 - 2354. DOI: 10.1109/TDEI.2014.003859

[23] Siada A, Hmood S, Islam S. A new fuzzy logic approach for consistent interpretation of dissolved gas-in-oil analysis. IEEE Transactions on Dielectrics and Electrical Insulation. 2013;20:2343-2349. DOI: 10.1109/TDEI.2013.6678888

[24] Tavner P, Ran L, Penman L, Sedding H. Condition Monitoring of Rotating Electrical Machines. The Institution of Engineering and Technology. London, United Kingdom; 2008. DOI: 10.1049/PBPO056E

[25] Hong Lan LT, et al. A new complex fuzzy inference system with fuzzy knowledge graph and extensions in decision making. IEEE Access. 2020;8:164899-164921. DOI: 10.1109/ACCESS.2020.3021097

[26] Kamthan S, Singh H. Hierarchical fuzzy logic for multi-input multi-output systems. IEEE Access. 2020;8:206966-206981. DOI: 10.1109/ACCESS.2020.3037901.

[27] Mei W. Formalization of fuzzy control in possibility theory via rule extraction. IEEE Access. 2019;7:90115-90124. DOI: 10.1109/ACCESS.2019.2928137.

Chapter 10

Health Monitoring of an Aircraft Fuel System Using Artificial Intelligence Techniques

Vijaylakshmi S. Jigajinni

Abstract

Aircraft is a non-linear complex system and is need of regular monitoring. Integrated Vehicle Health Management (IVHM) is a process of health management paradigm, which involves system parameter monitoring, assessment of current, future conditions through diagnostic and prognostic approaches by providing required maintenance activities. Deployment of diagnostic, prognostic and health management processes enable to improve the system reliability and reduces the operating cost of the aircraft. Health monitoring and management plays a vibrant role in safe operation and maintenance of aircraft. Soft computing methodologies such as Artificial Neural Networks (ANN) and Adaptive Neuro-Fuzzy Inference System (ANFIS) are used to estimate the health status of fuel system by developing model of a typical pump feed, twin-engine, four-tank small aircraft fuel system using Simulink in the laboratory environment. The controller is designed to generate the signals of the fuel tanks based on the fuel requirement of the engine. The ANFIS based management system helps to detect the faults existing in the fuel system and diagnose those faults using the expert's logical rules. During a fault ailment, the controller's performance is evaluated. The efficacy of this intelligent controller is verified with the present fuel control system and ANN controller.

Keywords: Aircraft fuel system, Fault detection and diagnosis, Prognosis, Integrated Vehicle Health Management, Artificial Intelligence, ANN and ANFIS

1. Introduction

1.1 Health monitoring of an aircraft fuel system

In recent years significant quantity of research work have been done using Artificial Intelligence (AI) and Expert Systems (ES) to predict and estimate the faults in the system in order to investigate the complete health status of the system [1]. Soft computing technique is an intelligent computing approach which integrates the human reasoning ability with learning capability of uncertainty and imprecision of the system. The prime importance of the soft computing technique is the non-requirement of the well-defined mathematical model of the system. Therefore, soft computing is becoming more popular in the fault diagnosis and prognosis applications. The powerful machine learning techniques used for the current work are Artificial Neural Network (ANN) and Adaptive Network-based Fuzzy Inference System (ANFIS). A brief description and implementation of these techniques for fuel system health management is presented in this chapter. The general steps to determine the system health status are described as below:

i. Detecting the symptoms of faults by monitoring and analyzing

ii. Identification of the cause of faults

iii. Manual insertion of faults within the model of the system

iv. Diagnosing the effect of the fault on the system and

v. Predicting the functional behavior of the system.

Figure 1 illustrates the health management architecture of the fuel system and its diagnostic and prognostic functioning process. This architecture comprises three main functions, such as monitoring, diagnosis and prognosis. The operation of the first function is to monitor the fuel flow rate and status of its co-operating components like pumps, the quantity of fuel in each tank, valve functioning, the rate of fuel consumption by the engine. Based on the initial fuel flow rate, the rate of fuel consumed by the engine is estimated by the monitoring function. Further, the required fuel flow rate from each tank is calculated by the prediction model. An

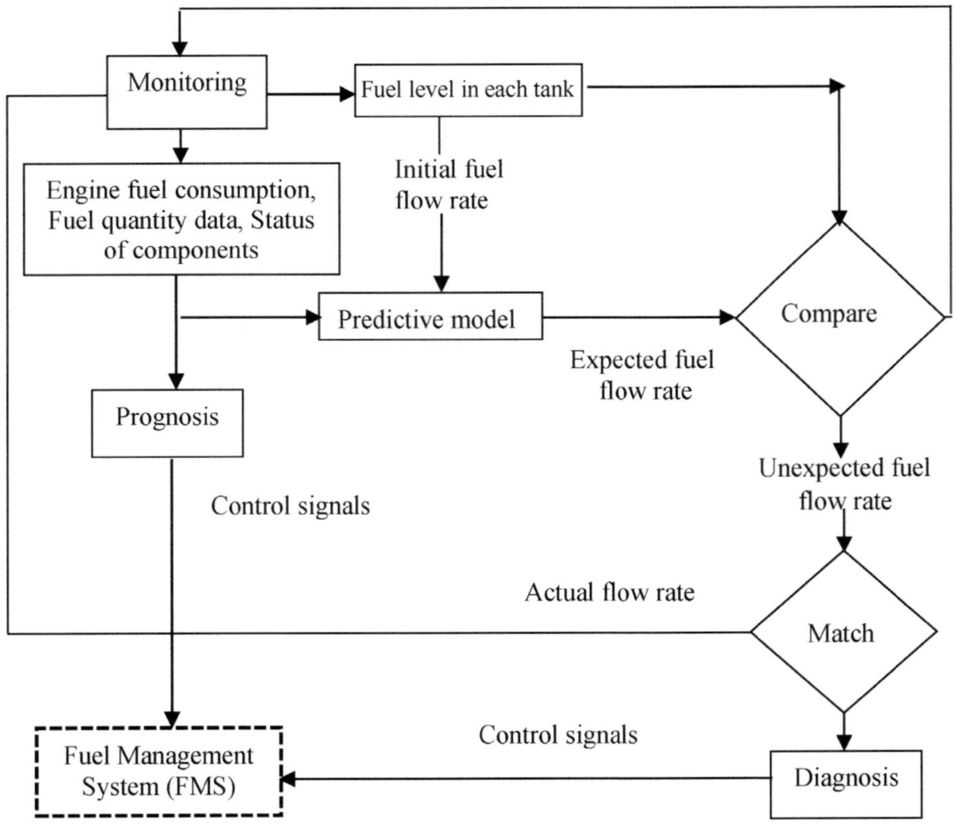

Figure 1.
Fuel system health management architecture.

artificial intelligence approach is implemented for the simulation of a typical small aircraft fuel system for fault analysis.

In the health management process initially, a comparison is done between the actual flow rate and the required fuel flow rate. If the flow rate is not in the desired range, an analysis is conducted to detect and determine the reason for this disagreement, which activates the fault analysis process. Further, it is used as the criterion of failure that leads to the diagnostic function. The prognostic function in the assumed fuel system is to forecast the fuel flow rate from each of the tanks. Thus, the monitoring function helps to predict the variations in the fuel flow rate and helps to assess the overall health status of the fuel system.

2. Literature review

Many different approaches of fault diagnosis and prognosis are demonstrated in the literature with their pros and cons and it is difficult to say which is a better approach. It depends on the type of the complex system considered to model, available expert knowledge and data. AI techniques are very effective in pattern classification, decision making, pattern recognition and problem solving *etc*. The hybrid (Neural Network with Fuzzy Systems) approach developed holds good when modeling the complex systems like aircraft and its subsystems for fault analysis and health management. The advantage of using fuzzy logic is to provide direct interpretation of human knowledge into the fault analysis process using the rules which are simple to comprehend. Further, the process of neuro fuzzy system is distributed parallel processing of data, knowledge inference matching conflict and distributed storage of diagnosis rules which help to overcome the black box limitation of a neural network. This combination provides a healthy solution for fault analysis by detecting the missed faults and decreasing the false alarms even in the presence of disturbance and changing environment of system during operation.

The combination of AI techniques with model-based approaches has excellent efficiency for fault diagnosis. ANFIS takes advantage of the learning ability of ANN and knowledge representation of Fuzzy logic systems. Implementation of ANFIS as fault diagnosis tool was reported by [2, 3]. A method for faults diagnosis of induction motors using ANFIS and regression techniques was described by [4, 5] used recurrent NN and ANFIS to forecast the crack propagation in rotating machine and provided the comparison results of both predictors. Application of prognosis in aerospace industries is an evolving arena with very a smaller number of real-time implementations. Thus, implementation of a model-based methodology with the integration of AI techniques is considered for providing better representation of the complex system like aircraft fuel system.

3. Fuel system health monitoring

The fuel level monitoring function is primarily used to manage and monitor the quantity of fuel in each tank. A platform for the study of the fuel system sequencing, a simulation of fuel management system was modeled by [6]. Fuel level sensors installed in each tank provides the quantity of fuel and also, transfer the signal to the fuel management system. The outer model of a typical small aircraft fuel system is shown in **Figure 2**. The Simulink model developed includes a tank model, tank parameters and fuel pump model as sub-models. The following are some assumptions made during the simulation using Simulink.

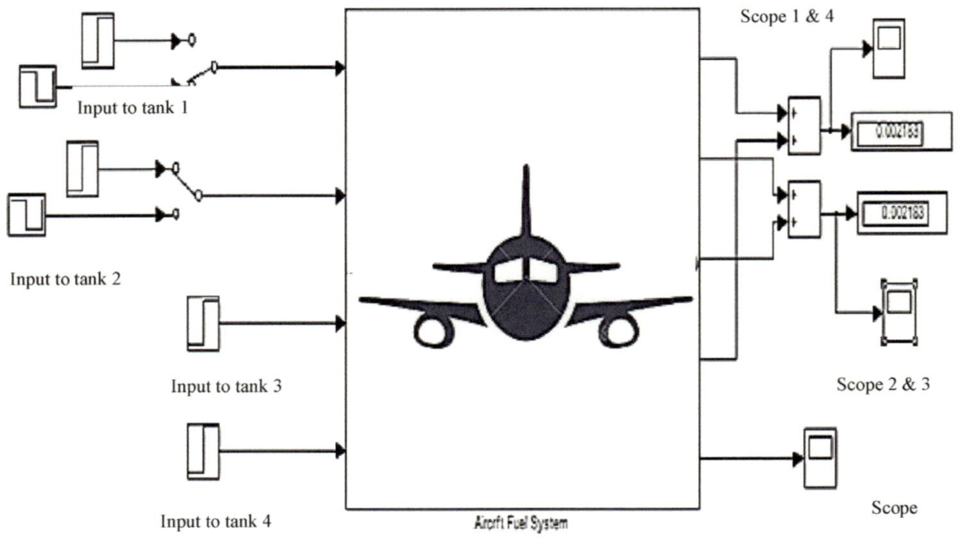

Figure 2.
Snapshot of outer model of the typical aircraft fuel system.

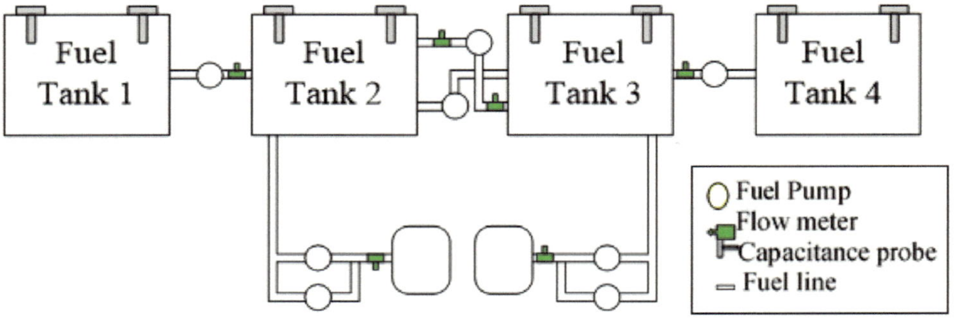

Figure 3.
Structure of a typical small aircraft fuel system.

i. In the aircraft fuel tank, the fuel and gas temperature are homogeneous

ii. The acceleration and deacceleration effect on the fuel is not considered while analyzing. Thus the sloshing is not interpreted for the simulation.

iii. Tank geometry of cubic meter is considered without any fixed reference frame

iv. The fuel properties are not affected by the changing ambient pressure and temperature

The assumed typical small aircraft fuel system contains four fuel tanks and a total of eight fuel pumps as depicted in **Figure 3**. It consists of four main pumps of which two pumps are used for backup and remaining two pumps for delivery of fuel between wings at emergency conditions [7]. The flow of fuel in this aircraft is managed and monitored by the designed adaptive intelligent controller. The primary objective is to manage the fuel flow to the engine without fail and to reach the required engine fuel consumption rate. If any fault arises in any of the fuel tanks,

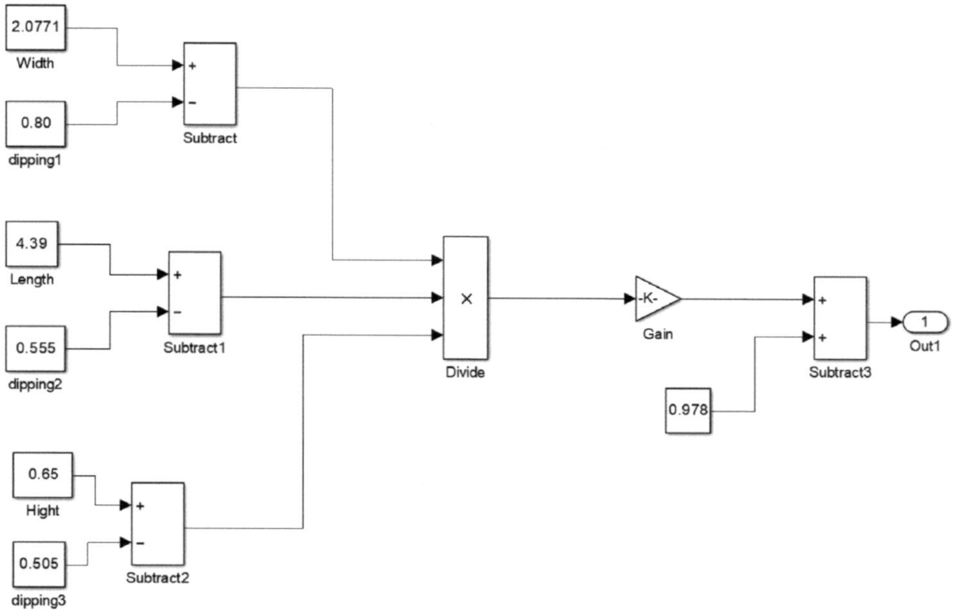

Figure 4.
Geometry of the aircraft fuel tank.

the controller model detects it and reacts as per the fuel requirement of the aircraft's engine. The typical fuel system consists of four tanks with a total fuel capacity of 2800 kg with each tank capacity of 700 kg respectively. The tanks are symmetrically located in both left and right wings.

The model of the aircraft fuel system is simulated based on the fuel system design [8], and the fuel system health management is analyzed using the hybrid fuzzy system. The efficiency ANN based controller is analyzed by comparing with ANFIS based intelligent tool. The total usable capacity of 2800 kg of fuel is sufficient for at least 30 minutes of operation of the assumed aircraft at full continuous power. Simulation of fuel pump is done using the properties of the Hyjet-4A fluid as fuel. The assumed temperature and the viscosity value of fuel are 22.72°C and 1, respectively. The fuel line is built and simulated in the same manner as in the practical fuel systems using the metal pipes that are firm and fixed. A pipe of length 500 mm is assumed which connects the internal wing-tip tanks. The length of pipe between the engines and the integral wing tanks is considered to be of 7200 mm. The pipes have an internal diameter of 10 mm, and the shape of pipes is decided with a geometry factor of 64. **Figure 4** shows the Simulink model of the geometry of the fuel tank. An integral fuel tank of height 0.65 m, length 4.39 m and width of 2.0771 m is modeled. An axial pump containing an electric motor drive is selected, with an angular velocity of 1770 rpm and the correction factor of 0.8. **Figure 5** shows the model of the fuel pump as well as the fuel line for the transfer of fuel between the wing tanks. In other words, the monitoring function helps to provide alert messages to the flight crew and provide data to the prediction model. **Table 1** provides the simulated values assumed at normal operating conditions for linear motion of the aircraft.

The flight range of simulated typical aircraft was calculated based on steady linear motion. Substituting assumed values of **Table 1** into the equation, for each tank the flight ranges are obtained. For a fuel system with four tanks each of 700 kg of fuel capacity, the flight range can be calculated as:

$$s = \left(\frac{V.v}{Q}\right) \tag{1}$$

$$= \frac{700 \times 81.9}{2800} = 20.475\, m/sec \cong 73.71 km/hr$$

Where, V is the quantity of fuel in each fuel tanks, v speed of the aircraft and Q is total quantity of the fuel.

Figures 6 and 7 shows the level of fuel of a healthy tank and a faulty tank. The faults are introduced intentionally in the first tank by decreasing the fuel level which may occur due to leakage in the tanks, pipelines, valve stuck or maybe because of the filters blocks or icing within the fuel system. **Figure** 7 demonstrates the leakage in the fuel system for 5 seconds, which affects the operation of the engine. The decrease in the level of fuel in the first tank is detected and diagnosed effectively by the developed health management methodologies. The level of fuel decreases after 6.5 sec during simulation for a simulation time of 25 seconds as shown in **Figure** 7.

3.1 Fuel system diagnosis and prognosis

This section discusses the diagnostic and prognostic function associated with the assumed fuel system. Prognosis is prediction of remaining useful life time of the system. Based on the available fuel quantity, engine fuel consumption and status of

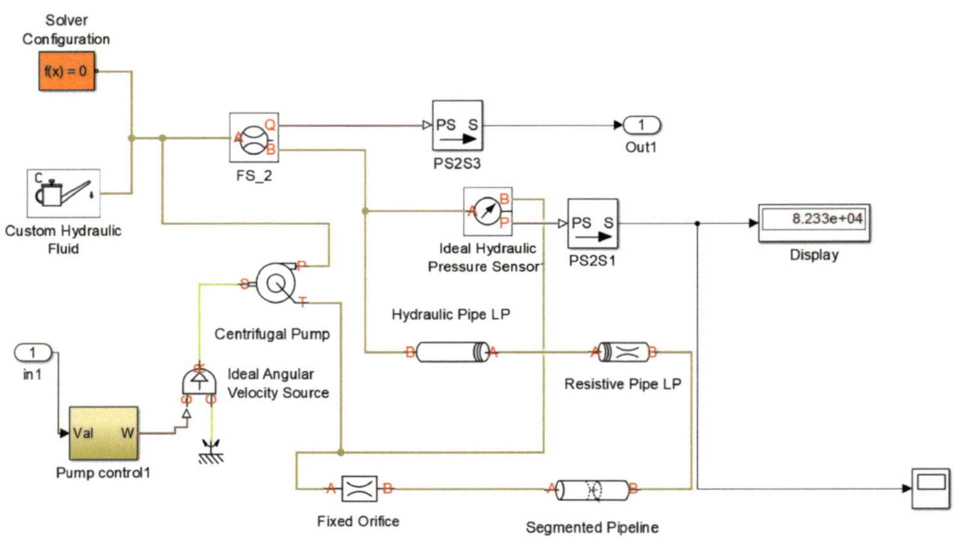

Figure 5.
Fuel pump and fuel line model.

Sl. No.	Parameter	Value
1	Mach number	0.26
2	Temperature of the ambient air	22.72°C
3	Speed of sound	314.8 m/sec
4	Aircraft speed	81.9 m/sec

Table 1.
Assumed simulated values of an aircraft.

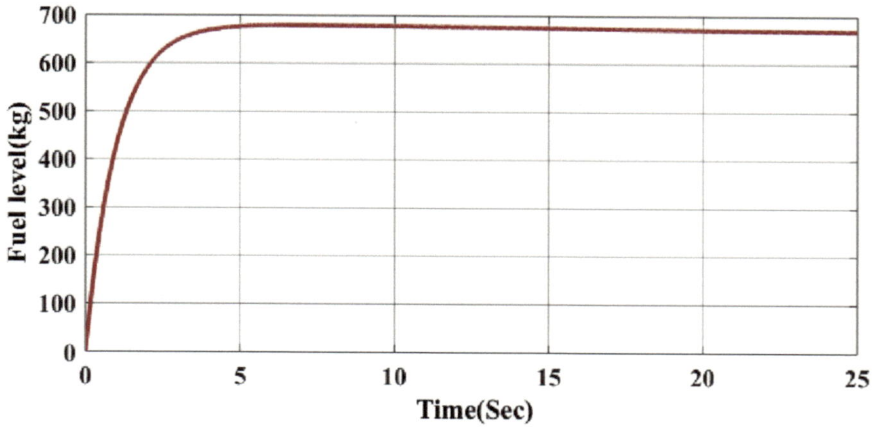

Figure 6.
Level of fuel in healthy tank.

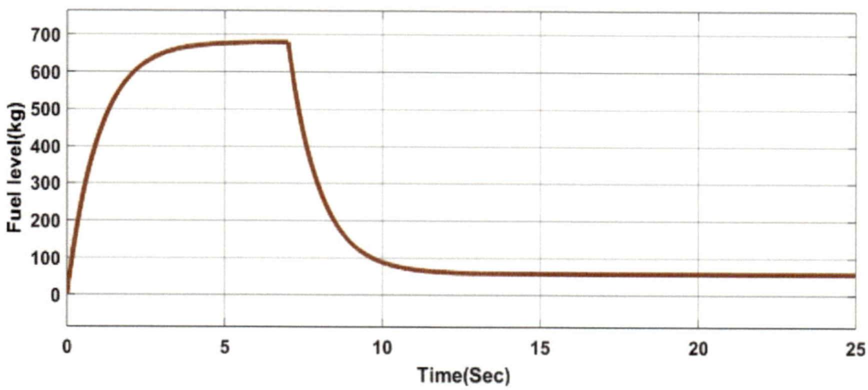

Figure 7.
Level of fuel in fault tank.

the components, the prognosis function generates the control signals to the fuel management system to predict and mitigate the faults within the system. Further, through the diagnosis function the obtained flow rate is matched with the control signals and corrective measures are taken to identify and detect the fault in the fuel system. In this way, using these functions health management of the fuel system can be achieved.

Soft computing techniques such as Artificial Neural Networks (ANN) and Adaptive Neuro-Fuzzy Inference System (ANFIS) are used for diagnostic and prognostic analysis of the aircraft fuel system. Health management process using ANN and ANFIS as a controller is illustrated in this chapter.

3.2 Health monitoring of fuel system using ANN

Backpropagation is considered to be the most efficient training algorithm among different artificial neural network algorithms [9]. Learning process in ANN is achieved by collecting the information and training them accordingly. For the assumed fuel system, ANN-based health management tool uses the fuel flow rate as input data and previous engine fuel consumption rate as the historical information. The ANN model has trained accordingly and generates output signals in relation to

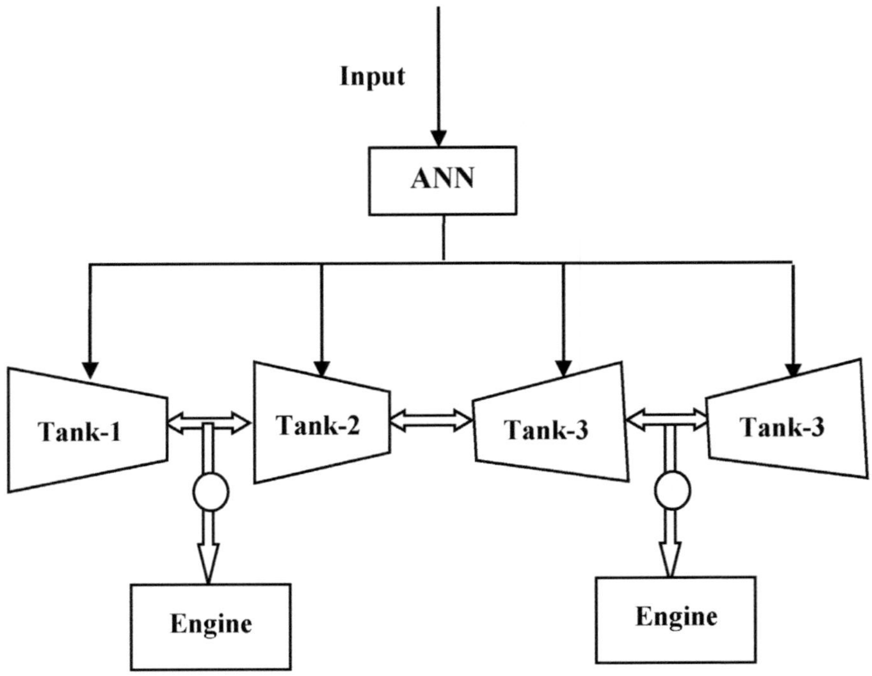

Figure 8.
Structure of the assumed fuel system with ANN as controller.

working condition of the fuel system. **Figure 8** shows the structure of the assumed fuel system with ANN controller. Based on the present and previous instance fuel flow rate, it generates the output signal in the presence of faults. The ANN health management tool continuously monitors the flow rate of fuel to the engines and provides the require d fuel flow to both the engines.

This section exploited the basics of ANN. Proper training of NN helps to perform fault analysis, detection, and diagnosis without the requirement of a complex mathematical model. The interpretation of the NN similar to human thinking and multiple parallel processing features of NN enhances the network performance [10].

4. Simulation of fuel system health monitoring using ANN

Feedforward neural network is implemented which regulates the input parameters of the assumed fuel system to obtain the desired outputs. A simple model of the neural network is shown in **Figure 9**. The 'n' number of input signals are denoted as $x'_1, x'_2 \ldots, x'_n$, and weights of each signal as $w'_{k1}, w'_{k2} \ldots, w'_{kn}$. These weighted input signals are summed and output signal Y' is found through activation function F.

Backpropagation (BP) is a supervised learning algorithm [4]. By gradient descent, the algorithm calculates the error function in relation to the weights of the neural networks. Learning process in ANN is achieved by collecting the information and training accordingly. The following equations describe the implementation of a feed-forward neural network.

$$v_k = \sum_{j=1}^{n} w'_{kj} x'_j \tag{2}$$

$$y'(k) = s(v_k) \tag{3}$$

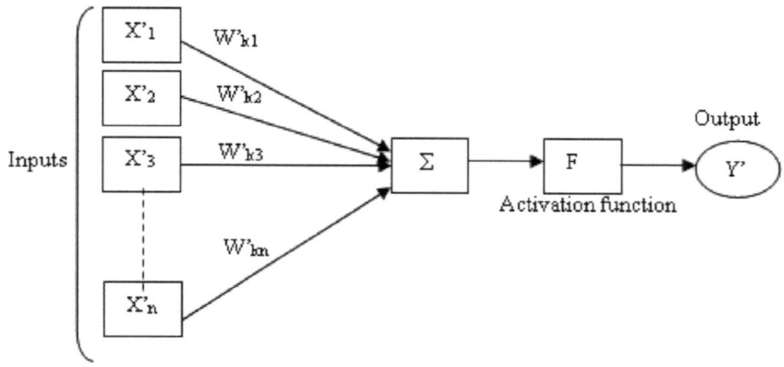

Figure 9.
A neural network model.

$$y = \theta\left(\sum_{k=1}^{n} y'(k)\right) \tag{4}$$

where, $x'_1, x'_2, \dots \dots \dots .x'_n$ are the inputs and $w'_{k1}, w'_{k2}, \dots \dots \dots .w'_{kn}$ are the weights with $y'(k)$ is the sigmoid function. $S(.)$ is the sigmoid activation function and $\theta(.)$ is the threshold activation function. y and n are the output of the NN and total number of neurons in the second layer of the network respectively. The output y obtained is the prediction based on the inputs of the NN and weights applied.

For hidden layers according to the BP algorithm, the derivation error of backpropagation is expressed as:

$$\delta_k = \left(1 - y'_k\right)\sum_k w_{kj}.\delta_p \tag{5}$$

$$\delta_p = y_p - t_p \tag{6}$$

Where, δ_p is the error derivation at pth neuron, y_p, pth unit activation output, δ_k derivation of error and t_p corresponding target. Based on the weights of the first layer and the next layer, the gradient of error is calculated. Further, the weights are updated accordingly.

For the assumed fuel system, ANN-based health management tool uses the fuel flow rate as input data and previous engine fuel consumption rate as the historical information. The ANN model has trained accordingly and generate the output signals in relation to the working state of the fuel system. The predictive output signal of the ANN model is obtained as:

$$v_k = f\left(w'_{k1}, w'_{k2}, \dots \dots \dots .w'_{kn}\right) \tag{7}$$

Where, v_k is the predicted output of the NN.

Training of ANN with backpropagation algorithm calculates the gradient error and regulates the weights by the required flow rate to be consumed by the engines. Continuous updating of ANN model helps to maintain the required flow rate of fuel irrespective of malfunctioning of the any of the component within the fuel system. The disadvantage of the Feed Forward Neural Network with BP algorithm is the necessity of complex mathematical calculation, slow rate of error convergence and more amount of time required for a non-linear system operating condition. A hybrid ANFIS methodology is utilized to overcome this drawback. This methodology provides better performance results for non-linear and changing operating condition for the assumed aircraft fuel system.

ANN controller monitors and manages the flow of fuel without any limitations to meet the required flow rate. Fault occurrence within the fuel tanks are detected, and corrective measures are taken by the ANN. Any change in the data is identified by the unique property of the ANN algorithm that differentiate the data points in noisy conditions. The ANN controller calculates the gradient error, and accordingly the target signals are generated by adjusting the weights of the BP algorithm. **Figure 10** shows the flowchart of the updation process of the ANN. Timely maintenance of the fuel tanks and fuel system helps to improve the performance of the aircraft. Faults in the tanks like leakage, filter blockage or pump failure during flight may degrade the operation of the aircraft. Under such circumstances, ANN algorithm continuously updates with the actual data and maintains the required flow rate.

4.1 Simulation results and discussion using ANN

The two hidden layers of ANN model using Simulink is depicted in **Figure 11**. In this neural network four input samples and four target samples are used. The hidden layer consists of 100 neurons and output layer of four nodes for the assumed four-tank fuel system. The neural network toolbox of Matlab/Simulink utilizes the data points in three stages. In the first stage, the training data points are used for training the neurons and the gradient error is calculated by using the weights of successive layers. The upadated weights of the network reduces the error for the given value. During second stage, validation fail check is done. Testing process of data points is carried out in the third stage. The performance of the developed NN is then analyzed.

The BP algorithm uses the Mean Square Error (MSE) procedure to compute the error in each step or iteration. MSE is the performance index of the BP algorithm. It is the error computed as variance between the network output and target output. The relation given in Eq. (8) computes the MSE in each step for each output,

$$MSE = \frac{1}{N}\sum_{1}^{N}\left[e_{i/p} - e_{o/p}\right]^2 \tag{8}$$

Where, $e_{i/p}$ is the actual input, $e_{o/p}$ is the output of the NN model and N is the number of iterations considered. The required number of iterations for converging the procedure and time taken for training depends on the size and structure of NN, learning methodology adopted, number of layers, and also on the length of data points including input data and output data.

During the training process of ANN after 4th iteration the performance of 4.33e-33 obtained which indicates the amount of minimized error. Gradient of 3.16e-12 indicates the variance occurred in the error rate, Mu of 1.00e-07 is the threshold value achieved after each iteration and validation check indicates reduced error after current (4th) iteration as compared to previous (3rd) iteration. The graph illustrated in **Figure 12** shows the performance and status of the training process. It is a curve showing the plot of MSE versus four epochs. The blue line plotted represents the training result, green color line denotes the result of validation check and red line indicates the test result. During the ANN training process, as the MSE decreases, consequently the network output reaches to target output. The performance using ANN as fault diagnosis and prognosis process is assessed by computing

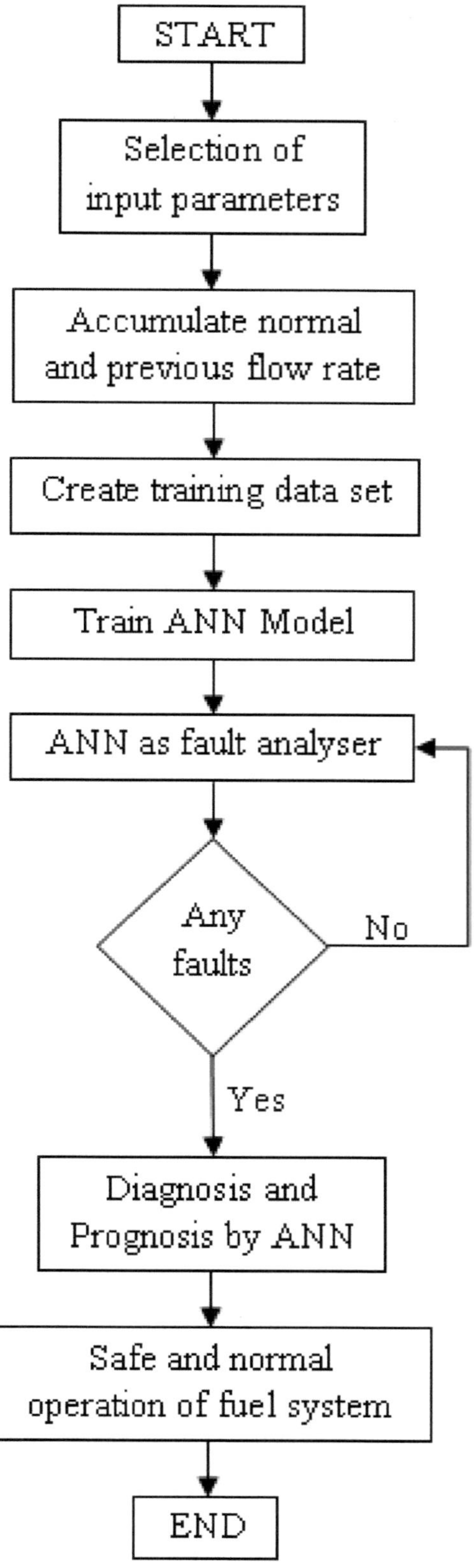

Figure 10.
Flowchart of ANN-based prognostic tool for the fuel system.

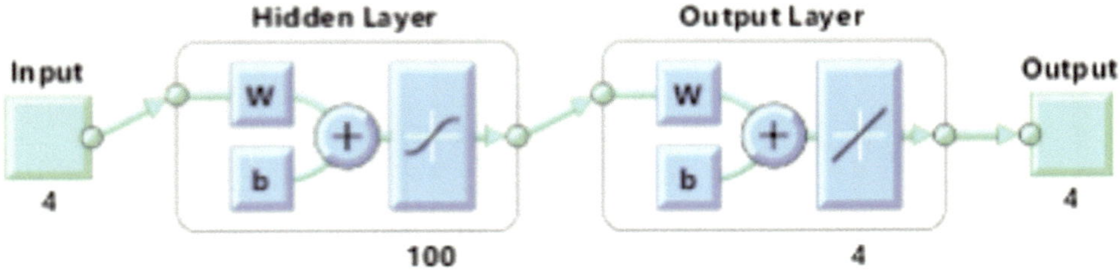

Figure 11.
Developed ANN Simulink model.

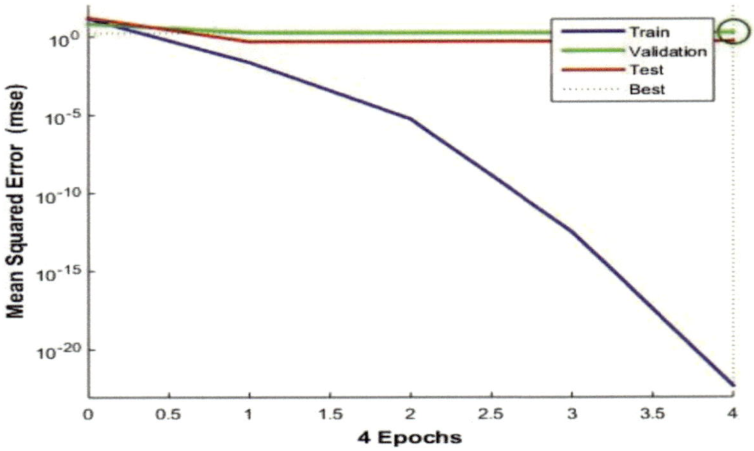

Figure 12.
Training performance of ANN.

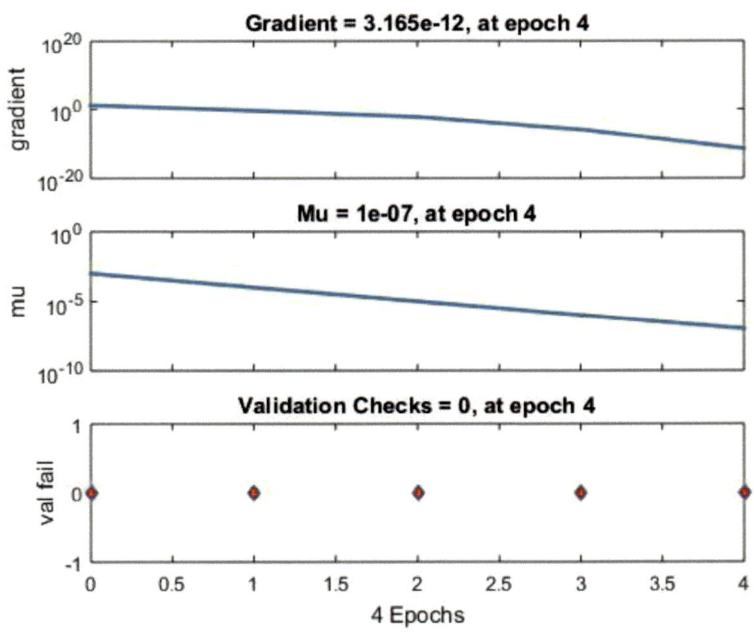

Figure 13.
Gradient and validation performance plot of the ANN model.

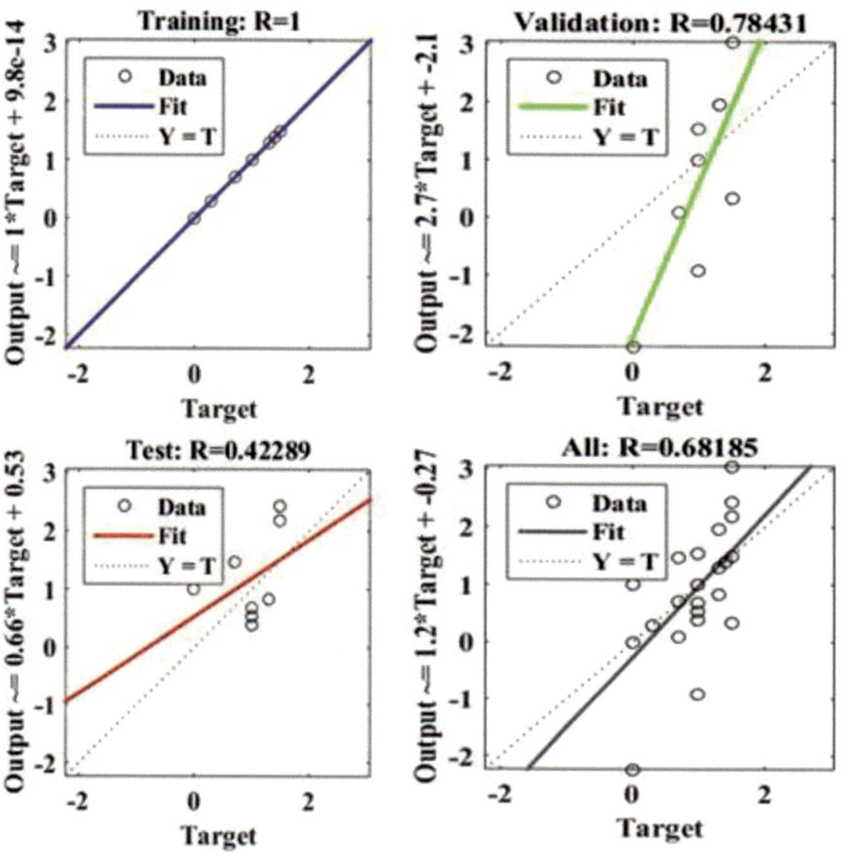

Figure 14.
Training performance plot of the ANN model.

Sl.No	Tank 1	Tank 2	Tank 3	Tank 4
1.	1	1	1	1
2.	0.7000	1	1	1
3.	0.3000	1	1	1
4.	0	1	1	1
5.	1	0.7000	1	1
6.	1	0.3000	1	1
7.	1	0	1	1
8.	1	1	0.7000	1
9.	1	1	0.3000	1
10.	1	1	0	1
11.	1	1	1	1
12.	1	1	1	1
13.	1	1	1	0

Table 2.
ANN training data.

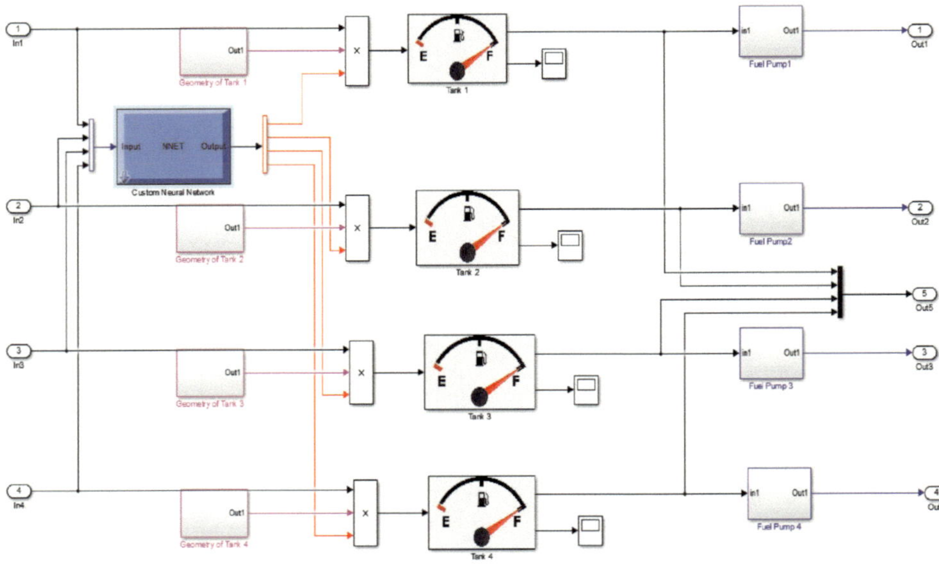

Figure 15.
The Simulink model of the aircraft fuel system with ANN as a controller.

Tank 1	Tank 2	Tank 3	Tank 4
1	1	1	1
0.7000	1.3000	1	1
0.3000	1.4000	1.3000	1
0	1.5000	1.5000	1
1	0.7000	1.3000	1
1.4000	0.3000	1.3000	1
1.5000	0	1.5000	1
1	1.3000	0.7000	1
1.3000	1.4000	0.3000	1
1.5000	1.5000	0	1
1	1	1.3000	0.7000
1	1.4000	1.3000	0.3000
1.3000	1.3000	1.4000	0

Table 3.
Target data of ANN.

the MSE based on the Eq. (8). The best validation performance achieved is of 1.6835 at four epochs. The graph with all the three results of training, testing and validation coincide is considered to be the best performance.

The gradient and validation plot determine a process of assessing the performance of the training and testing of ANN technique. The gradient of error and validation plot for four epochs obtained is as shown in **Figure 13**. It is observed that a smooth decrease in the gradient error at four epochs and maximum measures of validation check at four iteration fails to zero.

Another important factor considered while evaluating the performance of the ANN is the correlation coefficient obtained for training, testing and validation.

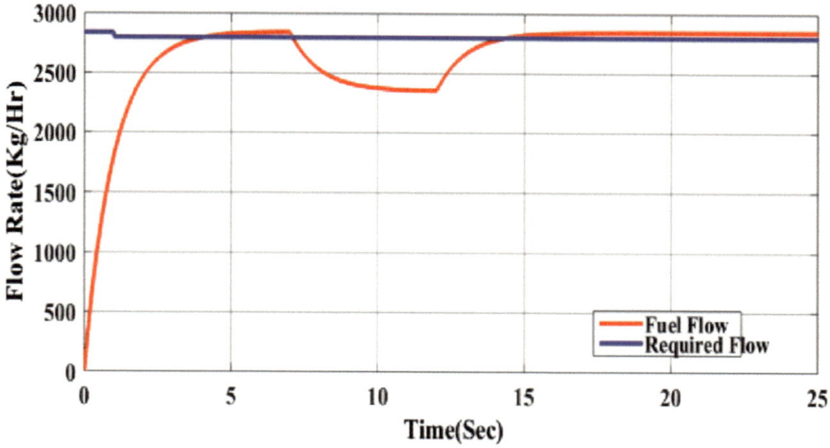

Figure 16.
Fuel consumption by the engine of the aircraft fuel system without a controller.

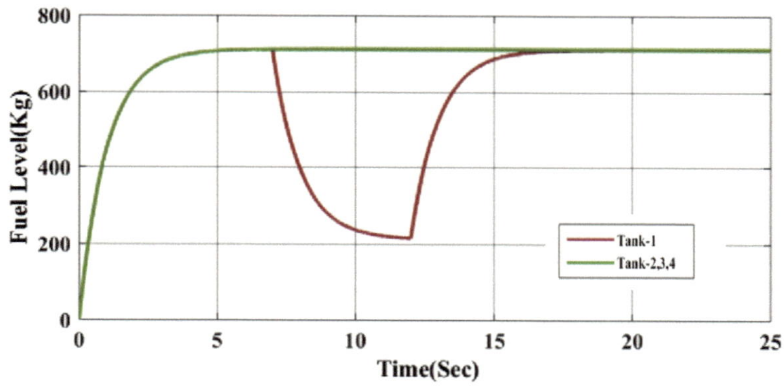

Figure 17.
Fuel management test without a controller.

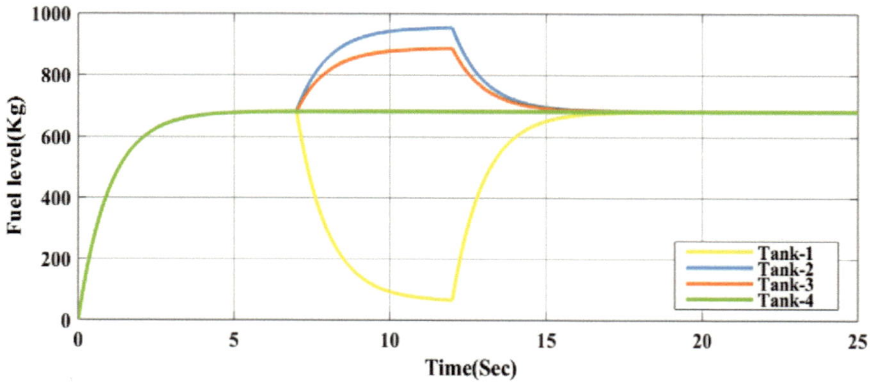

Figure 18.
Fuel management test with the ANN model as a controller.

From the **Figure 14**, it is seen that the overall correlation coefficient is 0.6818. **Table 2** provides the details of the ANN training data where intentional faults are inserted into the tanks and trained accordingly. **Figure 15** shows the Simulink model of four tank fuel system with ANN as a controller. Due to faults, the input

Tank-1	**70 kg**	**Required Fuel**
Tank-2	960 kg	2800 kg
Tank-3	870 kg	
Tank-4	700 kg	
Total	2600 kg	

Table 4.
Fuel fetched from each tank using ANN as controller.

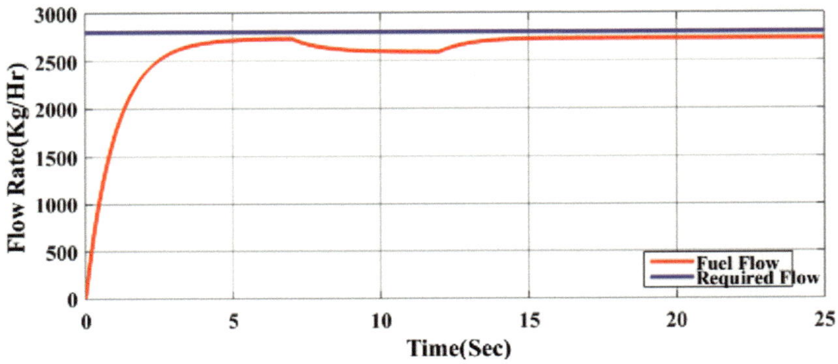

Figure 19.
Fuel consumption by the engine using ANN as a controller.

data points vary and such variations are trained consequently by updating the weights of the neural network. The ANN-based health management process is implemented by taking corrective measures in the presence of faults (**Table 3**).

The simulated model of the fuel system can operate continuously for 30 minutes with full control and power. Fuel consumption test and fuel management test performed on the fuel system without a controller is shown in **Figures 16** and **17** respectively. In the plot blue line indicates the required fuel flow rate and red line indicates the obtained flow rate without controller within the fuel system.

As observed in **Figure 17** for a simulation time of 20 seconds, within four seconds engine(s) get the required amount of fuel. The blue line indicates the required flow rate of 2800 kg/hr., which is fulfilled by delivering the fuel from all the four tanks with capacity of 700 kg per tank. It was obseverd that after a time period of 4 seconds, the level of the fuel in the first tank decreases indicating the presence of faults in the fuel tank. Because of faults the level of fuel in first tank decerases. Thus, when the quantity of fuel and level of fuel in the tank varies, it affects the flow rate of fuel to the engine. This may lead to the fuel starvation that causes the failure of the aircraft engine because of insufficient fuel supply to the engine. The existing aircrafts fuel system with automatic and programmed fuel management system may not identify sudden decrease in the level of fuel properly due to which the performance of the system gets degraded.

Figure 18 shows the fuel management test using ANN as a controller which identifies, detects the change in the fuel quantity and level of fuel in the first tank and correct it by fetching the required fuel from the second and third tanks. The ANN as controller predicts the required fuel by fetching the fuel quantity from the other tanks as per the **Table 4**.

The weight updating process of the BP algorithm during training process classifies the fault data, and testing process identifies the fault occurrence and mitigate it by predicting and fetching the required fuel from other tanks. Compared to the existing fuel management system, ANN as a controller in the presence of faults manages the fuel quantity and flow rate more efficiently. It detects the time of the fault, diagnoses it and mitigates it providing the complete health management of the fuel system. Minimizing the gradient of error for four iterations, the ANN model could be able to fetch the fuel flow rate of 2600 kg/hr. as shown in **Figure 19**. Using ANN as contorller during fuel managemnt test able to fetch 2600 kg/hr. of fuel where as the required flow rate by the engine is 2800 kg/hr. This variation in the flow rate is because of the drawbacks such as slow rate of error convergence and necessity of intensive calculation of BP algorithm for assumed operating conditions. Thus, an intelligent control model ANFIS as a health management tool is used to fulfill the engine fuel consumption requirement.

5. Health monitoring of fuel system using ANFIS

5.1 Adaptive neural fuzzy inference system

ANFIS is a supervised gradient descent algorithm. In this, fuzzy rules configured upon the NN structure provide a qualitative description for the fault analysis of aircraft fuel system. In the resultant hybrid model, the NN recognize the fault pattern and adapt to the changing atmosphere. On the contrary, the FIS integrates the data and performs inferencing and decision making. The dynamic performance of the system can be represented by modeling the neuro-fuzzy method by extracting the numerical data from the model. Based on this approach, the system modeling serves two purposes. They are: the functional behavior of the assumed system can be predicted from the derived model, and the design of a controller is done using the resultant model. The ANFIS model is built first by initializing the input variables with the rules extracted from the input–output data of the assumed system. Later, the NN is utilized to fine tune the rules of the fuzzy model. The flow chart for ANFIS training procedure is as shown in **Figure 20**. In this work, ANFIS is used to detect and identify the presence of faults in the aircraft fuel system.

5.2 Implementation of ANFIS algorithm

The ANFIS structure developed is based on the model developed by [11]. The ANFIS network is mapping of input and output variables in a multi-layer network with a single target output [12]. The operating model of the ANFIS controller is depicted in **Figure 21**. The ANFIS methodology as a fault diagnosis and prognosis process for aircraft fuel system is briefly described in the following sub-sections. ANFIS is a structural plan that links expert's knowledge and the knowledge capability of the neural networks. ANFIS builds a FIS whose membership function parameters are obtained by training appropriately. Consider FIS with two inputs 'x' and 'y', two connected Membership Functions (MFs) and one output 'z'. The fuzzy if-then rules for the present work based on the model [13] are framed as follows. If 'x' is A_1 and 'y' is A_2, then the target 'z' is $f(x,y)$, where A_1 and A_2 are the sets in the antecedents, and $z = f(x,y)$ is a crisp function in the consequent. $f(x,y)$ is a polynomial for x and y input variables. A zero-order Sugeno fuzzy model is formed when $f(x,y)$ is zero or constant which is a Mamdani FIS [14]. A first-order Sugeno fuzzy system model is formed if $f(x,y)$ is first order polynomial. **Figure 22** shows a five-layer ANFIS architecture.

ANFIS is a five-layered feed-forward neural network, *viz* the input layer, product layer, defuzzy layer, normalized layer, and an output layer. The nodes may be

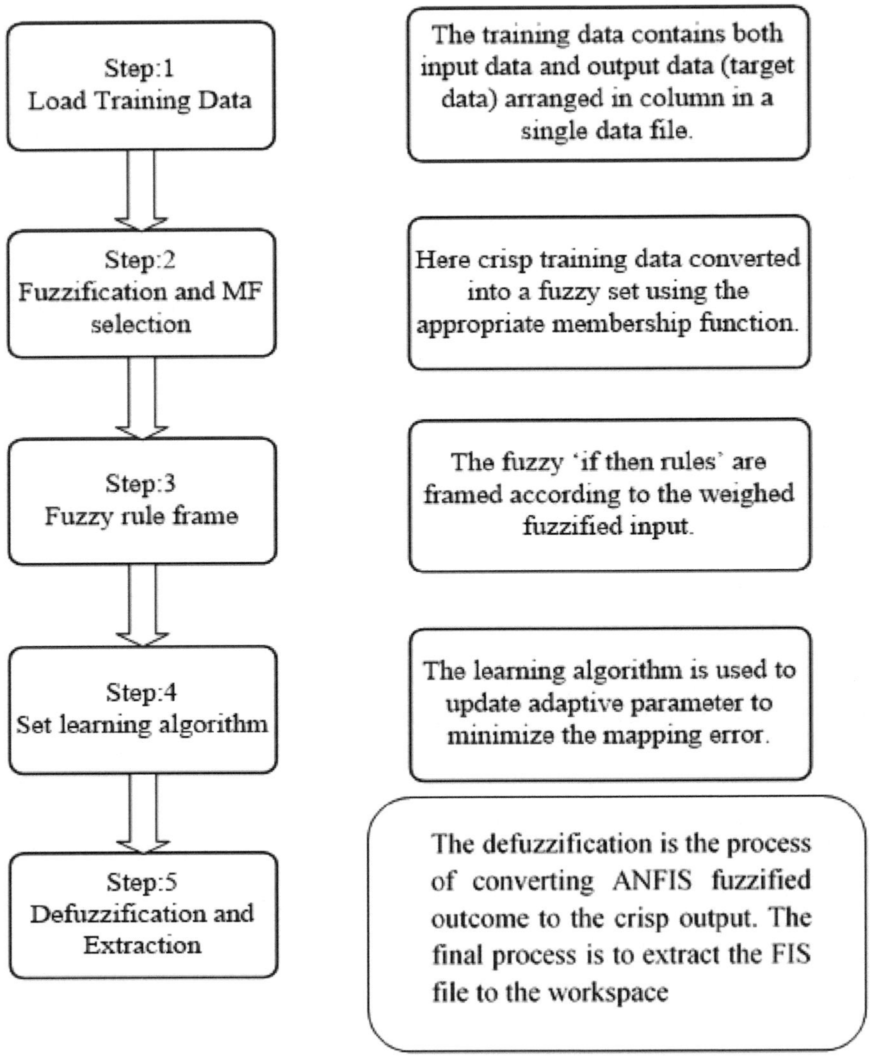

Figure 20.
Flowchart for ANFIS training procedure.

adaptive or fixed. The nodes in a square shape are adaptive and the nodes in the form of a circle are fixed. In this case, the inputs to the ANFIS considered are the flow of fuel at prior instant 'x' and engine's fuel consumption 'y'. The output signals of the fuel tank 'z' are measured as the target output. These parameters aid the ANFIS in formulating the rules as well as realizing a better tuning performance. Every rule covers the unity weight, and the learning procedure of ANFIS is achived on the classified signals. In the ANFIS architecture, two if then rules based on first order Takagi and Sugeno [14] are considered as below:

Rule 1: If x_i is A_1^1 and y_i is A_2^1 then $f_1 = P_1x + Q_1y + C_1$.

Rule 2: If x_i is A_1^2 and y_i is A_2^2 then $f_2 = P_2x + Q_2y + C_2$.

where, P_1, P_2, Q_1, Q_2, C_1 *and* C_2 are the linear parameters, A_1^1, A_2^1, A_1^2 *and* A_2^2 are the nonlinear parameters. Activation levels of the fuzzy rules are considered using the realtion of Eq. (9),

$$W_i = P_i(x) * Q_i(y), \; i = 1, 2, \ldots n \tag{9}$$

where the logical operator “and” is modeled as continuous term an in this case it is stated as a product. The individual o/p of all rules is obtained as a linear combination among parameters of the antecedents of every rule as signified by Eq. (10).

$$Z_i = P_i * x + Q_i * y + C_i,\ i = 1, 2, \ldots n \tag{10}$$

The output of the model Z_0 is found by multiplying the standardized activation degree of the rules by the individual output of all rule, and it is stated by subsequent Eq. (11).

$$Z_0 = \frac{W_1 Z_1 + W_2 Z_2}{W_1 + W_2} \Rightarrow Z_0 = \overline{W}_1 \overline{Z}_1 + \overline{W}_2 \overline{Z}_2 \tag{11}$$

where, $\overline{W}_1$ and $\overline{W}_2$ are the normalized values of W_1 and W_2.

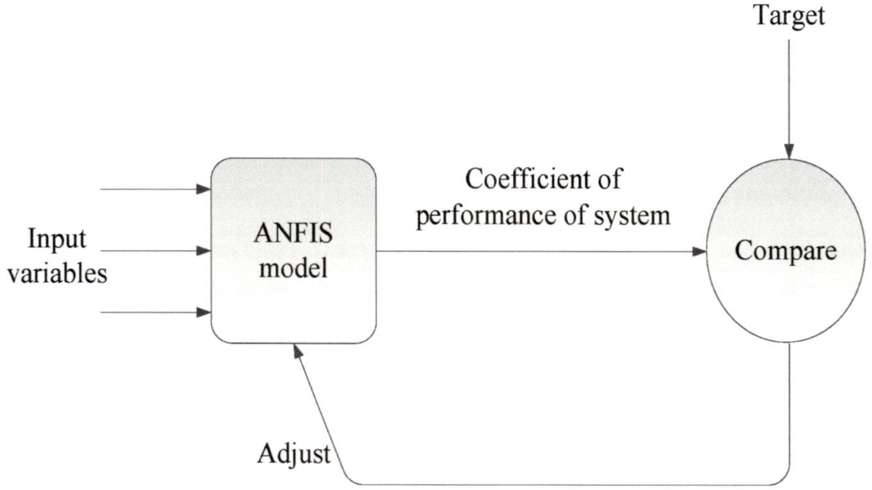

Figure 21.
Working model of the ANFIS controller.

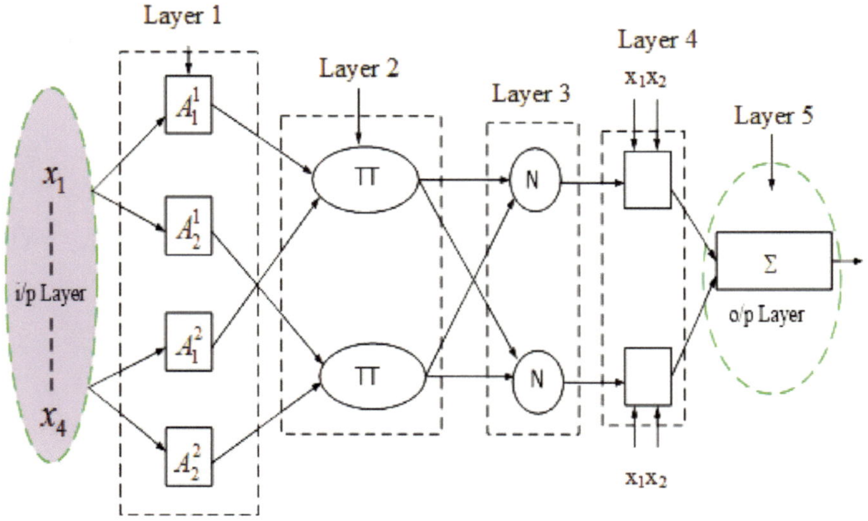

Figure 22.
The equivalent ANFIS architecture.

The hybrid ANN signifying this inference is an adaptable network with five layers. All layers indicating the operation of the Fuzzy Inference System of the ANFIS is examined as follows.

5.2.1 Fuzzification layer

In this method, every input layer is represented as an input variable, and it refers to the fuzzification layer. The input parameters x_i and y_i have the nodes A_1^1, A_2^1, A_1^2 *and* A_2^2 which are the linguistic labels of fuzzy system for isolating the membership performances. The output of the fuzzy layer is given by,

$$F_{L1,i} = \mu A^1{}_i(x) \ , i = 1, 2; \tag{12}$$

$$F_{L1,j} = \mu A^2{}_j(y) \ , j = 1, 2; \tag{13}$$

where, $F_{L1,i}$ and $F_{L1,j}$ are the outputs of the fuzzy layer, '*x*' and '*y*' are the input to nodes *i* and *j*. $\mu A^1{}_i(x)$ and $\mu A^2{}_j(y)$ are the membership performance of the fuzzy layer.

5.2.2 Product layer

This layer may be identified as the π that performs logical "AND" operation, *i.e.*, the multiplication of the input membership functions. In this process, the output is the weighted input function of the next node which is symbolized by W_1 and W_2. The output is described by,

$$Z_1 = F_{L2,i} = \mu A^1{}_i(x).\mu A^2{}_i(y), \ \ i = 1, 2 \tag{14}$$

$$Z_2 = F_{L2,j} = \mu A^1{}_j(x).\mu A^2{}_j(y), \ \ j = 1, 2 \tag{15}$$

5.2.3 Normalization layer

In this layer each node of this layer is fixed which represents the "if" part of a fuzzy rule. It is process of normalization of the input weights that can complete the fuzzy "and" operation. In this layer, each node computes the ratio of the ith rule firing strength to the total firing strength of all rules. The normalized firing strength of the ith node is given by,

$$\overline{Z_1} = F_{L3,i} = \frac{Z_i}{Z_1 + Z_2}, \quad i = 1, 2 \tag{16}$$

$$\overline{Z_2} = F_{L3,j} = \frac{Z_j}{Z_1 + Z_2}, \quad j = 1, 2 \tag{17}$$

where, $\overline{Z_1}$ *and* $\overline{Z_2}$ are the outputs of this layer.

5.2.4 Defuzzification layer

This is an adaptive layer that gives output membership function based on predetermined fuzzy rules. The node function is given by Eqs. (18) and (19).

$$\overline{Z_1} f_i = F_{L4,i} = \frac{Z_i}{Z_1 + Z_2} \left[A_1^1(x) + A_2^1(y) + C_1\right] \tag{18}$$

$$\overline{Z_2}f_j = F_{L4,j} = \frac{Z_j}{Z_1 + Z_2}\left[A_1^2(x) + A_2^2(y) + C_2\right] \tag{19}$$

where, $\overline{Z_i}$ is the o/p of the third layer and $\{P_i, Q_i, C_i\}$ is the consequent parameters set.

5.2.5 Output layer

The output layer is symbolizing the THEN part of the fuzzy rule. This consists of one fixed node that computes the total output which is the summation of the input signals given by the following Eq. (20).

$$f = F_{L5,i} = \sum \overline{Z_l}f_i = \frac{\sum \overline{Z_i}f_i}{\sum Z_i} \tag{20}$$

Where, f is the total output and the function of ANFIS is verified by considering a higher number of signals. Training the ANFIS model for the given inputs generate the control signals which help to maintain the fuel flow rate within the aircraft fuel system.

The important benefits of ANFIS are improved learning capacity, the ability to incorporate the non-linear structure of the system and rapid adaption capability. ANFIS can achieve exceptionally nonlinear mapping, far better than other techniques. Some of the drawbacks of this technique are: there are no standard methodologies to incorporate the changing human learning or experience into the base of a FIS and also there is a need for agent techniques used for tuning the membership functions to diminish or minimize the error during execution.

5.3 Simulation of fuel system using ANFIS methodology

This section defines the simulation procedure of health management of aircraft fuel system using ANFIS as a controller. **Figure 23** shows the model of aircraft fuel system with ANFIS as controller. It includes fuel tanks, pumps, pipelines that connects the tanks and pumps with the engines. The fuel system function is to distribute clean fuel at the required pressure and fuel flow rate to the engines in different operating conditions. The diagnostic and prognostic process of small aircraft fuel system is regulated by the ANFIS intelligent control model, that provides better fuel flow rate compared to ANN methodology. That is because of ANFIS's significant features of significant reasoning ability and the low level of computational power during training process. The main purpose of the ANFIS control model is to direct the fuel flow to the engine and to access the essential engine fuel consumption rate. If a fault arises in any of the fuel tanks, the controller model detects the fault and activates the necessary actions as per the fuel requirement of fuel engine.

The ANFIS controller is concentrated on the optimization of parameters of the aircraft fuel system. Similar to the ANN controller, the ANFIS controller is assessed with the previous instance fuel flow and the engine fuel consumption value of the fuel system.

5.4 Simulation results and discussions using ANFIS

ANFIS's learning ability carried out through the five-layer structure of the Fuzzy Logic system helps to approximate the non-linear functions which depends on the

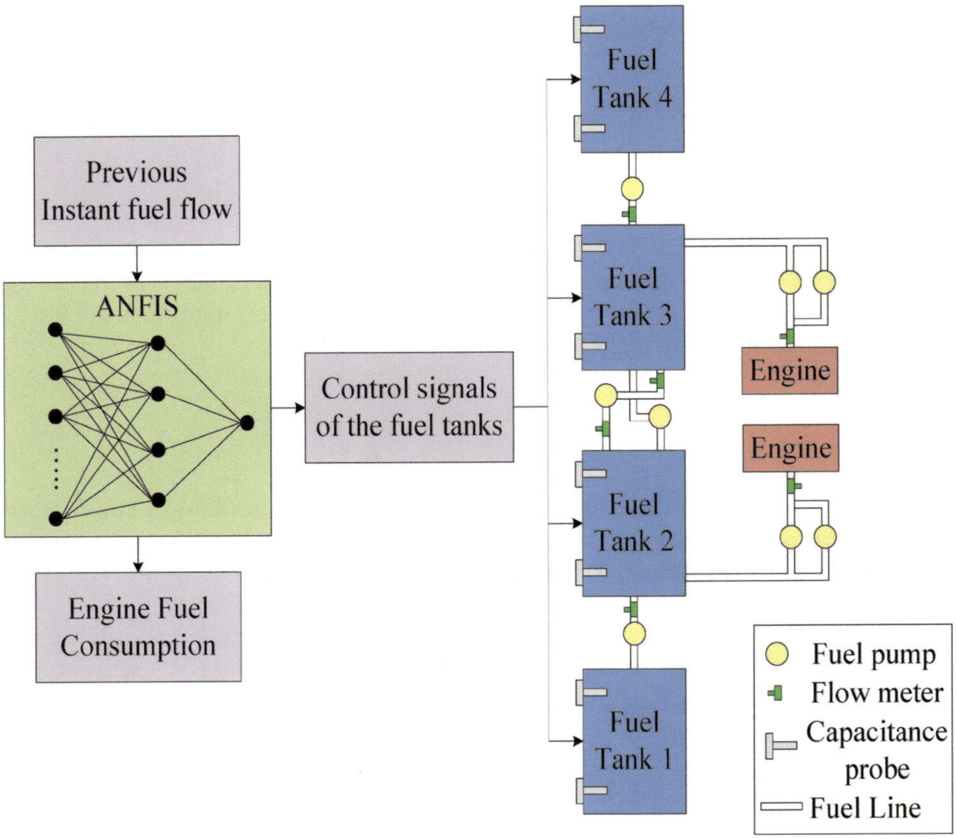

Figure 23.
Structure of the aircraft fuel system with ANFIS as controller.

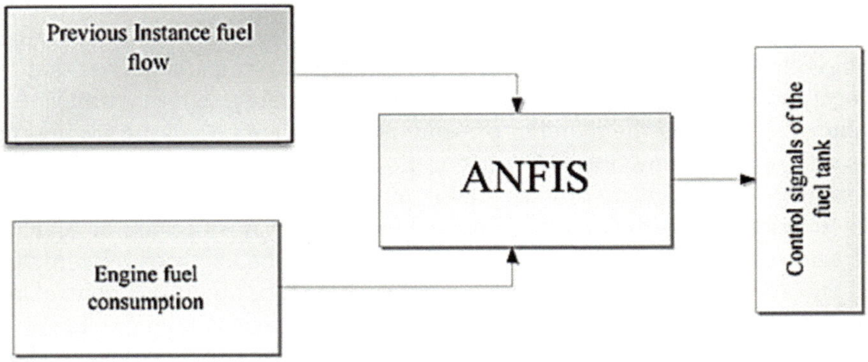

Figure 24.
The structure of the proposed controller.

antecedent and consequent parameters. ANFIS is more robust and has better performance compared to conventional computing methods. These unique properties of the ANFIS such as improved computational power and high reasoning ability, permit it to be used in the fault diagnosis and prognosis of the fuel system to manage fuel flow rate as per engine consumption. The structure of the ANFIS process implemented for the health management of the fuel system is as depicted in **Figure 24**. The control signals generated are decided based upon the input parameters such as engine fuel consumption and previous instance fuel flow to the engine.

The rule structure of ANFIS is determined by the interpretation of the features of the variables of the fuel system model. ANFIS learn details of the input data points, calculate the membership function that best suits to track the input data and output data. The parameters related to the membership functions varies with the learning process of the fuzzy system which depends on the gradient vector. This gradient provides the measure to check the ability of the FIS for the given set of system parameters. The performance is evaluated by considering the error the difference between the actual and desired outputs. **Table 5** shows the fuzzy inference rules framed for the four-tank fuel system. Based on these seven logical rules the learning of the system parameters related to the fuel system with the fault data is carried out. The training data points of input data and output data sets applied to the ANFIS scheme is configured similarly as applied to the ANN.

During the evaluation, the ANFIS structure enables the change in the rules of the FIS. This property of ANFIS helps to optimize itself for the given number of iterations by changing the shape of the membership function, rules and also removes the unnecessary rules during training. A suitable ANFIS Simulink model is designed and developed for the health management purpose of the aircraft fuel system. ANFIS as a controller is designed for the aircraft fuel system with two input

	Input				Output			
Rules	**Tank 1**	**Tank 2**	**Tank 3**	**Tank4**	**Tank 1**	**Tank 2**	**Tank 3**	**Tank4**
1	Normal	Normal	Normal	Normal	Normal	Normal	Normal	Normal
2	Medium	Normal	Normal	Normal	Medium	High	Normal	Normal
3	Low	Normal	Normal	Normal	Low	High	High	Normal
4	Very low	Normal	Normal	Normal	Very low	Very high	Very high	Normal
5	Normal	Medium	Normal	Normal	Medium	High	Normal	Normal
6	Normal	Low	Normal	Normal	High	Low	High	Normal
7	Normal	Very low	Normal	Normal	Very High	Very low	Very high	Normal

Table 5.
Fuzzy inference rules.

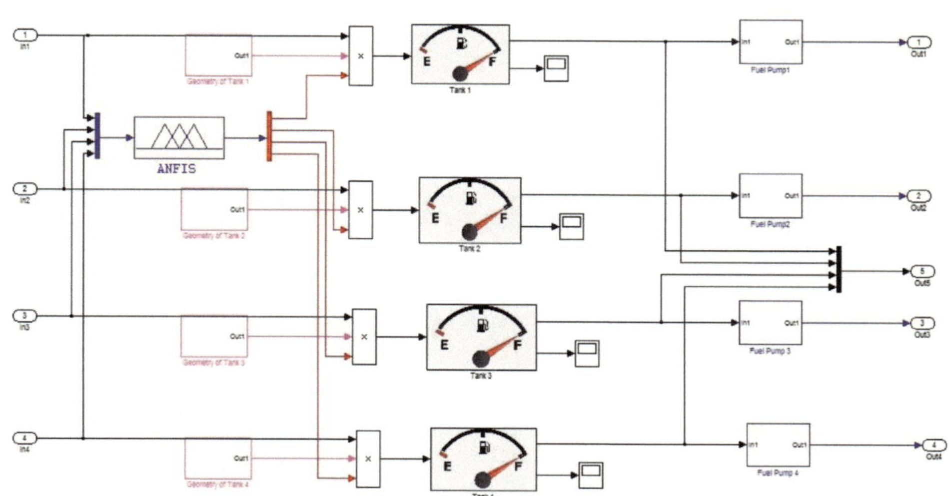

Figure 25.
Snapshot of the Simulink model of a fuel system with ANFIS as a controller.

Tank 1	Tank 2	Tank 3	Tank 4
1	1	1	1
0.7500	1.3300	1	1
0.3500	1.4200	1.3300	1
0	1.5500	1.5500	1
1	0.7500	1.3500	1
1.4000	0.3500	1.3500	1
1.5500	0	1.5500	1
1	1.3500	0.7500	1
1.3000	1.4500	0.3500	1
1.5500	1.5500	0	1
1	1	1.3500	0.7000

Table 6.
ANFIS target data.

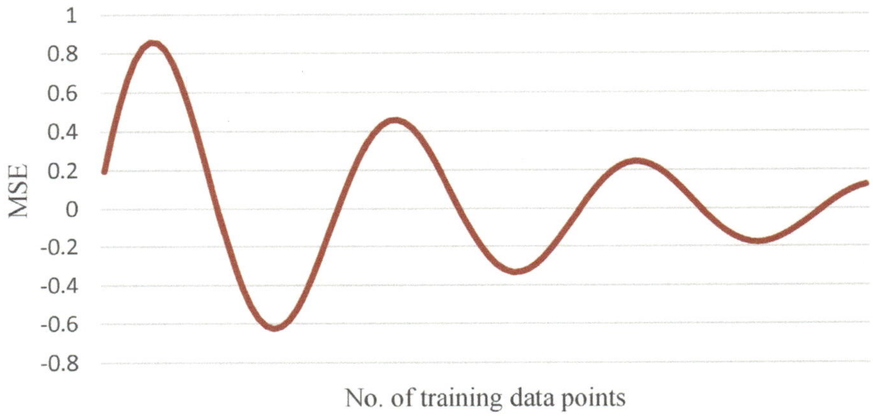

Figure 26.
Training process of ANFIS.

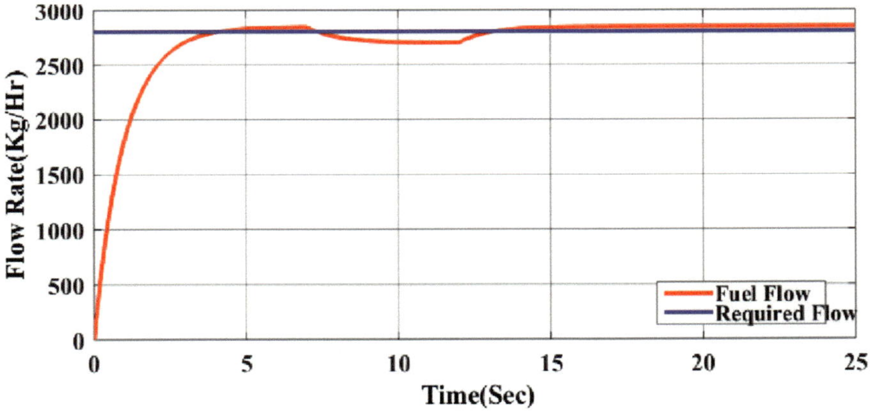

Figure 27.
Fuel consumption by the engine using ANFIS as a controller.

parameters and five bell membership functions for each input unit. **Figure 25** shows Simulink model of aircraft fuel system with ANFIS controller.

Table 6 gives the details of the target output generated by the ANFIS. Similar to the ANN, ANFIS performance is evaluated based on the MSE value. **Figure 26** depicts the curve of convergence of the training data with ANFIS target data indicated as reduction of MSE error. As the number of iterations are increased the MSE reduces indicating the fulfillment of desired target data from the training process of the ANFIS.

Figures 27 and **28** shows the rate of fuel consumption and management test using ANFIS as a controller. The target output generated as control signal identifies the presence of a fault and provide 2700 kg/hr. of fuel flow rate which is almost

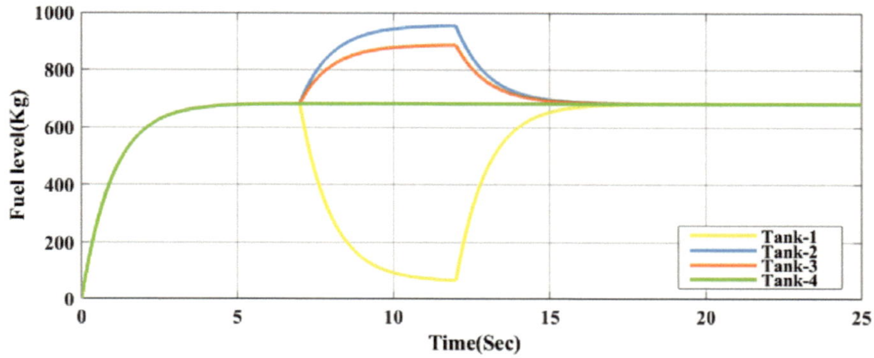

Figure 28.
Fuel management test with the ANFIS controller.

Tank-1	70 kg	Required Fuel
Tank-2	990 kg	2800 kg
Tank-3	940 kg	
Tank-4	700 kg	
Total	2700 kg	

Table 7.
Fuel fetched from each tank using ANFIS as controller.

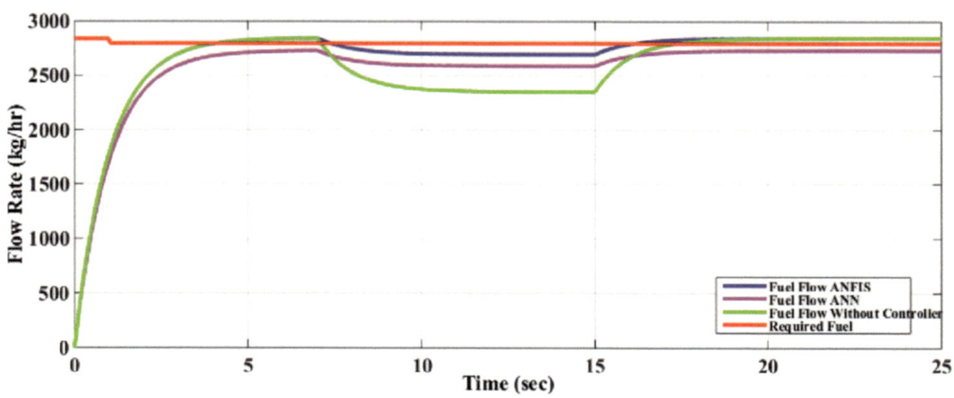

Figure 29.
Comparisons of fuel consumption.

near to the engine requirement 2800 kg/hr. **Figure 27** also illustrates that the learning ability of the ANFIS reproduces accurately the desired output as compared to the ANN process. Thus, the error difference between the actual output value and the obtained output value is very small which is of 100 kg/hr. of flow rate. The ANFIS controller predicts the required fuel by fetching the fuel from the other tanks as per the **Table 7**.

The effectiveness of the ANFIS health management scheme is evaluated by comparing with ANN and fuel system without a controller. The details of the ANFIS based fault diagnosis process is presented in work of [15]. Both techniques uses similar fault conditions. In terms of comparison of training process using ANFIS and ANN techniques, it is clear from the **Figure 29** that ANFIS provides better results. ANN and ANFIS methods detect the time of the fault, diagnose and predict the required flow rate by injecting the additional fuel from other tanks. However, the weight updation process of ANN is based on the historical dataset, which gives the mismatching results during the testing time. Due to this reason, the fuel flow rate obtained by the ANN method is 2600 kg/hr. and is not the desired requirement of fuel by the engine. Hence, it shows that ANFIS technique manage the health status of the aircraft fuel system by monitoring and managing the accurate fuel flow to the engine.

6. Conclusion

Soft computing methodologies like ANN and ANFIS are described in this chapter. A comparison study is made in [16]. All the simulation done is considered for the laboratory conditions only. Based on the theory of NN and FIS, the concept of hybrid five-layer ANFIS structure is implemented and simulated for the health management of the fuel system. Both the techniques help to monitor and manage the rate of fuel flow as required by the aircraft's engine by generating the control signals. Further, based on an adaptive algorithm fault analysis is carried by the author in paper [17]. Diagnostic and prognostic process are carried out through managing the previous fuel flow and fuel consumption by the aircraft engine using ANFIS. ANFIS is a hybrid computational tool, which helps to tune and explain past data and predict future behavior of the system. The fuzzy inference rules that are created in ANFIS rely on both the input and the target output. Tuning can be accomplished with the learning ability of NN. To achieve flawless performance possible faults in the fuel system are detected and corrected by generating the appropriate control signals before the occurrence of massive damages in terms of economy and human life. In this scenario, health management tools have been encouraged.

Acknowledgements

First of all, I would like to thank the Supreme power, the Almighty, being with me and always guiding me to work on the right path of life. Without His grace, this would not have been possible. I would like to express my sincere thanks to my Research Guide Dr. Vanam Upendranath, Senior Principal Scientist, Aerospace Electronics and Systems Division, CSIR-National Aerospace Laboratories (NAL) Bangalore, Karnataka, India and Co-guide Dr. D. S. Jangamshetti, Prof. Dept. of EEE, Basaveshwar Engineering College, Bagalkot, Karnataka. This work would not have been possible without their support, encouragement and able guidance.

Nomenclature

$S(.)$	Sigmoid activation function
$y'(k)$	Sigmoid function
$\theta(.)$	Threshold activation function
δ_p	Error derivation
δ_k	Derivation of error
t_p	Corresponding target
v_k	Predicted outputs
$e_{i/p}$	Actual input
$e_{o/p}$	Output of the NN model
N	Number of iterations
$F_{L1,i}$ and $F_{L1,j}$	Outputs of the fuzzy layer
$\mu A^1{}_i(x)$ and $\mu A^2{}_j(y)$	Membership functions
$\overline{Z_i}$	Output of the third layer
$\{P_i, Q_i, C_i\}$	Consequent parameters set
f	Total output
W	Regular hyperplane
R	Highest distance value
$\|\|w\|\|$	Geometry distance
Φ	Kernel function
$f_T^1(t)$	Time taken by the fuel tank
$f_T^{fuzz}(t)$	Time taken by the fuel tank after fuzzification
$T_{ar}{}^{fuzz}(t)$	Target time of fuel tank

Author details

Vijaylakshmi S. Jigajinni
Department of Electronics and Communication Engineering, Basaveshwar Engineering College (A), Bagalkot, India

*Address all correspondence to: talk2vijusj@gmail.com

References

[1] G. Vachtsevanos, F. Lewis, and M. Roeme, "*Intelligent Fault Diagnosis and Prognosis for Engineering System*", John Wiley & Sons, 2006.

[2] Shukri M., Khalid M., Yusuf R. and Shafawi M., "Induction Machine Diagnostic using Adaptive Neuron Fuzzy Inference System", KES, pp. 380-387, 2004.

[3] Ye Z., Sadeghian A., and Wu B., "Mechanical fault diagnosis for an induction motor with variable speed drives using adaptive neuro-fuzzy inference system", Electric Power Systems Research, Vol. 76, pp. 742-752, 2006.

[4] Tran V.T., Yang B.S., Oh M.S. and Tan A.C.C., "Fault diagnosis of induction motor based on decision trees and adaptive neuro-fuzzy inference", Expert Systems with Applications, Vol. 36(2), pp. 1840-1849, 2009.

[5] Jimenez Juan F., Jose M. Giron Sierra, C. Insaurralde and M. Seminario, "A simulation of aircraft fuel management system", Simulation Modelling Practice and Theory, Vol. 15(5), pp. 544-564, 2007.

[6] Wang W.Q., Golnaraghi M.F., and Ismail F., "Prognosis of machine health condition using the neuro-fuzzy system", Mechanical Systems and Signal Processing, Vol. 18, pp. 813-831, 2004.

[7] Skybrary, "Aircraft Fuel System", Skybrary Available: http://www.skybrary.aero/index.php/Aircraft_Fuel_Systems.

[8] Robert Breda and Vladimir Beno, "Modeling of control circuit of aircraft fuel system", Przegląd Elektrotechniczny, Vol. 89, pp. 172-175, 2013.

[9] S. J. Hong and G. S. May, "Neural Network based sensor fusion of optical emission and mass spectroscopy data for real time fault detection in reactive ion etching", IEEE Trans. Ind. Electron., Vol. 52(4), pp. 1063-1072, 2005.

[10] Haykin S, Neural networks. A comprehensive foundation. Macmillan Collage Publishing Company, New York, 1994.

[11] M. Sugeno and G. T. Kang, Structure identification of the fuzzy model, Fuzzy Sets and Syst.,vol. 28, pp. 15–33, 1988.

[12] Azimi, Hamed, Hossein Bonakdari, Isa Ebtehaj, Seyed Hamed Ashraf Talesh, David G. Michelson, and Ali Jamali, Evolutionary Pareto optimization of an ANFIS network for modeling scour at pile groups in clear water condition, Fuzzy Sets and Systems, Vol.319, pp.50-69, 2017.

[13] T. Takagi and M. Sugeno, derivation of fuzzy control rules from hu an operators control action, in Proc. IFAC symp. Fuzzy Inform., knowledge Represenstation, and Decision Analysis, July 1983, pp. 55-60.

[14] E. H. Mamdani and S. Assilian, An experiment in linguistic synthesis with a fuzzy logic controller, Int. J. Man-Machine Studies, vol. 7, no. 1, pp. 1–13, 1975

[15] Vijaylakshmi S. Jigajinni and Vanam Upendranath, ANFIS based fault diagnosis tool for a typical small aircraft fuel system, in the proceeding of International Conference on Intelligent Communication, Control and Devices, 2016.

[16] Vijaylakshmi S. Jigajinni and Vanam Upendranath, Simulation and modeling of ANN-based prognosis tool for an aircraft fuel system health management, Proc. of 7th Int. Conf. of AI & Soft Comp, Chennai, CS & IT-CSCP, July 14-15, 2018, pp 83-93.

[17] Vijaylakshmi Jigajinni and Vanam Upendranath, Health management of a typical small aircraft fuel system using an adaptive technique, in Proc. of 25th IEEE Int. Conf. on High Performance Computing, Data and Analytics held at Bangalore, India Dec. 17-20th , 2018, pp 104-112.

Section 6

Control Systems

Chapter 11

Fuzzy Logic Control with PSO Tuning

Jeydson Lopes da Silva

Abstract

Several applications of artificial intelligence in the area of control of dynamic systems have proven to be an efficient tool for process improvement. In this context, control systems based on fuzzy logic - Fuzzy Logic Control (FLC) are part of a series of advances in the areas of control systems. Fuzzy control is based on natural language and therefore has the ability to make approximations closer to the real nature of the problems. The use of metaheuristic algorithms such as the particle swarm optimization (PSO) allows it to provide adequate adjustments to the fuzzy controller in an optimized manner. This technique allows to adjust the FLC in a simple way according to the performance desired by the designer, without the need for a long time of conventional tests.

Keywords: FLC, PSO, artificial intelligence, controller, optimization

1. Introduction

FLC represents a family of intelligent controllers with a lot of potential for use in the world for industrial control systems. Its popularity is mainly due to its performance being robust in several operating conditions and its functional simplicity, in addition to its ease of implementation, allowing engineers to operate them in a simple and direct way. Even with the emergence of new control techniques, FLC controllers will still be on the market for a long time in industrial plants [1].

A good parameterization of the FLC inference functions is essential to allow a good performance of this type of closed loop controller. The tuning of the controller is a persistent problem in the area of control and automation, from the initial approach of this topic to the present day; a definitive solution has not yet been reached, being a subject constantly addressed in several works in the field of control engineering. However, it is important to note that despite the various techniques that produce adjustments in the FLC parameters, it is still necessary to assess the designer regarding the result of the parameterization of this controller [2].

In recent years, the computational capacity available allows optimization techniques developed in the field of artificial intelligence to gain space in the solution of several engineering problems [3, 4]. In this context, algorithms based on metaheuristics can provide adequate solutions to the FLC parameterization problem. Since the parameters necessary for the proper functioning of the FLC can be numerous and often complex, techniques based on intelligent computing provide an alternative solution to this type of problem.

2. Particle swarm optimization

Particle Swarm Optimization (PSO) algorithm is a population metaheuristics created from models of the collective and social behavior of animals, in their coordination of movements in the tasks of searching and obtaining food. These models were simplified, losing the requirement to maintain a minimum distance between its neighbors. In addition, the communication architecture was transformed, which was initially inspired by spatial proximity and was changed to use a topology defined by a graph. Therefore, PSO ends up having, nowadays, more similarities with models of mutual influence between human beings in their ways of thinking and acting [5].

In the PSO there is a fixed amount of agents, called particles. This set of agents is called a swarm, designed in such a way that each agent is able to communicate with its neighbor, which is a subset of its peer, and can be defined in a static or dynamic way. Each of the particles moves in the solution space with a certain speed, always evaluating the solution corresponding to that occupied position in each iteration. The speed of the particles must be influenced by your own experience (cognitive factor), and also influenced by your neighbors (social factor). Such influences are implemented as two attractors, the first located in the best position already evaluated by the particle itself and the second located in the best position visited by the neighboring particles [6, 7].

In general, the position of a particle i at a time k is represented by a vector x_k^i and its respective speed as a vector v_k^i. Both vectors are stored during the learning process in generation k and used for updating in the next generation. In addition to the vectors x_k^i and v_k^i, PSO uses the best position of the particle i (p_k^i), as well as the position of the best particle of the whole swarm (p_k^g), both evaluated throughout the process until the moment k evaluated.

The equations that govern the PSO can be defined as

$$v_{i+1} = \omega v_i + c_1 r_1 (p_i - x_i) + c_2 r_2 \left(p_g - x_i\right) \tag{1}$$

$$x_{i+1} = x_i + v_{i+1} \tag{2}$$

where x is the position of the particle; v is the velocity of the particle; w is the inertial weight, controlling the impact of the previous speed at the current speed; c_1 and c_2 are positive constants, controlling the social and individual behavior of each particle; r_1 and r_2 are random numbers in the interval [0.1], contributing to diversify the exploration of the problem search space.

The basic PSO algorithm can be defined as follows.

Randomly initialize the particle positions x_k^i and speeds v_k^i within the search space at k = 0.

Algorithm 1: PSO algorithm

Randomly initialize the particle positions x_k^i and speeds v_k^i within the search space at k= 0;
while *Unsatisfied stop conditions* **do**
  Evaluate the objective function of each of the population particles;
  Update the best position of each particle individually p_k^i and the best position of the swarm p_k^g;
  Update the position of each particle in time *k+1* based on position and speed in time *k*, based on equations (1) and (2);
end

---

3. Fuzzy logic controller

From the concepts of fuzzy logic, which include fuzzification, defuzzification, inference system and compositional rules, it was possible to build a fuzzy logic controller, where control actions are generated from a set of linguistic knowledge. In general, the FLC is conceptually easy to understand. Despite its easy understanding, its design requires a greater amount of parameterization than normal in other controllers [8].

Several previous works attest that the application of FLC can overcome the effectiveness of traditional controllers in several tasks. In fact, the FLC works perfectly in a dynamic control environment, its steps can be similar to other controllers. In the fuzzification stage, the signals from the plant's sensors and any other information that is declared important are mapped to the pertinence functions and appropriate truth values. Then, the processing step makes use of each appropriate rule, thereby generating a particular result individually and then combines the results of those rules. Finally, the result of the operation in the previous stage is then converted into the defuzzification to an output value according to the system in which the FLC is inserted [9, 10].

Regarding the FLC memberships functions (MFs), it should be noted that the most applicable functions are the triangular, trapezoidal and Gaussian MFs type. This is due to the fact that. In the FLC project there is no exact number of how many curves, or what types of curves must be inserted in order to improve the FLCs performance, this work is part of the specialist's commitment.

A fuzzy controller can contain dozens of pertinent functions and associated rules. The rules have an empirical character, since they are elaborated from the knowledge of a process specialist. This fact is especially important and makes the FLC different and advantageous compared to other controller architectures. The reason for this is that empirical rules are especially useful in plants that contain complex processes or even with insufficient information that would decrease the effectiveness of more conventional control techniques.

3.1 FLC designer

Similar to what happens in fuzzy logic applications in other areas, the FLC project requires several operational requirements that must be adequately dimensioned for the correct functioning of this type of control. An important choice as to how the final control signal will take place is the decision between the TSK or Mamdani FLC. The general rule model is given as:

$$\textbf{IF}\, x\ is\ A\ \textbf{AND}\, y\ is\ B\ \textbf{THEN}\, u\ is\ C, \tag{3}$$

where, A is a fuzzy set of X, B is one of the fuzzy set of Y and C is fuzzy set U (signal of controller).

In the case of the FLC of the Mamdani type, each rule is a conditional fuzzy proposition, and different fuzzy relationships in A x B x C can be derived from it. The implementation of each rule is done by defining operators to process the rule's antecedent and the implication function that will define its consequent. In this case, the action of the FLC is defined based on the aggregation of rules that make up the algorithm. This aggregation results in the fuzzy set C, which defines the output of controller C. The effective output of the controller is then obtained through a defuzing process applied to set C.

The TSK FLC is a simplified model of the Mamdani controller, where the consequent of each rule is defined as a function of the linguistic input variables.

Each rule results in no longer a numerical value but a fuzzy set. The weight assumes the pertinence value resulting from processing the rule's antecedent. The value of **u** can be defined as a constant (single value with relevance equal to one) - **singleton**. From a singleton it is possible define rules with output values that represent a classification of the controller response, without changing the simplified way of determining the final controller response.

In many practical situations of real controllers trapezoidal MFs are used for the function inputs. The reason is that several sensors installed in the system can present different noise in the measurement and as a consequence change some conditional rule in the operation of the FLC.

Regarding the design of the basic FLC rules, the expert needs to decide the total number of rules and how these operational conditions are related.

3.2 FLC optimization

The central idea of using metaheuristics in the pursuit of optimizing FLC performance is in the design characteristic of this controller. The various parameters and variables that are necessary for the good functioning of the FLC do not present general rules, making the role of the operator essential in the analysis and dimensioning of these terms [11].

Despite the specialist's knowledge, the resulting FLC may not perform as optimally. Metaheuristics, on the other hand, can function as an intelligent search engine for the various possible architectures for the FLC, without the need for the exhaustive work of trial and error on the part of an expert control engineer [11–13].

The use of a metaheuristic for the optimization of the FLC needs to be correctly dimensioned for its proper and effective functioning. There are different architectures to be explored when using these techniques despite an FLC [13]. **Figure 1** presents an FLC architecture using metaheuristics for its optimization.

From the observation of **Figure 1**, it is possible to notice that metaheuristics can help in the solutions of different areas of the FLC. There are several options for the parametric optimization of the FLC, therefore, the designer needs to define what one wants to obtain or which are the important parameters to be defined in his control project.

3.2.1 FLC membership function optimization

An application of FLC parametric optimization with metaheuristics is based on determining the positions and characteristics of MFs. In this case, algorithm

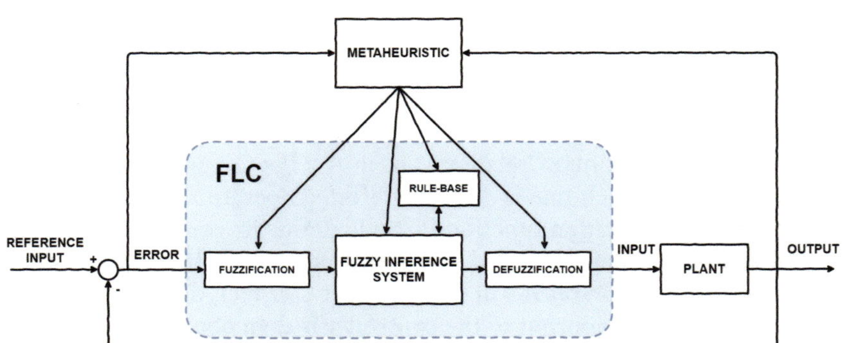

Figure 1.
FLC optimization scheme.

objective is to find the values of each MF individually that allows to obtain a better performance of this controller [14].

Assume that you have determined a set of rules and chosen all the MFs in your project. A common application of a metaheuristic algorithm in this case is to determine the ideal positioning of the MFs. In this particular context, the MF designer is a time-consuming exercise and it is usually possible to achieve good results. For this reason, some techniques offer benefits to develop such functions. In this case, it is necessary to turn the positions of each MF into optimization variables of the algorithm. **Figure 2** presents a representation of this application.

Figure 2 presents a possible model for the application of metaheuristic optimization of MFs. From a set of previously selected MFs, it is assumed that the positions of these functions are the variables (x_i) of the optimization problem. In this way, the algorithm seeks to select such parameters, increasing or decreasing the space of each one in the possible set. The position and size of each MF is important since this has a direct influence on the output of the FLC system.

In this FLC optimization application, it is necessary to define which MFs are involved and the respective definition limits. In practice, it is necessary to limit the range of values for each variable (x_i), thus avoiding that the optimization does not result in inadequate results.

3.2.2 FLC rules optimization

The optimization of the FLC rules is also an interesting application on this topic. Determining the rules of an FLC is naturally an empirical process, since trial and error are very necessary in order to be successful in these rules [15].

The application of metaheuristics in this case is configured to determine the best rules that, when combined in the logic of the FLC, optimize its operation. The variables in this situation are the defining set of MFs and can be combined in different ways to compose the rules. In this case, based on a previous set of information, it is possible to determine which variables will be part of a given rule and even how many rules should drive the FLC. **Figure 3** shows a metaheuristic scheme for the determination of a specific rule of the **IF-THEN** type.

Figure 3 shows an optimization of rules based on IF - THEN. From the set of MFs (NB, NS, NM, Z, PS, PM, PB), the subsequent terms $y_1(i)$ (de/dt) and $y_2(j)$ (e), as well as the consequent term $z(k)$ can be defined appropriately for the formation of FLC rules. In this example, i = 7 and j = 7, totaling 49 possible rules to be determined by metaheuristic optimization.

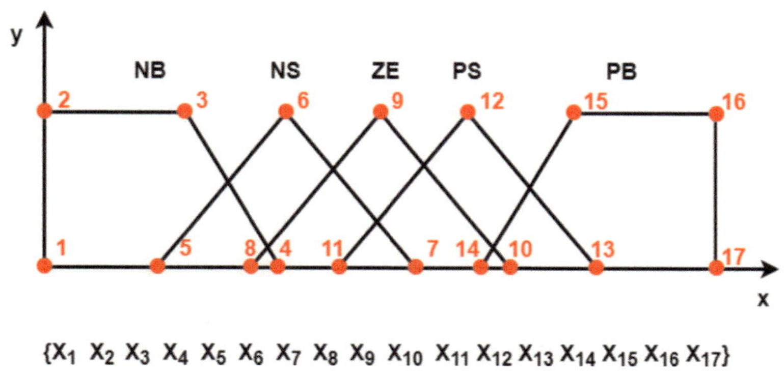

Figure 2.
MF optimization scheme.

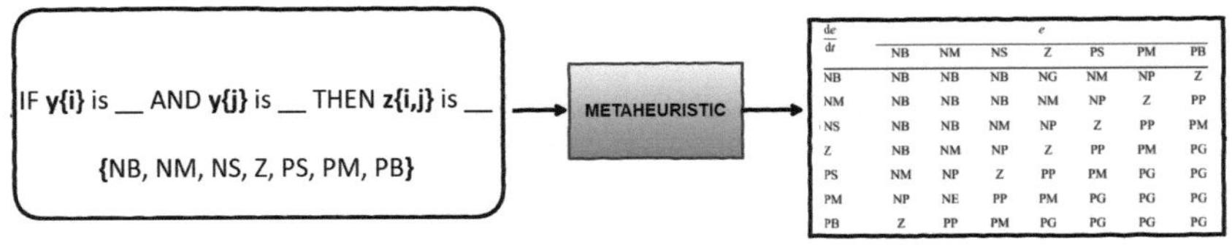

Figure 3.
Optimization scheme for fuzzy rule conditionals.

Metaheuristic algorithm tests all possible combinations of the IF-THEN condition simultaneously, thus allowing to speed up this process of determining the FLC rules.

In fact, there are several possibilities for the elaboration of FLC rules, which makes this problem complex for the use of traditional optimization methods.

4. Fitness function modeling

One or more performance requirements can be used to compose the performance index of the system, which will serve as the objective function (fitness) of the optimization problem used to obtain the tuning of the controller. The performance and robustness requirements act as restrictions on the tuning of the controller, that is, they allow imposing limits on the behavior of the control loop. The performance index to be useful to the problem must be composed of system parameters, in addition to that, it must be computed with some ease, analytically or experimentally [16].

In order to make the optimization functional, the definition of a fitness function is extremely necessary. This function aims to assess the quality and behavior of the control system, allowing the designer to check how well his project is performing. In general, dynamic performance quality indices are adopted as variables of an objective function in control systems.

The composition of the objective function is based on the dynamic characteristics that the operator wants the FLC to achieve in the optimization. This definition happens empirically based on the observation of the system and execution tests.

If the designer wishes the controller response to have a low overshooot value, this variable should be added with some gain in the composition of the fitness function. This gain depends on how much this variable is intended to influence the optimization goal. Moreover, if there is a desire to minimize some error rate in general, ITAE, ISE or IAE can be adopted in the function.

In such cases, the goal of metaheuristics is to minimize the objective function and, since this function is directly associated with a specific parameter, the specific objective can be achieved.

In order to demonstrate the construction of the objective function, consider the following general formulation

$$J = C_1X_1 + C_2X_2 + \ldots + C_nX_n, \tag{4}$$

where C_i represents the constants that are associated with the variables of interest X_i. The designer must determine how many variables one wishes to compose the function such that it achieves its final FLC performance objective. However, some variables in the function can be redundant and not affect the overall performance of the optimization. In this way, the designer's expertise is essential to obtain a function that allows an effective optimization of the FLC.

5. FLC optimization example

In order to present a practical application of FLC optimization via metaheuristic algorithm, this section presents an example of using a metaheuristic (PSO) for positional optimization of MFs in an FLC.

An air flow system is presented as a transfer function and an FLC is connected to it, which is parameterized manually and also by the PSO. Since the PSO algorithm is relatively simple and easy to apply, consider it for use in this application example.

The FLC is of the TSK type (control output is a specific value). In addition, the error and its derivative are considered as input to the system. The reason for this is that in addition to the error, it is also necessary to understand the speed and direction of the error variation.

In order to facilitate application, an example in MATLAB is used. **Figure 4** shows the FLC scheme associated with a metaheuristic.

From **Figure 4**, it can be seen that the FLC has the error and the error derivative as inputs. The error signal is used to calculate the ITAE and IAE indices. The FLC model has two outputs, one for the control signal and one for the rules. The control signal output passes through an integrating signal to compose the control signal. Moreover, the other output of the FLC is being used for the optimization of this controller. In this example five MFs are used (NB, NS, Z0, PS, PB) for the input error *e* (triangular), input error derived *d/dt* (triangular) and for the controller *output*. The optimization focuses on determining the position of the MFs (section 3.2.1) for the inputs and output.

First, the optimization focuses on the triangular functions, choosing one of the three positions that make up the triangle to carry out the modification. This is done by adding a variable X_1 and X_2 to this position while the others remain with a fixed value. Second, we seek to find the value of the output using the optimization variables X_3 and X_4. In addition, the rules have already been pre-established and are not targets for optimization. The source code for the definition of the FLC, including the MFs and the respective rules are presented in **Figure 5**.

Regarding the fitness function to be used in this example, the system uses the ITAE and IAE error indexes to formulate this function (**Figure 6**).

$$J = 0.5 * ITAE + 0.5 * IAE \tag{5}$$

For each new population resulting from the PSO swarm, the variables from the algorithm are stored and inserted in the FLC and, from that, the result of the fitness function (J) is estimated. This is done until reaching the stopping criterion. In this example, add the value of $J < 10^2$ as the main stop criterion for the algorithm. The code that presents the implementation of the objective function is shown below.

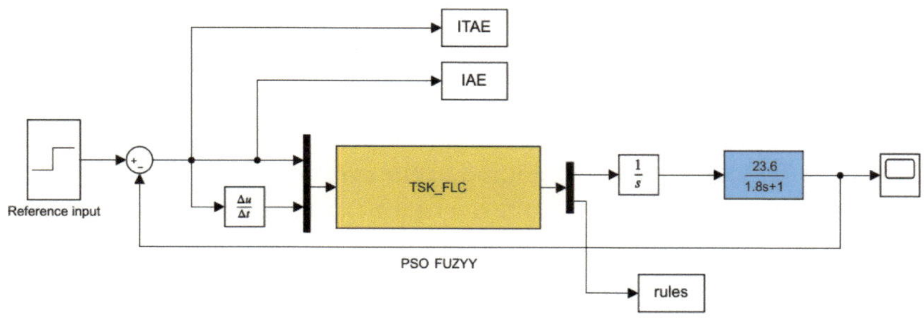

Figure 4.
FLC scheme for optimization by PSO in MATLAB.

```
function out=functions(input)
X=evalin('base','X');
warning off;
error=input(1);
d_error=input(2);
fuz=newfis('fuz','sugeno');          %Sugeno Type FLC

Sa1=[0.5,0.5];        %Input and Output membership functions are equallt distributed
Sb1=[0.5,0.5];        %Input and Output membership functions are equallt distributed
Sc1=[0.5];            %Input and Output membership functions are equallt distributed

fuz = addvar(fuz,'input','erro',X(1)*[-1 1]);
fuz = addvar(fuz,'input','d_erro',X(2)*[-1 1]);
fuz = addvar(fuz,'output','u1',[-1 1]);
% MF for input1
fuz = addmf(fuz,'input',1,'NB','trimf',[-1 -1     -Sa1(2)]);        %Five Triangular Membership functions
fuz = addmf(fuz,'input',1,'NS','trimf',X(1)*[-1 -Sa1(1) 0 ]);
fuz = addmf(fuz,'input',1,'ZO','trimf',X(1)*[-Sa1(1) 0   Sa1(1)]);
fuz = addmf(fuz,'input',1,'PS','trimf',X(1)*[0    Sa1(1) 1]);
fuz = addmf(fuz,'input',1,'PB','trimf',X(1)*[Sa1(2) 1     1   ]);
% MF for input2
fuz = addmf(fuz,'input',2,'NB','trimf',X(2)*[-1 -1     -Sb1(2)]);    %Five Triangular Membership functions
fuz = addmf(fuz,'input',2,'NS','trimf',X(2)*[-1 -Sb1(1) 0 ]);
fuz = addmf(fuz,'input',2,'ZO','trimf',X(2)*[-Sb1(1) 0   Sb1(1)]);
fuz = addmf(fuz,'input',2,'PS','trimf',X(2)*[0    Sb1(1) 1]);
fuz = addmf(fuz,'input',2,'PB','trimf',X(2)*[Sb1(2) 1     1   ]);
% MF for output1
fuz = addmf(fuz,'output',1,'NB','consXnt',[-1]);                     %Five Triangular Membership functions
fuz = addmf(fuz,'output',1,'NS','consXnt',[-Sc1(1)]);
fuz = addmf(fuz,'output',1,'ZO','consXnt',[0]);
fuz = addmf(fuz,'output',1,'PS','consXnt',[Sc1(1)]);
fuz = addmf(fuz,'output',1,'PB','consXnt',[1]);

matrix_rules =[  1    1    1    1    1; 1    2    1    1    1; 1    3    1    1    1; 1    4    2    1    1;
                 1    5    3    1    1; 2    1    1    1    1; 2    2    1    1    1; 2    3    2    1    1;
                 2    4    3    1    1; 2    5    4    1    1; 3    1    1    1    1; 3    2    2    1    1;
                 3    3    3    1    1; 3    4    4    1    1; 3    5    5    1    1; 4    1    2    1    1;
                 4    2    3    1    1; 4    3    4    1    1; 4    4    5    1    1; 4    5    5    1    1;
                 5    1    3    1    1; 5    2    4    1    1; 5    3    5    1    1; 5    4    5    1    1;
                 5    4    5    1    1;];

fuz=addrule(fuz,matrix_rules);
out= evalfis([error d_error],fuz);
```

Figure 5.
FLC code in MATLAB.

```
function OptFUZZY =OptmiFUZZY(x1,x2,x3,x4)
%clear all
%close all

%Objective function calculation - partially calculated inside of the model

assignin('base', 'X', [ x1 x2 x3 x4]);

sim('Fuzzy_PSO.mdl');
OptFUZZY =0.5*max((Sys_Out1.Data(end))+0.5*max((Sys_Out2.Data(end))); %ITAE + IAE


end
```

Figure 6.
Fitness function code in MATLAB.

For the application of the PSO, it is first necessary to initialize the algorithm parameters. In this case, initially the value of the particle positions in the swarm is randomly defined. The position of all particles is usually started with some speed, in this example all speeds are started at zero. Finally, it is defined as a minimum value of the best initial particle in 1000. The particles in the swarm must reach and exceed at least this value of the "best initial particle". **Figure 7** shows the code used.

The operation of the PSO is simple, the population is evaluated and its position is updated based on its position and previous speeds. The swarm positions are then entered in the *X* variables, which in turn are the inputs for the "*OptFuzzy*" optimization function. The *X* values are actually the values that will adjust the positions of the FLC's input MFs. If the value obtained in the optimization is less than the current value of the objective function, the best individual particle values and the best global value are increased. This is done until reaching the stopping criterion. **Figure 8** shows the PSO operation code in this FLC optimization example.

In order to test the PSO in the optimization of the FLC MFs of the present example, we try to test the algorithm with 10, 20, 50, 100, 200 and 500 particles in the swarm. **Figure 9** shows the behavior of the objective function "*OptFuzzy*" over 50 iterations.

In general, the greater the amount of swarms in the PSO, the faster the minimization of the objective function will occur, since this way there will be a greater exploration of the search space of the algorithm.

```
%% Particle Swarm Optimization Simulation

iterations = 5000;
inertia = 1.0;
correction_factor = 1.0;
swarm_size =500;

% ---- initial swarm position -----
%index = 1;
for index = 1 : iterations

        swarm(index, 1, 1) = randi([0,10]);
        swarm(index, 1, 2) = randi([0,10]);
        swarm(index, 1, 3) = randi([0,10]);
        swarm(index, 1, 4) = randi([0,10]);
        %index = index + 1;
end

swarm(:, 4, 1) = 1000;          % best value so far
swarm(:, 2, :) = 0;             % initial velocity
```

Figure 7.
PSO parameters.

```
%% Iterations
for iter = 1 : iterations

    %-- evaluating position & quality ---
    for i = 1 : swarm_size
        swarm(i, 1, 1) = swarm(i, 1, 1) + swarm(i, 2, 1)/1.3;     %update x1 position
        swarm(i, 1, 2) = swarm(i, 1, 2) + swarm(i, 2, 2)/1.3;     %update x2 position
        swarm(i, 1, 3) = swarm(i, 1, 3) + swarm(i, 2, 3)/1.3;     %update x3 position
        swarm(i, 1, 4) = swarm(i, 1, 4) + swarm(i, 2, 4)/1.3;     %update x4 position
        x1 = swarm(i, 1, 1);
        x2 = swarm(i, 1, 2);
        x3 = swarm(i, 1, 3);
        x4 = swarm(i, 1, 4);

       value = OptFUZZY(x1,x2,x3,x4);
       %(x1,x2,x3,x4);

       %OptFUZZY =0.5*max((Sys_Out1.Data(end)))+0.5*max((Sys_Out2.Data(end)))
       % fitness evaluation (you may replace this objective function with any function having a global minima)
       Fitness = val;
       if val < swarm(i, 4, 1)                  % if new position is X2_ter (fitness value of this particle)
           swarm(i, 3, 1) = swarm(i, 1, 1);     % update best x,
           swarm(i, 3, 2) = swarm(i, 1, 2);     % best y postions
           swarm(i, 3, 3) = swarm(i, 1, 3);
           swarm(i, 3, 4) = swarm(i, 1, 4);
           swarm(i, 4, 1) = val;                % and best value
       end
    end
```

Figure 8.
PSO code in MATLAB.

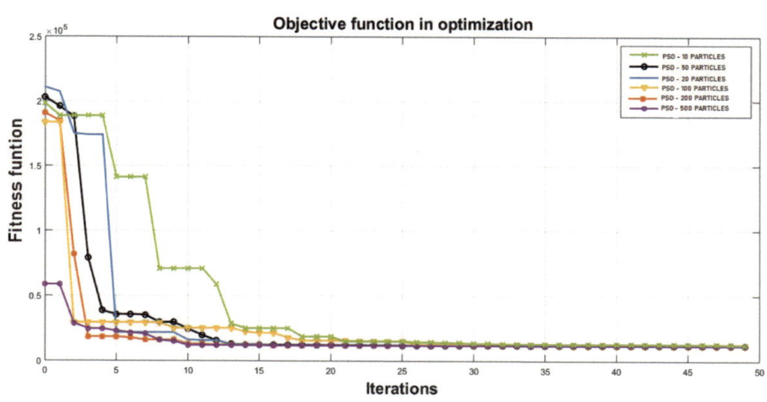

Figure 9.
Fitness function optimization.

In addition to optimizing the FLC's input (*e* and *de/dt*) and output via PSO, manual tuning is also applied to these MFs. The manual tuning basically consists of determining "*manually*" the positional values of the triangular functions of the MFs

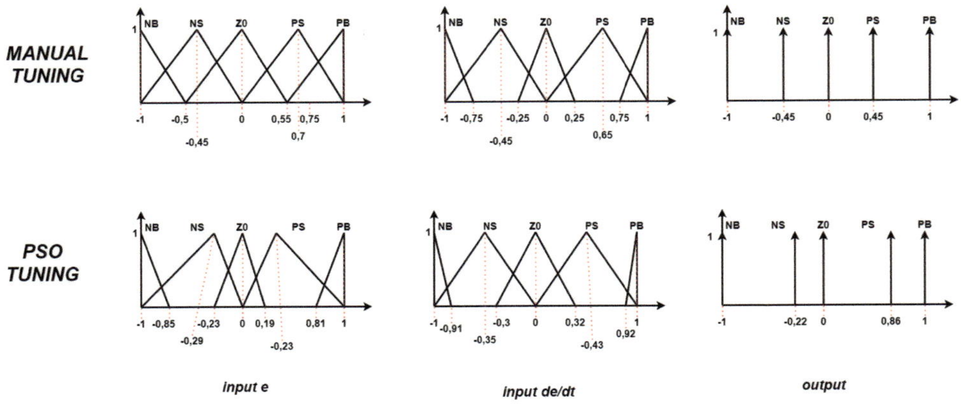

Figure 10.
MFs manual tuning vs. MFs PSO tuning.

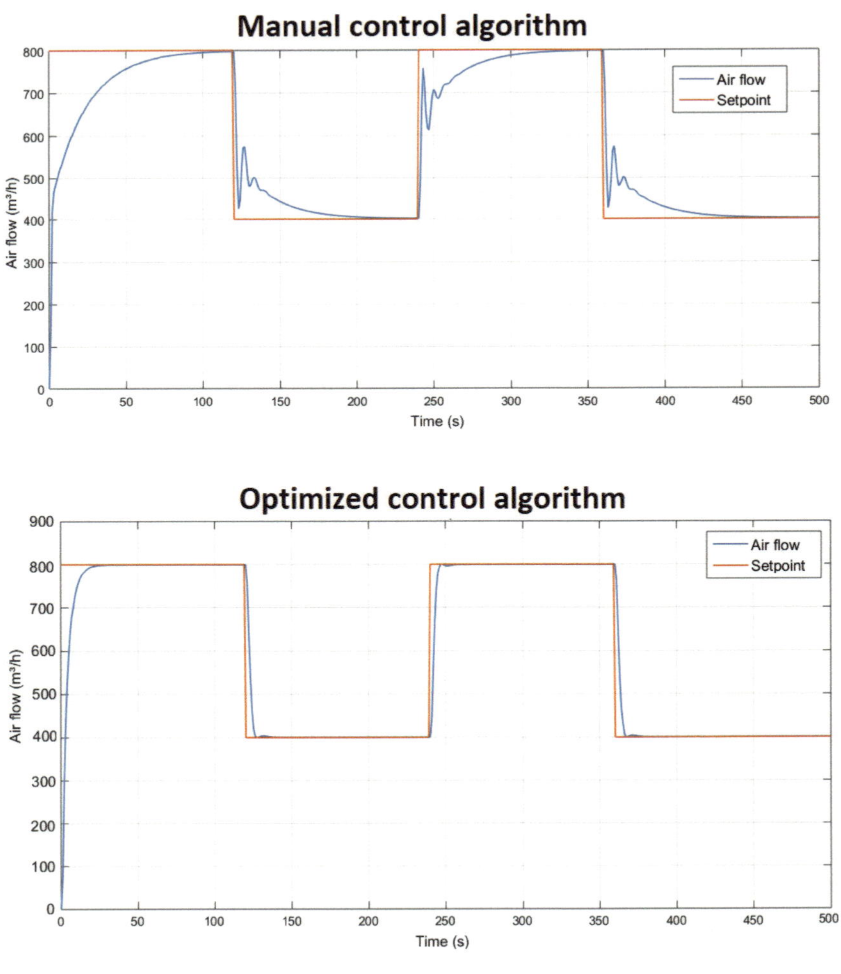

Figure 11.
FLC manual tuning vs. FLC PSO tuning.

FLC Tunning	Rise Time (s)	Settling Time (s)
Manual	33,94	72,10
Metaheurístic	3,56	6,99

Table 1.
Comparison between the different tunings of the FLC.

and the value of the MFs at the output. This choice is empirical and is based on the operator's knowledge and experience regarding the control system. The result of the MFs tuning process in different ways is shown in **Figure 10**.

From the MFs resulting from the tuning processes discussed above (**Figure 4**), the FLC controller is simulated in both tuning situations (Manual and PSO) by applying different steps to the control system input. **Figure 11** shows the result of the behavior of the FLC resulting from the different synotnias mentioned above.

From **Figure 11** it is possible to notice that the FLC obtained from the PSO presented a much more satisfactory performance in terms of dynamic behavior. **Table 1** presents detailed values of the performance indices for both processes.

6. Conclusions

The FLC is an important controller in the area of intelligent systems of control engineering, however its design aspects represent a challenge to the specialist, as its operational requirements require a characteristic knowledge based on linguistic resources.

An alternative to the development of the FLC is the use of optimization resources to aid the development of this type of controller. Metaheuristics can clearly contribute to this issue, as they represent an effective way of exploring solutions to complex problems. The use of this feature involves the definition of an optimization problem, that is, the definition of a fitness function and which variables are associated in the process.

An FLC optimized by a metaheuristic algorithm represents a viable alternative for several applications, as the solutions can be adapted to the optimization criteria of these algorithms, thus allowing the specialist to develop an FLC to meet their specific performance needs.

Abbreviations

FLC	fuzzy logic control
MF	membership function
PSO	particle swarm optimization
GA	genetic algorithm
TSK	Takagi–Sugeno–Kang

Author details

Jeydson Lopes da Silva[†]
Federal University of Pernambuco, Recife, Brazil

*Address all correspondence to: jeydson.lopes@ufpe.br

† These authors contributed equally.

References

[1] Chuen L. Fuzzy logic in control systems: fuzzy logic controller. I. IEEE Transactions on Systems, Man, and Cybernetics, vol. 20, no. 2, pp. 404–418. DOI: 10.1109/21.52551

[2] Yahaya, S Z, Hussain Z, Boudville R, Ahmad F, Taib M N. Optimization of FLC parameters for optimal control of FES-assisted elliptical stepping exercise using GA and PSO. 2014 IEEE International Conference on Control System, Computing and Engineering (ICCSCE 2014), Batu Ferringhi, 2014, pp. 663–667. DOI: 10.1109/ICCSCE.2014.7072801

[3] Ronaldo A, Luis C, Davidson M, Jeydson S, Camila B, Andre A. An Emotional Controller PLC Implementation for an Industrial Fan System. Neural Networks (IJCNN), 2016 International Joint Conference on, 2016.

[4] Arrofiq M, Saad N. A PLC-based Self-tuning PI-Fuzzy Controller for Linear and Non-linear Drives Control. 2nd IEEE International Conference on Power and Energy (PECon 08), Malaysia, 2008.

[5] Kennedy J, Eberhart R. Particle swarm optimization. Proceedings of IEEE International Conference on Neural Networks, p. 1942–1948, 1995.

[6] Saptarshi S, Sanchita B, Richard A P. Particle Swarm Optimization: A Survey of Historical and Recent Developments with Hybridization Perspectives. Journal of Machine Learning and Knowledge Extraction. DOI: 10.3390/make1010010.

[7] Juan L F M. A Brief Historical Review of Particle Swarm Optimization (PSO). Journal of Bioinformatics and Intelligent Control 1(1):3–16 DOI: 10.1166/jbic.2012.1002.

[8] C. C. Lee, "Fuzzy logic in control systems: fuzzy logic controller. I," in IEEE Transactions on Systems, Man, and Cybernetics, vol. 20, no. 2, pp. 404–418, March–April 1990, doi: 10.1109/21.52551.

[9] Zhou, et al.Optimized Fuzzy Logic Control Strategy for Parallel Hybrid Electric Vehicle Based on Genetic Algorithm. Applied Mechanics and Materials, 274, 345–349. https://doi.org/10.4028/www.scientific.net/amm.274.345

[10] P. Rajakumar, R. Saravanakumar. Design of Fuzzy Logic Controller for DSTATCOM. Advanced Materials Research, vol. 905, Trans Tech Publications, Ltd., Apr. 2014, pp. 401–405. Crossref, doi:10.4028/www.scientific.net/amr.905.401.

[11] Lamkhade P, Parvat B, Kadu C. Design and implementation of fuzzy logic controller for level control. 2015 International Conference on Energy Systems and Applications, Pune, 2015, pp. 475–479, DOI: 10.1109/ICESA.2015.7503395.

[12] Katbab A. Fuzzy logic and controller design-a review. Proceedings IEEE Southeastcon '95. Visualize the Future, Raleigh, NC, USA, 1995, pp. 443–449. DOI: 10.1109/SECON.1995.513133.

[13] Shahraki A, Shahraki M, Mosavi M. Design and simulation of a fuzzy controller for a busy intersection. 2013 International Conference on Computer Applications Technology (ICCAT), Sousse, 2013, pp. 1–6. DOI: 10.1109/ICCAT.2013.6521986.

[14] Rahma A, Khemliche M. Combined approach between FLC and PSO to find the best MFs to improve the performance of PV system. 2014 International Conference on Electrical Sciences and Technologies in Maghreb (CISTEM). doi:10.1109/cistem.2014.7077038

[15] Li P, Du X. A GA Optimization for FLC with Its Rule Base and Scaling Factors Adjustment. In: Huang DS., Li K., Irwin G.W. (eds) Computational Intelligence. ICIC 2006. Lecture Notes in Computer Science, vol 4114. Springer, Berlin, Heidelberg. https://doi.org/10.1007/978-3-540-37275-2_1

[16] Ogata K. Modern Control Engineering, Prentice-Hall, 2003. , Prentice-Hall, 2003.

Chapter 12

Functioning Fuzzy Logic in Optimizing the Solar Systems Work

Daoud Raid, Ahmed Omer and Al-khashab Yaareb

Abstract

Fuzzy logic has been used in many fields, either to control a specific movement, improve the productivity of a machine, or monitor the work of an electrical or mechanical system or the like. In this chapter, we will discuss what are the basic factors that must be taken to use the fuzzy logic in the aforementioned matters in general, and then focus on its employment in the field of renewable energy. Three main axes for renewable energy are solar panels, a wind turbine and finally, solar collectors. The key to working and the basis of the static system is the mechanism for selecting the inputs that directly affect the output in addition to the methods and activation functions of the fuzzy logic.

Keywords: renewable, energy, solar panals, effeciency and optimization

1. Introduction

In this chapter a complete review of how the fuzzy dominant operates in renewable energy, analyzes, and control. Usually many of the words which are used arbitrarily in our lifestyle have a simple meening and big work. When representing or discus a system or any think, words were used such as big, small, long, short, cold, warm, hot, sunny, cloudy, fast, slow, etc., which are vague in nature. Humans are use unguaranti, catchy and muddy words when showing something or report a decisions to produce a certain actions. According the age, call were individual old, mid-age, young, old plus, and new young. Applying gas or stop pressure according to road situation, whether dry, slippery, sloping or flat. If the light level in the classroom is low, we increase the brightness with one touch, otherwise, we decrease it. These examples illustrate how our brain behaves and makes decisions during uncertain and ambiguous situations.

Studies of systems with unconfirmed and disinformation have reached the era of substitution with the submitting the article "Fuzzy Groups" by Lotfi Zadeh [1]. Although this text was first published in 1965, the use of Symbolic Logic (FL) increased after the latter half of the 1970s when Lotfi A. Zadeh two additional articles [2, 3], in which pure fuzzy mathematics was used for uncertain systems and decision making. FL apps have been gaining fast speed since the Japanese started using them in commercially available devices. Nowadays, it is possible to search for ambiguous applications in almost every region [4]. Depending The sustainable sources are containes more parameters and variables which difficult to control, but using the artificial intellignt make the contol simple and easy to use. FL works in many areas of use for automatic control systems and monitoring issues [4–7]. It can be used in database control tool to manage the data flow and knowledge, huge data, innovative method, and smart work for the motors. Graphical works, signal prob., and body-motor simulation are also another applications where FL is function. Additionally, fuzzy is utilized as a mathmatics processor in some cases such as equation optimization, selecting, figures smoothing, etc. [4].

In this chapter, the mechanism of programming and designing mechanisms for improving the work of solar energy systems will be explained and how to make the most of their work. As for solar cells, as it is known, they have several uses, including: converting solar radiation into electrical energy, distillation, storage, heating water and so on from these applications. The initial design of these systems may be with a certain efficiency or at a specific degree of use with specific physical or material matters, here comes the role of methods for improvement, searching for weaknesses and addressing them in systems in general and solar systems in particular. Artificial intelligence, it has been used by many researchers in improving the work of solar systems in terms of increasing efficiency and choosing the values of the best variables at different times of the day in which the intensity of solar radiation, inclination angle, humidity, and the like changes.

The optimization process for the operation in general is based on the mathematical model reached in the initial design of the system. In an AI environment, the default system inputs are identified with the expected output of these inputs. When the ideal result for the system is available, the proposed controller, which is built on the basis of artificial intelligence, changes the internal factors of the system and adjusts the initial weights to make the general system work according to the expected standards and the productivity, as far as possible, is ideal.

2. Fuzzy sets

Fuzzy groups are the basic elements of FL. Fuzzy groups are characterized by organic functions. In fact, these organic functions are just kind of mysterious numbers. One must know the meaning of ambiguity in order to know the terms FL, membership function, and ambiguous number. For example, two specific colors are mixed in the color world and they appear in **Figure 1**. First, it's white, and then it's

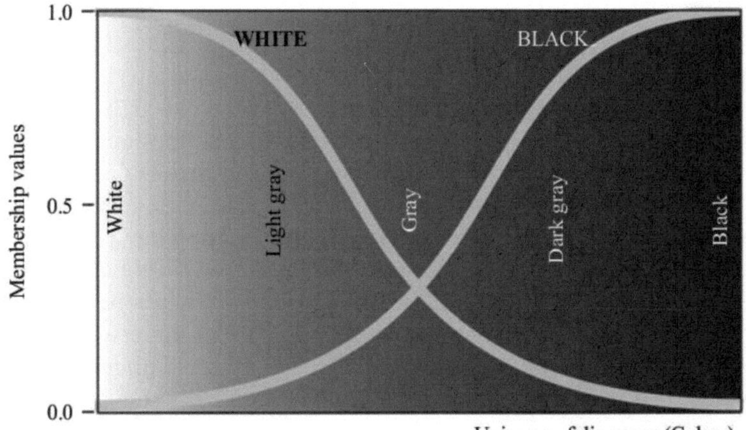

Figure 1.
Blended colors in the universe of colors.

changed to black over a transition area so that it becomes light gray, gray, taupe, and black as we move from left to right. It is not a single color that is transparent in transition. Includes a black and white reminder and no one can distinguish one color from the other because the transitional part is blurred. Colors within the transition region are often highlighted as white, light gray, gray, dark gray, and black shown in **Figure 1**.

There are two color classes in **Figure 1**, black and white, called fuzzy subsets or fuzzy organic functions. The Fuzzy subset WHITE in **Figure 1** shows the tones that are white while the blur The BLACK subset represents the black hue. The prolonged changing zone, the tow different colors. The color is different from the two origin. The color range along the line is blurred. FL is used in the search for solutions in the common areas between ideal solutions and wrong solutions, and for this, the search process may produce unexpected and unexpected solutions. Most of the researchers used fuzzy logic to improve the productivity of a machine or improve the efficiency of a specific system or the like, due to the rule that this type of research is characterized by, in foggy areas, which allows the system to use solutions closer to reality than it was in the past. In mathematical equations, an example of the above, there is one or two roots in most possibilities. When using fuzzy logic, it will find a set of roots that give a much lower error rate which depends on the permittivity specified in the software.

2.1 Fuzzy membership functions

The functions of FMF can be seen as a tunnel between unconfirmed bits and a hazy outland. The fragile home of muddy information is subdivided and shown by unclear organic form, see **Figure 2**. Shades of gray become darker as we move from left to right or from white to black. Semi-color of gray were re-arranged into subsections as high, little high, low, very low and zero refered by the triple-kind mysterious organic method [1].

The organic function is the basis for the system modeling process in order to improve its operation or reduce the error rate. Where the number of functions must be determined with the quality of each function and according to the data that will be dealt with. There are linear and nonlinear models, each of them has a function quality that differs from the other, in order to reach the goal more quickly and accurately than is recognized. The organic functions that distinguish fuzzy groups and the groupings with which they are performed are the idea of fuzzy sets and systems of symbolic logic. Therefore, understanding ambiguous groups and their groupings is vital to understanding what is often done with fuzzy sets and symbolic logic. Therefore, this chapter is reserved for introducing ambiguous groups and

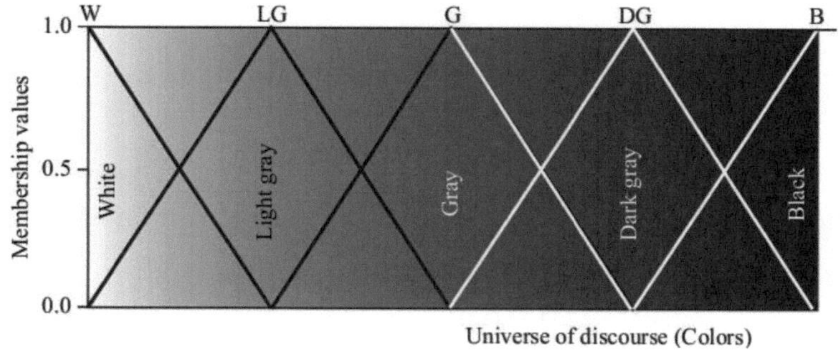

Figure 2.
Blended colors in the universe of gray colors.

analyzing their properties from a control application point of view. Known membership functions will be reviewed to represent fuzzy groups one by one, and MATLAB functions for each will be written as a neighborhood to develop a symbolic, logical, user-defined toolbox.

3. Photovoltaic

Photovoltaics has become the most cost-effective source of electrical energy in regions with high solar potential, with bids as low as $ 0.01567/kWh [8] in 2020. Panel prices have decreased 10 times in a decade. This competitiveness opens the way for a global transition to sustainable energy, which may be needed to help mitigate heating. Emissions of carbon dioxide allow the achievement of a target of 1.5 degrees that will be spent in 2028 if emissions remain at the current level. However, using photovoltaic energy as a first supplier requires a power saver or main divider by a big energy DC busses which adds fees.

Solar PV has some impoertant benifet as an power supply: when first operate, its response produces no infection and no emissions of cleanhouse gases, it is additionally viewed in terms of energy is widely available within the shell. Photovoltaic systems have always been used in specialized applications where standalone installations and grid-connected PV systems have been used since the 1990s [9] PV modules were first produced in large quantities in 2000, when German environmental scientists, and thus Eurosolar, obtained government funding for the Ten Thousand Roofs program [10].

4. Application of fuzzy logic in solar system optimization

Before starting with the details of this application, we will mention the basics followed to represent the solar system and what are the most important variables that must be followed to improve work and make the most of it.

The solar radiation which received by the solar still was calculated using ASHRIE model depending on the environmental conditions in Iraq (35.33No, 44.5 Eo) due to lack of Meteorological data.

The intensity of total solar radiation incident on inclined plane by δ angle was calculated using the following formula:

$$I_o = I_{DN} * \left[\cos(\theta) + C * \frac{(1 + \cos\delta)}{2} + s * (C + \sin\beta) * \frac{(1 - \cos\delta)}{2} \right] \quad (1)$$

where: I_{DN} represents the amount of direct radiation incident on the perpendicular surface and calculated from the following equation:

$$I_{DN} = A_1 \exp\left(-\frac{B}{\sin(\beta)}\right) \quad (2)$$

A1 is the intensity factor of solar radiation and calculated from the following relationship:

$$A_1 = 1158 * [1 + 0.066 * \cos(360 * ND/370)] \quad (3)$$

B is the atmospheric extinction coefficient calculated according to the following equation:

$$\begin{aligned}\mathrm{B} = 0.175 * [1\text{-}0.2 * \cos(0.93 * \mathrm{ND})] \\ -0.0045 * [1 - \cos(1.86 * ND)]\end{aligned} \tag{4}$$

s the quantity of reflective ground referred to as (Alpedo) with an estimated value adequate to 0.25 within the current research.

β is an angle calculated due to sun radiation.

C is a factor can determined by:

$$\begin{aligned}\mathrm{C} = 96.5 * 10^{-3} * [1\text{-}0.42 * \cos((0.97) * \text{number of days})] \\ -7.5 * 10^{-3} * [1 - \cos(1.95 * \text{number of days})]\end{aligned} \tag{5}$$

θ is an angle determined by:

$$\cos\theta = (\sin(beta) * \cos(segma) \pm \cos(beta) * \cos(phai) * \sin(segma)) \tag{6}$$

φ is the angle between projection of rays on the surface and vertical line on the surface, called the surface azimuth angle and takes a negative signal if the surface tilted away from the sun.

ND Number of the day in the year.

To facilitate the mathematical analysis in question a set of assumptions were adopted, namely:

1. The heat transfer is one dimension through the transparent cover.

2. The temperature of the glass cover is constant.

3. Transparent cover reflects the infrared radiation.

4. Heat transfer is one dimension through the insulating layer.

5. The properties of various material not dependent on temperature.

According to the assumptions above, we can write the heat balance equation of solar still as:

$$C_s \frac{dT_w}{dt} = (\alpha_g + \alpha_w \tau_g) I_o - q_{ga} - q_b - q_f \tag{7}$$

where C s is the heat capacity of the solar still, including the heat capacity of water-gravel layer, the glass cover and the structure of still. While the water and gravel in the basin occupy a certain space of each other and changing in so-called porosity factor that has the symbol (ϕ) Which represents the volume of water relative to the total volume, the water mass is (mw) and the gravel mass (ms) under the following equations:

$$\begin{aligned} m_w &= V_t \phi \rho_w \\ m_s &= V_t (1 - \phi) \rho_s \end{aligned} \tag{8}$$

The intensity of gravel used 2560 kg/m 3 and specific heat capacity is equal to 900 J/kg.K [11].

Thermal equilibrium of still glass cover was given by the following relationship:

$$q_{ga} = q_r + q_c + q_e + \alpha_g I_o \tag{9}$$

Heat is transported across the bulk of the humid air inside the sill by free convection of air between the water surface and the cold glass cover. Khalil and Al-Jibouri [12] estimated the transmitted heat (watts per square meter of glass surface) using the following formula [12]:

$$q_c = 0.8831\left[(T_w - T_g) + \frac{P_w - P_{wg}}{0.265 - P_w}(T_w + 273)\right]^{1/3}(T_w - T_g) \tag{10}$$

where T w and T g is the average temperature for the mixture of water - gravel and glass cover used, respectively.

P_w and P_{wg} the vapor pressure (Absolute MPa) at temperature Tw and Tg, respectively, calculated from the following formula:

$$\begin{aligned}\log_{10}P_w &= -3.2154 + 3.13619\times10^{-2}T_w - 1.22512\times10^{-4}T_w^2 \\ &+3.6384\times10^{-7}T_w^3 - 5.67607\times10^{-10}T_w^4\end{aligned} \tag{11}$$

A hot air carrying water vapor from the warm water surface on the base of the still to the inner surface of the cold glass covered causing condensation on it, the estimated internal rate of heat transfer intensification (qe) (watts per square meter) of the following equation [13]:

$$q_e = 0.0061\left[(T_w - T_g) + \frac{P_w - P_{wg}}{0.265 - P_w}(T_w + 273)\right]^{1/3}(P_w - P_{wg})L_w \tag{12}$$

Where L_w is the latent heat of evaporation at water temperature (Tw) of still calculated by the following formula:

$$L_w = 2501.67 * 10^3 - 2389T_w \tag{13}$$

And the rate of condensate water (per square meter of glass cover) calculated from the following formula:

$$D_e = \frac{q_e}{L_{w,g}} \tag{14}$$

$L_{w,g}$ is the latent heat of water vaporization at glass temperature (Tg.).

Also the heat is transferred by radiation between water surface and the glass cover an estimated rate of transmission in this way is as follows:

$$q_r = F\sigma\left[(T_w + 273)^4 - (T_g + 273)^4\right] \tag{15}$$

where σ is Stefan-Boltzmann constant and the supposition of the small distance between the water surface and glass cover, compared with the length and width of still, the surfaces can be considered as an infinite and the shape factor computes from the following formula:

$$F = \frac{1}{\frac{1}{\varepsilon_g} + \frac{1}{\varepsilon_w} - 1}. \tag{16}$$

where ε_g and ε_w the emissivity of the inner surface of glass cover and water, respectively.

The glass cover loses heat to the outer surrounding by convection and radiation together and the lost heat calculated from the following formula:

$$q_{ga} = h_{ga}(T_g - T_a) + \varepsilon_g \sigma \left[(T_g + 273)^4 - (T_{sky} + 273)^4 \right] \tag{17}$$

where h_{ga} is the convection heat transfer coefficient between the outer surface of the glass cover and ambient air is highly dependent on wind velocity using the following formula to calculate this factor [10]:

$$h_{ga} = 5.61 + 1.09v \text{ where } v < 18. \tag{18}$$

$$h_{ga} = 2.64 * v^{0.78} \text{ where } 18 \le v \le 110 \tag{19}$$

where v is wind velocity, and given in units of (km/hour).

T_{sky} is the apparent sky temperature and calculated from the following equation [14]:

$$T_{sky} = 0.0552 * (T_g + 273)^{1.5} - 273 \tag{20}$$

The bottom and sides of solar still were insulated to minimize heat loss to the external environment and an estimated rate of heat loss of the solar still base per square meter of glass cover from the following relationship:

$$q_b = U_b(T_w - T_a) \tag{21}$$

where U_b is the overall coefficient of the transfer of rear heat through the base of still, calculated by the following formula:

$$\frac{1}{U_b} = \frac{t}{k_b} + \frac{1}{h_b} \tag{22}$$

where t is the thickness of insulation, and k is thermal conductivity of insulation material, h_b the transfer coefficient of the lower side of still.

If the feeding system continues to feed the water into the solar still, heat amount depleted as a result of this feeding calculated by the following formula:

$$q_f = G_f c_p (T_w - T_{f,i}) \tag{23}$$

where G_f feeding rate per square meter of the glass cover and c_p the heat capacity of feeding water, $T_{f,i}$ the temperature of feeding water interring the still usually taken equal to the temperature of the atmosphere in the first hours of the day.

In this sectio, tow implemented works were fixed to show how the fuzzy optimiz the overal system performance.

4.1 Hybrid solar collector

In this application example, we will explain how to take advantage of the fuzzy controller to improve the efficiency of a solar collector with three thermal complexes, as the necessary parameters for each complex were determined and then installed on a software model to be in harmony with what is logical and ideal.

In **Figure 3**, with the aim of comparing the thermal and electrical performance of a hybrid solar panel, each model consists of most of the parts:

Hybrid collector base (Iron Stand)

The wooden structure

Solar cells

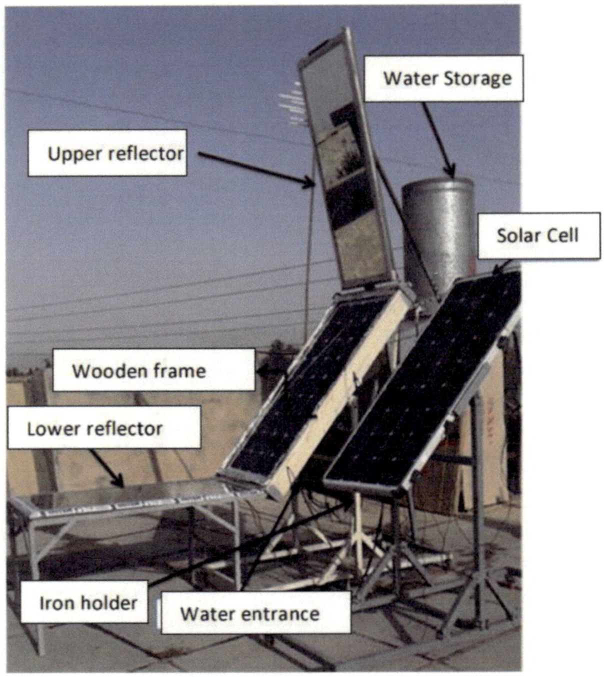

Figure 3.
A photograph of the hybrid solar collector used.

Reflective mirrors
Measuring devices

The first model consisted of a monocrystalline photovoltaic cell, described in **Table 1** inside the wooden structure and the photovoltaic cell housed within a wooden structure, isolated from the back and sides by a solid insulation. The device shown in **Figure 3** is behind the photovoltaic cell, fix the glass cover over the PV cell 2.5 cm away. Use a lower mirror with dimensions (length 120 cm, width 50 cm) on the top of a solar panel. Iron holder is for this bottom mirror. This stand is mounted on the iron stand of the solar dish. They are often moved by different angles of the solar thermal collector and thus have the same dimensions as the bottom mirror and the reflectivity of those mirrors estimated at 0.95. Mirror edges surround an aluminum frame to secure it and make sure not to break it. The second model might be a photovoltaic cell without refrigeration (it has no device below) and has no solar mirrors installed on it. The aim of constructing a symmetric model is to achieve an accurate comparison of the effect of design and operating variables affecting the performance of the hybrid solar panel to succeed in the acceptable range of this case. Many kind of recording devices have been fixed to determine temp. a variety of thermocouples were created to live the temp. in several regions of validation.

These parameters are fixed within devices, hence FL is used for control issue. Three entrances are defined as PV / T corner, mirror (upper corner) and lower

V.\I/P	No. of atte.	Temp. (°C)	I (W/m^2)
Simple	S-u	L-t	L-i
Med	M-u	M-t	M-i
Big	L-u	H-t	H-i

Table 1
The linguistics term for the FL input.

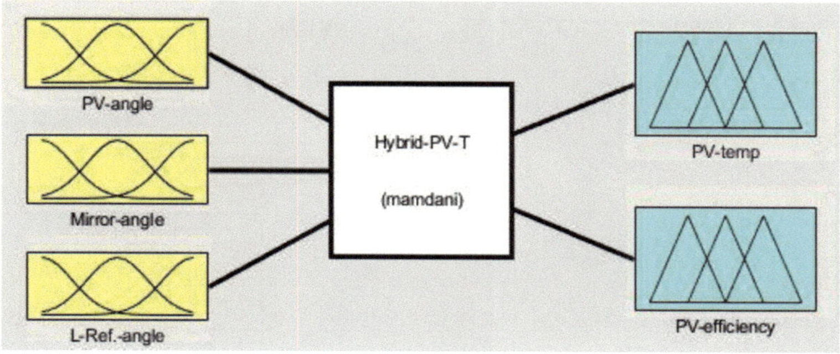

Figure 4.
The FL system architecure.

reflector angle. Draw outputs are monitored to optimize system efficiency, PV/T temperature and power generation, see **Figure 4**. Mamdani model was used to optimizing the proposed system with three inputs and two pairs of outputs. However, FL controls the sun's radiation and thus the amount of water needed by the collector.

The function of the MF member vessel is determined according to the nature of the information for each part of the system. Therefore, the amount of MFs is fixed to three for both input and output, in addition to the present, the amount of bases has reached approximately 3 * 3 * 3 27. Use the solar density meter (solar meter SM206) to accurately live the radiation intensity (± 10 W/M2), the wind velocity was measured accurately (± 0.1%) so the instrument (MT- 1280) to live the voltage and current accurately (± 0.5 use 4 21 watts (DC)) lamps as the load of the photoelectric cell and a flowmeter for live water flow when entering the heat exchanger Behind the photovoltaic cell, **Figures 5** and **6** represent a diagram of the device used.

It is known that the solar cell produces electricity in the form of DC. When solar energy falls on the solar cell, electricity is produced by the positive and negative electrodes outside the cell. Its main advantages are that it does not contain moving parts that are subject to breakdown. The energy generated from the cell In this study according to the following equation.

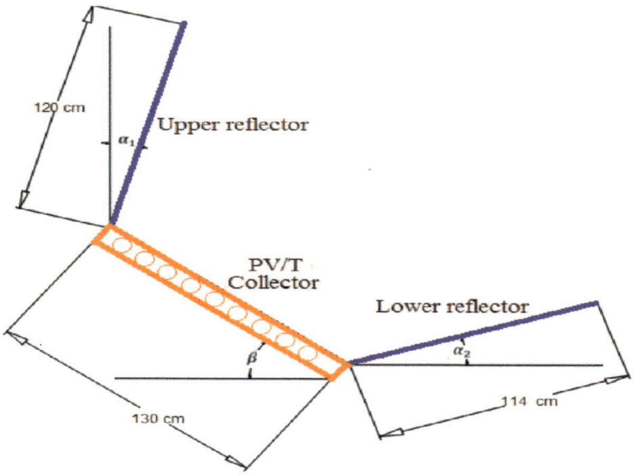

Figure 5.
Dimensions and angles of the mirrors used in the experiment.

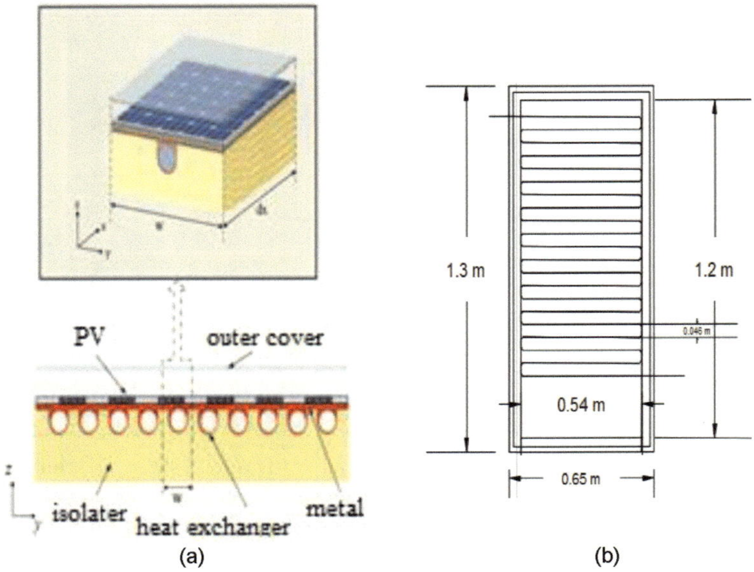

Figure 6.
A diagram of the parts of the solar collector (PV/T collector). (a) Side clip of solar cell and heat exchange. (b) Dimensions of heat exchanger.

$$P_{pv} = V * I \tag{24}$$

Where the represents current I and the voltage of the unit of the solar cell, each of which is measured using a device (Digital Scale), which is characterized by high accuracy of the measure of electricity (DC), and it is necessary that the measuring devices are accurate in order to calculate the efficiency and electrical power out of the solar cell to be correct And close compared to standard conditions specifications, see **Figure 7**.

The proposed of the solar dish fixed envolve of a photovoltaic and a device on second end photovoltaic cell. The proposed bowl is roofed with a single trnsparency layer. In addition, the top mirror is attached to the top and thus the below inverter is attached to the underside of the solar hybrid dish. **Figure 6** is a diagram of a hybrid solar dish.

The total radiation absorbed by the hybrid system connected to the upper and lower reflectors can be found from the following equation [15].

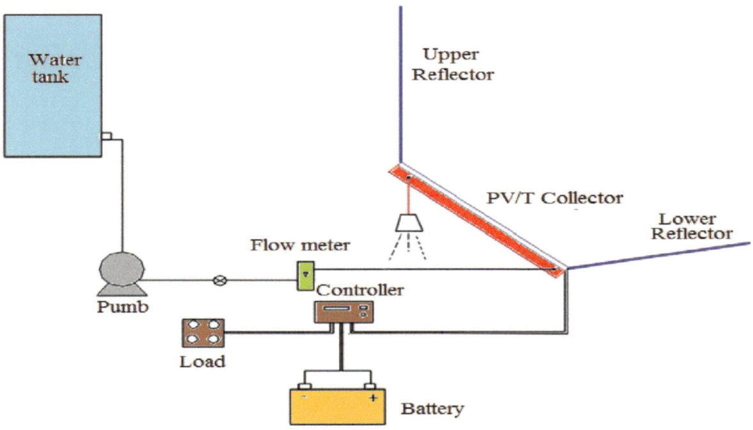

Figure 7.
The diagram of the device used.

$$A_{total} = A_b + A_d + A_g + A_{refr\,1} + A_{refr\,2} \quad (25)$$

Each part of the equation is found as follows [9]:

$$A_b = I_b R_b\,(\tau\alpha)_b \quad (26)$$

The below text can be reached (**Figure 8**):

The most applicable point during this thik is that the optimization, and thus the percentage became 84% after 66% in the old paper. Power production typically solar hybrid plate optimized with high value in the radiation by the reflecting plates, with the increase in the fluid flow without a transperant cover to prolong the cooling effect. Smart setting of system parameters with FL makes the system more installable and operating in optimum mode. Having both upper and lower external reflectors has a greater effect than having only one on the performance of a hybrid solar plate. The presence of dust negatively affects the performance of the hybrid solar dish and outdoor reflectors [12].

4.2 Special solar collector parameter determination

In this section, the symbolic logic of the angle organization and the total reservoir required for a particular place is also proposed due to the traditional tools of fresh water. Three inputs are defined for FL: indoor users, temp., and hence sun power; which varies from place to place and should be well fixed. The values of the input variables were transferred to the Linguistics model for better FL addition; This process is called fuzzification. Most of the parameters to be regulated within the wedge storage complex are the angle and size of the tank; These two things will be the FL output of (10°-85°) for angle, and (500-10000 L) for tank volume. The system was validated and tested by a combination of random value inputs and the output response monitored at the end of the system, when the users were around 20° C, so the density was 450 W/m2, and the output was 3560 liters for tank volume and 47.7° for angle. The control of the proposed system is straightforward to monitor and smooth transition within the output resulting in specification of parameters covering the full values available [16].

The proposed method for determining most of the stabilization parameters for storing the wedge assembly is based on FL. Throughout this paper, three method control inputs were defined to work on parameters defined for the new tank

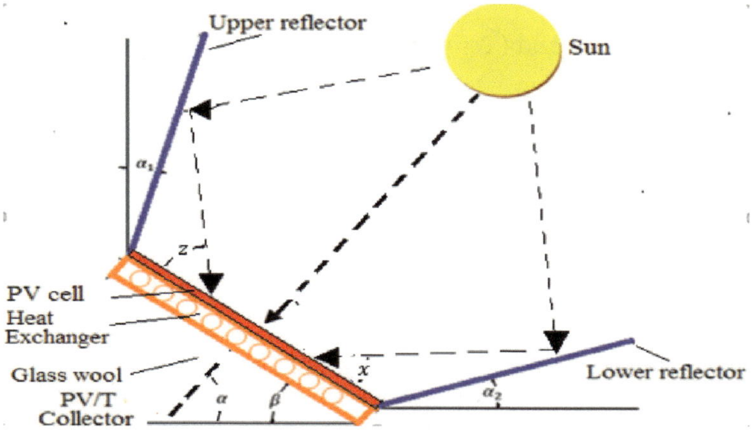

Figure 8.
Schematic of hybrid solar collector (PV/T).

geometry type. The input variables are: users inside the building, temperature, and thus daylight intensity, which are most of the FL inputs that are used to predict the response of the output. The range of the input data is: users inside the building (1–00), and the temperature inside the surrounding area is (10–60) degrees Celsius, so the intensity of daylight is (0–1000) W/m2 2. The specified input has a special range of values. And its unit which is one among FL facility points that may drive multiple types of knowledge. The basic step in FL is that of fuzzification, which proposes converting numeric data into linguistic variables for each input. 3- Membership function (MF) for each input variant and each Gaussian MF to overcome nonlinear changes within the input file. **Table 1** shows linguistic terms derived from the names of the input variables.

Due to these language words; The fussy will derive the monitor signal to the machine through a Mamdani process. Values within the input range will be validated, not normal or rearranged by alternate model, because they are within real model to meet optimal work with very small error ratio. When the information is set up through a process of scrambling, the inference engine will drive the control flow through well-established rules that support the state of the input values. The rule-base is often represented by an unclear factor to provide the specific response consistent with the mathematical model of system control. The amount of rules for each ambiguous project is often determined by keeping the MF amount to the input count power; Eq. (2) is often used for this purpose [17, 18]:

$$\boldsymbol{rules_{project}} = \left(\boldsymbol{n_{inputs}}\right)^{n_{MF}} \tag{27}$$

In this work 3-MF is installed on the input and there are 3 input variables; This indicates that 27 is the maximum number that is often reached to take care of the control process. The operator drives things to meet the optimum system response and control the conditions available to the input variables. A list of some derivative rules for the current issue throughout this paper is explained below, clear to ascertain the fuzzy operator and to force all three input values available in the current data to work on the specified parameters of the wedge complex:

1. If (UserNum is S-u) and (Temp. is L-t) and (I is L-i) then (TankS is M-T) (CutAn is L-A)

2. If (UserNum is S-u) and (Temp. is M-t) and (I is M-i) then (TankS is S-T) (CutAn is M-A)

3. If (UserNum is S-u) and (Temp. is H-t) and (I is H-i) then (TankS is S-T) (CutAn is S-A)

4. If (UserNum is S-u) and (Temp. is L-t) and (I is M-i) then (TankS is S-T) (CutAn is M-A)

5. If (UserNum is S-u) and (Temp. is H-t) and (I is H-i) then (TankS is S-T) (CutAn is S-A)

MF is 3 for all parameters and Gauss for non-linear process and is closest to the active response. **Figure 2** shows the curves of the member-ship for each proposed FL model with the general block of the manager board (**Figure 9**).

FL response is often examined either during curves or by adjusting the input value through principles and observing the output response. Throughout this paper, both types of results are fixed to further validate the proposed work and

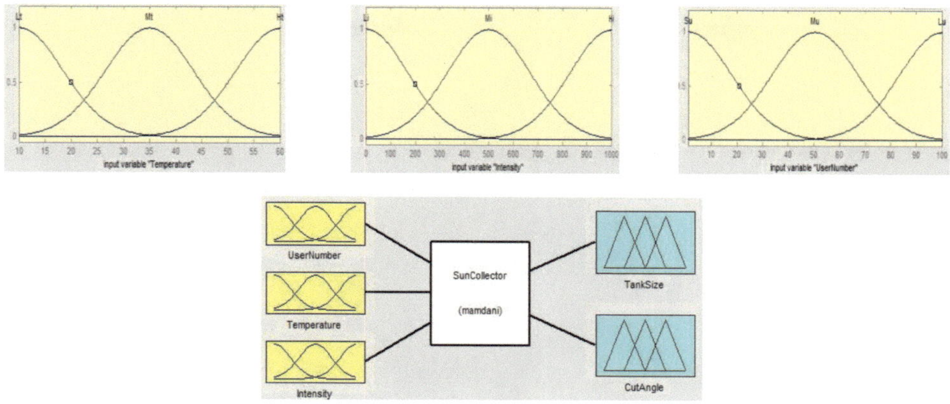

Figure 9.
The MF for the i/p points with the rest.

demonstrate how the input affects the specific parameters of the system. **Figure 10** shows the determined surface response to the number of users and temperature with an indication of the collector volume in liters.

In this figure, it is evident to ascertain the transition response when the input varies from low to high values and thus the output increases or there is still change in both cases. The surface corresponding to the temperature and intensity with reference to the volume of the tank is shown in **Figure 11**.

The tank volume is particularly related to the number of users inside the building, and the opposite input variables have little influence on it, so the values are arranged between 5498 and 5499 liters. The reason for the agile transition within the output response in **Figures 4** and **5** is to facilitate the proposed control method that deals with the smallest change within the data entry. For example; When the number of users is low (7), the temperature is medium (27° C) and the intensity is low (200 W/m2); Expected parameters for FL are: 2,240 liters for tank volume and 74.7 liters for angle. The size of the tank mainly depends on the number of users, and the other has a touch effect on it.

In **Figure 12**, the relationship between users and density is shown with respect to tank volume. Clear to ensure the transmission of values is consistent with the state of the entry.

When cutting is the most respected variable, the response will be more realistic and have little change as data entry increases or decreases. The angle varied

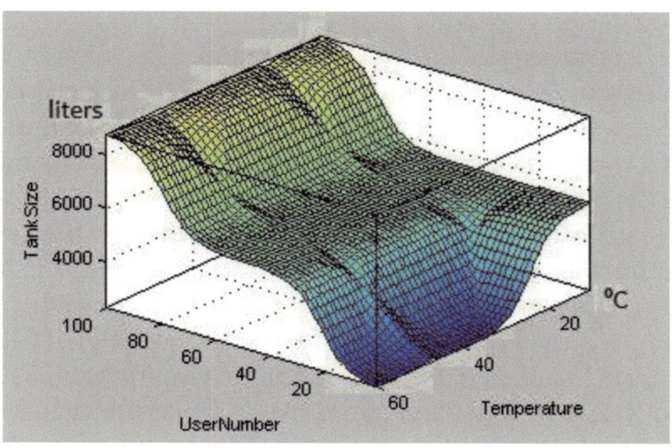

Figure 10.
The response of the proposed control system for the users and temps.

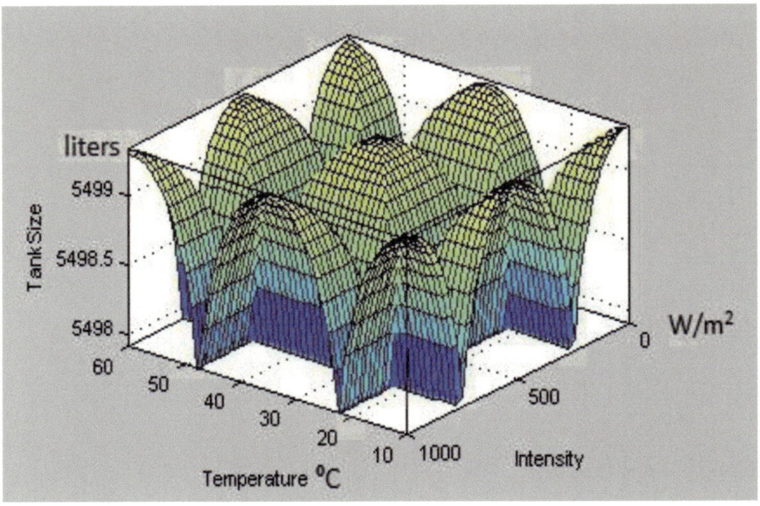

Figure 11.
The output surface response for the temperature and intensity with respect to tank size.

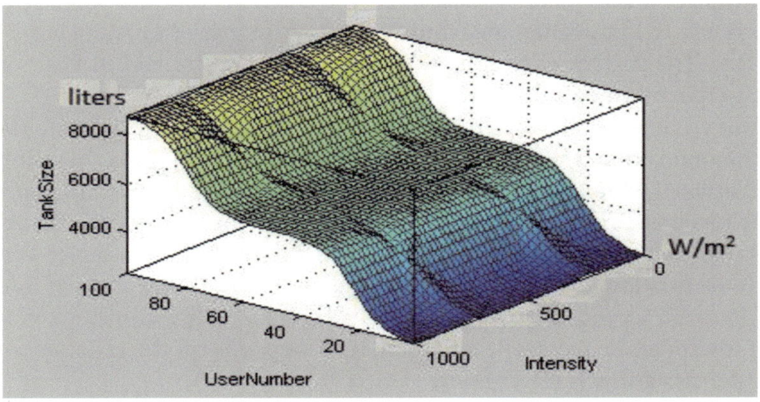

Figure 12.
Tank size response according to resent user ans intensity numbers.

according to the strength of the candle and the temperature values which would reflect the good design of the proposed method. When the temperature and intensity are at the maximum values, the angle will be 45 degrees; Otherwise, the angle begins to expand to collect optimum sunlight.

FL may be an instrument that simplifies the control problem by means of a simple and real response with very slight actuators [17]. During this work, FL identifies and compares different types of temperatures and the number of users for their service with appropriate storage of the wedge complex. The input variables are: users inside the building, temperature, and thus daylight intensity, which are most of the FL inputs that are used to predict the response of the output. Whereas there are two output variables: tank size and hence cut angle; Which are the wedge storage complexes. During a particular situation; When the number of users is low (7), the temperature is medium (27° C) and the intensity is low (200 W/m2); Expected parameters for FL are: 2,240 liters for tank volume and 74.7 liters for angle. FL gives different values when input file variables are changed or a new building must be constructed in a new location. The control of the proposed system is straightforward to monitor and smooth transition within the output which results in specification of parameters covering the full available values. When the specified values are applied to the actual design of the collector store, there is approximately 25% improvement share for the water system. The proposed control satisfies the corner rule for storing the collector which has an inverse relationship to the radiation.

Author details

Daoud Raid*, Ahmed Omer and Al-khashab Yaareb
Northern Technical University, Mosul, Iraq

*Address all correspondence to: raid.daoud@gmail.com

References

[1] Lotfi A. Zadeh: Fuzzy sets. Information and Control. 1965; 8: 338–353.

[2] Lotfi A. Zadeh: Outline of a new approach to the analysis of complex systems and decision processes. IEEE Transactions on Systems, Man, and Cyber-netics. 1973; 3(1): 28–44.

[3] Lotfi A. Zadeh: The concept of linguistic variable and its application to approximate reasoning. Part I, II, and III. Information Sciences. 1975; 8(3): 199–249, 8(4):301–357, 9(1):43–80.

[4] Maiers J, and Sherif Y,: Applications of fuzzy set theory. IEEE Transac-tions on Systems, Man, and Cybernetics. 1985; 15(1): 175–189.

[5] Mamdani E, and Assilian S,: An experiment in linguistic synthesis with a fuzzy logic controller. International Journal Man-Machine Studies. 1975; 7 (1):1–13.

[6] Mamdani E,: Application of fuzzy algorithms for control of simple dynamic plant. IEE Proceedings. 1974; 121(12): 1585–1588.

[7] Mamdani E,: Advances in the linguistic synthesis of fuzzy controllers'. International Journal Man-Machine Studies. 1976; 8(6): 669–678.

[8] "KAHRAMAA and Siraj Energy Sign Agreements for Al-Kharsaah Solar PV Power Plant". Qatar General Electricity & Water Corporation "KAHRAMAA". Retrieved 26 January 2020.

[9] Bazilian, M.; Onyeji, I.; Liebreich, M.; MacGill, I.; Chase, J.; Shah, J.; Gielen, D.; Arent, D.; Landfear, D.; Zhengrong, S. (2013). "Re-considering the economics of photovoltaic power" (PDF). Renewable Energy. 53: 329–338. CiteSeerX 10.1.1.692.1880. doi:10.1016/j.renene.2012.11.029. Archived from the original (PDF) on 31 August 2014. Retrieved 4 September 2015.

[10] Palz, Wolfgang (2013). Solar Power for the World: What You Wanted to Know about Photovoltaics. CRC Press. pp. 131–. ISBN 978-981-4411-87-5.

[11] Elsayed M. M. et.al: Design of solar thermal systems. Thermal engineering dept., *King Abdulaziz university*, Jeddah, 1994.

[12] Omer Khalil Ahmad Al-Jibouri,: A Cheap Way to Improve the Performance of Simple Solar Still. Energy Science and Technology. 2014;7 (1), 1-7, DOI:10.3968/j.est.1923847920140701.3915.

[13] Dunkle R.V,: Solar water distillation: the roof type still and a multiple effect diffusion still, International heat transfer conference. University of Colorado, U.S.A, Part 5, 1961; 895-902.

[14] Duffie J.A. and Beckman,: Solar engineering of thermal process, *John Wiley and sons*, New York,U.S.A, 1980.

[15] Lo Piano, Samuele; Mayumi, Kozo (2017). "Toward an integrated assessment of the performance of photovoltaic systems for electricity generation". Applied Energy. 186 (2): 167–74. doi: 10.1016/j.apenergy.2016.05.102.

[16] Omer Kh. Ahmed, Raid W. Daoud,: Determining the Coefficients of the Wedge Storage Solar Collector Based on Fuzzy Logic, International Conference on Engineering Technologies and their Applications (ICETA), Islamic University – ALNajaf – IRAQ, 978-1-5386-7858-9/18/$31.00 ©2018 IEEE

[17] Volosencu, Constantin, On Some Properties of Fuzzy Systems, Proceedings of the 8th WSEAS International Conference on Signal Processing, Robotics and Automation

(ISPRA09), Cambridge, UK, Feb. 21-23, 2009, pag. 178-186, ISBN 978-960-474-054-3, ISSN 1790-5117.

[18] Volosencu, Constantin, Pseudo-Equivalence of Fuzzy PI Controllers, Proceedings of the 8th WSEAS International Conference on Signal Processing, Robotics and Automation (ISPRA09), Cambridge, UK, Feb. 21-23, 2009, pag. 162-168, ISBN 978-960-474-054-3, ISSN 1790-5117.

Section 7

Green Infrastructure

Chapter 13

Location Selection for Smog Towers Using Zadeh's Z-Numbers Integrated with WASPAS

Janani Bharatraj

Abstract

Fuzzy sets have been extensively researched and results have been developed based on the extensions of fuzzy sets. In this chapter, fuzzy sets and its extensions are discussed. Z-numbers along with weighted sum product assessment method is used to obtain a feasible solution to the location selection problem for installation of smog towers in a densely populated locality. The degrees of freedom namely degree of membership, degree of non-membership and the degree of hesitancy have been expressed as Zadeh's Z-number with probability quotient for the degrees. Further, ranking of the alternatives based on Z-numbers and WASPAS to allocate smog towers to residential areas stricken by air pollution.

Keywords: Z-numbers, WASPAS, fuzzy set, Generalized fuzzy set, smog tower, air pollution

1. Introduction

Mathematics is a study of quantity, structure, space and change. Patterns are observed to understand the structures and reasoning is provided to real time phenomena. Mathematics can be subdivided into arithmetic, algebra, geometry and analysis. Further, there are subdivisions linking the core of mathematics to other fields like logic, set theory, empirical mathematics and more recently to the rigorous study of uncertainty, imprecision and vagueness. Set theory is a branch of mathematical logic that studies 'sets', a collection of well-defined objects. An object under consideration either 'belongs to' a set or 'does not belong' to a set. Thus classical set theory could answer membership of an element in terms of 0's and 1's. This binary logic could not translate the imprecision prevailing in the real world. The need to bridge the precise classical mathematics with the imprecise world gave birth to the concept of fuzzy sets. These sets were introduced independently by Zadeh [1] and Dieter Klaua in 1965 as an extension of classical set theory. In contrast to binary terms, fuzzy set theory allowed gradual assessment of the membership of elements in a set described by a membership function valued in the interval [0,1]. Zadeh went on to propose new operations in fuzzy logic and proved that fuzzy logic was a generalization of classical Boolean logic. He also proposed the concept of fuzzy numbers which were special case of fuzzy sets. The mathematical operations were also defined and thus fuzzy sets paved the way for many extensions, whose edifice stands strongly on the mathematics concept.

1.1 The extensions

Interval-valued fuzzy sets (IVFS) were introduced as an extension to fuzzy sets in which the membership degrees are represented by an interval value reflecting the uncertainty in assigning membership degrees. Larger the interval, more uncertainty is seen in assigning membership degrees.

Intuitionistic Fuzzy sets (IFS) is also an extension of fuzzy set introduced by Atanassov [2]. The addition of 'degree of non-membership' of an element improved the efficiency of modeling uncertainty.

Interval-Valued Intuitionistic Fuzzy Set (IVIFS):

Atanassov [3] combined IVFS and IFS to form IVIFS where in are all intervals in [0,1].

Neutrosophic Sets:

Having defined fuzzy sets, IVFS, IFS and IVIFS, researchers still could not handle uncertainty efficiently. The question of, "what if I had a neutral opinion?" had to be answered. Thus, Smarandache broke free the inter-dependencies of all three membership functions. Thus neutrosophic sets were defined

$$A = \{(x, \mu_A(x), \upsilon_A(x))/x \in X\}$$
$$\mu_A(x) : X \to [0,1], \upsilon_A(x) : X \to [0,1], \Gamma$$

such that,

$$0 \le \mu_A(x) + \upsilon_A(x) + \Pi_A(x) \le 3.$$

Neutrosophic sets, thus, generalized all the sets with classic set theory as the foundation.

In a nutshell, the holy trinity were introduced as a trigger for astounding research all over the world.

1.2 Recent extensions

Picture Fuzzy sets (PFS):

These sets were introduced by Cuong [4] to model situations where in human opinions involved refusal towards a particular event. For instance, voting in an election could have four categories of people; people wanting to vote for a particular party, people abstaining from voting, people not wanting to vote for a party and people refusing to vote. Thus degree of refusal membership is given by $\eta_A(x)$, with

$$A = \{(x, \mu_A(x), \upsilon_A(x))/x \in X\}$$
$$\mu_A(x) : X \to [0,1], \upsilon_A(x) : X \to [0,1], \Pi_A(x) : X \to [0,1]^{\cdot}$$
$$0 \le \mu_A(x) + \upsilon_A(x) + \Pi_A(x) < 1.$$

Pythagorean Fuzzy Sets (PyFS):

These sets were introduced as a generalization for IFS by Yager [5]. The main feature of PyFS is that it is characterized by the degrees in which the sum of the square of each of the parameters equal to 1.

Let X be a universal set. Then a Pythagorean fuzzy set A, which is a set of ordered pairs over X, is defined by the following;

$$A = \{(x, \mu_A(x), \upsilon_A(x))/x \in X\}$$

$$\mu_A(x) : X \to [0,1], \upsilon_A(x) : X \to [0,1]$$
$$0 \le (\mu_A(x))^2 + (\upsilon_A(x))^2 \le 1,$$
$$\Pi_A(x) = \sqrt{1 - \left((\mu_A(x))^2 + (\upsilon_A(x))^2\right)}$$

Hesitant Fuzzy Set (HFS):

HFS were introduced by Torra [6]. Hesitant fuzzy sets were defined in terms of a function that returns a set of membership values for each element in the domain.

HFS, to a large extent, were able to model uncertainty, but with in-depth research, a significant drawback appeared, namely, loss of information. To overcome this drawback, Zhu and Xu [7] proposed the concept of Probabilistic Hesitant Fuzzy Set (PHFS) which incorporates distribution information in HFS. PHFS depicts not only the hesitancy of decision-makers when they are irresolute for one thing or the other, but also hesitant distribution information.

Spherical Fuzzy Sets (SFS):

These sets were introduced as an extension to Picture fuzzy sets and Pythagorean fuzzy sets [8] (**Figure 1**).

A spherical fuzzy set $\tilde{A}_S$ of the universe of discourse U is given by

$$\mu_A(x) : X \to [0,1], \upsilon_A(x) : X \to [0,1], \Pi_A(x) : X \to [0,1]$$

With $0 \le (\mu_A(x))^2 + (\upsilon_A(x))^2 + (\Pi_A(x))^2 \le 1$, for any x in the universal set U.

The spherical fuzzy sets extend PFS and PyFS, but however, these sets are nothing but a particular case of Neutrosophic sets.

1.3 Decision-making techniques

Decision-making is process which involves problem-solving yielding a solution deemed to be "optimal" or satisfactory to an extent. A major part of decision-making involves analysis of finite set of alternatives with respect to a given set of criteria. The task involves ranking these alternatives based on feasibility when all

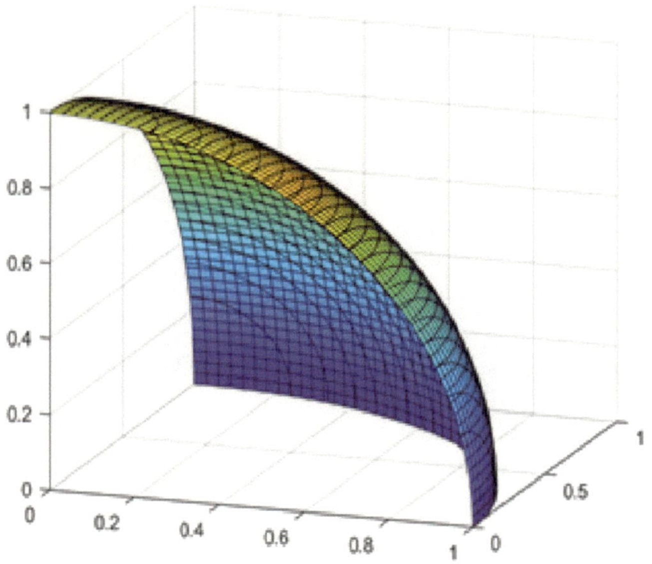

Figure 1.
Spherical fuzzy set.

the criteria are considered simultaneously. This area of decision-making has always attracted researchers and is still a highly debatable concept as there are many such methods which yield different results when applied to the same set of data. Emotion also appears to aid the decision-making process. Decision-making often occurs in uncertainty about whether one's choices will lead to benefit or harm.

Under fuzzy environment:

Decision making under uncertainty means a decision process in which the constraints or goals are fuzzy in nature, but the system need not be fuzzy. As per Bellman and Zadeh [9], fuzzy goals and constraints can be precisely defined as fuzzy sets in the space of alternatives. A fuzzy decision is then viewed as an intersection of the given goals and constraints. Decision-making under uncertainty basically translates to taking decisions in which the goals or constraints are fuzzy in nature. This implies that the constraints consists of alternatives whose boundaries are not sharply defined. An example of a fuzzy goal is "x should be in the vicinity of y", where y is a constant. Here vicinity is a source of fuzziness.

We thus divide the decision-making process into seven steps:

1. Outline the goal and outcome

2. Gather data

3. Develop alternatives

4. List pros and cons of each alternative

5. Identify the best alternative

6. Evaluate and monitor the solution

7. Examine feedback when necessary.

1.4 Zadeh's Z-numbers

A Z-number is an ordered pair of fuzzy numbers (A,B). Z-number is associated with a real-valued uncertain variable X, with the first component A, playing the role of a fuzzy restriction R(X), on the values of which X can take, written X is A. $R(X) : X \, is \, A \rightarrow \mu_A(u)$.

Here $\mu_A(u)$ is the degree to which u satisfies the constraint.

The second component B, is referred to as certainty or reliability or probability or strength of belief related to the component A.

For example, (finding an enclosed space of 900 s.m. in a densely populated area, low, not sure) [10].

1.5 Weighted aggregated sum product assessment (WASPAS)

WASPAS method was introduced by Zavadskas et al. [11]. This MCDM method is a combination of two simple decision-making techniques; Weighted Sum Model (WSM) and Weighted Product Model (WPM).

The total relative importance is given by

$$Q_i = \lambda \sum_{j=1}^{n} \overline{x}_{ij} w_j + (1 - \lambda) \prod_{j=1}^{n} \left(\overline{x}_{ij}\right)^{w_j}, \lambda = 0, 0.1, 0.2, \dots, 1 \tag{1}$$

Here,
x_{ij} is the performance of ith alternative with respect to the jth criterion.
$\overline{x}_{ij}$ is the normalized value of x_{ij} evaluated as follows;

$$\overline{x}_{ij} = \frac{x_{ij}}{\max_i x_{ij}} \text{ for beneficial criteria} \tag{2}$$

$$\overline{x}_{ij} = \frac{\min_i x_{ij}}{x_{ij}} \text{ for non - beneficial criteria} \tag{3}$$

In the next section a literature review of the various aspects of decision-making using Z-numbers will be discussed.

2. Review of literature

Qiao et al. [12] proposed a simple computational method for ranking Z-numbers inspired by the concept of possibility degree of interval numbers. Outranking relations along with a weight acquisition algorithm relative to the possibility degree were developed. Finally, an extended PROMETHEE II based on the proposed ideas were developed. The same was applied to selection of travel plans.

Aliev et al. [13] suggested human-like fundamental approach for ranking Z-numbers. The approach was based on two ideas;

a. The optimality degree of Z-numbers was computed, and

b. The obtained degrees were adjusted using human opinion formalized by a degree of pessimism.

The concept was then used to solve a real time decision-making problem and results were obtained.

Peng and Wang [14] developed an innovative method for addressing MCGDM problems with Z-numbers with unknown weight information about the criteria. Cloud model was employed to analyze Z-numbers. Power aggregation operations of normal Z + -value was proposed using ratio analysis and full multiplicative form. This model was used to evaluate potential air pollution concerns.

Wang and Mao [15] developed a novel approach based on power plant location selection problem with Z-fuzzy based AHP model and successfully modeled the location selection problem.

Chatterjee and Kar [16] proposed COPRAS-Z methodology for Z-numbers. They modeled the fuzzy numbers with reliability degree to represent imprecise judgment of decision-makers in evaluating weights of criteria and selection of renewable energy alternatives.

3. Lacuna and new definition

Chakraborty et al. [17] validated the applicability of WASPAS under five real time manufacturing related problems which resulted in acceptable solutions.

Kahraman and Otay [18] used Z-numbers with AHP to select a location for solar energy PV plant using a 4-level hierarchy. Several criteria and sub-criteria were considered to understand the location selection problem with a Z-fuzzy based AHP method with a real life case study from Turkey.

Decision making techniques have been used mainly in supply chain management which covers the processes from the initial materials provision to the ultimate consumption of finished product linking all the supplier-user entities. Zarandi and Zarandi [19] proposed a modular architecture for the information agent which uses nine different modules, each of which is responsible for one or more functions for the information agent. This automated supply chain is adaptable to an ever changing business environment.

Another area which requires decision making techniques is site selection. Landfill site selection should take into account a wide range of alternative and evaluation criteria in order to reduce negative impacts on the environment. Aydi [20] presented a geographic information systems based multicriteria site selection of municipal solid waste landfill in Tunisia. The methodology involved integrating fuzzy logic and AHP to rank the best suitable landfill sites. The landfill suitability was accomplished by applying weighted linear combination that uses a comparison matrix to aggregate various scenarios associated with environmental and socio economic objectives. The study led to two candidate landfill sites best suited for the procedure.

Sadollah [21] clearly explained the role of membership functions and how to choose an appropriate membership function based on the data available. Computational time is also an important factor that decides the need of a particular type of membership function for decision making methods.

The literature review paves the way for some new definitions to be introduced. Thus, in this chapter, the degrees of freedom are combined with Z-numbers and applied to decision-making problem related to selection of location of smog towers in a densely populated area to combat air pollution problem faced by residents [22].

3.1 Definition

Let U be the universal set. Then a fuzzy subset S can be defined as

$$S = \{(x, (\mu_S(x), p), (\upsilon_S(x), q), (\Pi_S(x), r))\}$$

Where $(\mu_S(x), p), (\upsilon_S(x), q), (\Pi_S(x), r)$ are all Z-numbers with p, q and r the respective probabilities.

$$\text{Here}, \mu_S(x) \in [0, 1], \upsilon_S(x) \in [0, 1], \Pi_S(x) \in [0, 1]$$
$$\text{and } 0 \leq \mu_S(x) + \upsilon_S(x) + \Pi_S(x) \leq 1$$

$0 \leq \mu_S^n(x) + \upsilon_S^n(x) + \Pi_S^n(x) \leq 1$, n is an integer $n > 1$.

Figures 2 and **3** justify the above definition.

4. Application to location of smog towers in the capital city of Tamilnadu

The first of its kind smog towers were designed by Studio Roosegarde as a long term campaign for clean air. The seven metre tall smog-free tower uses patented positive ionization technology to let out smog free air into atmosphere. The tower is designed to clean 30,000 cubic meter of air per hour and is supposed to use small amount of green energy. A similar kind of tower was installed in New Delhi, the capital of India in a busy place called Lajpat Nagar. This tower could trap particulate matter of all sizes suspended in the air. It is capable of treating 2,50,000 to

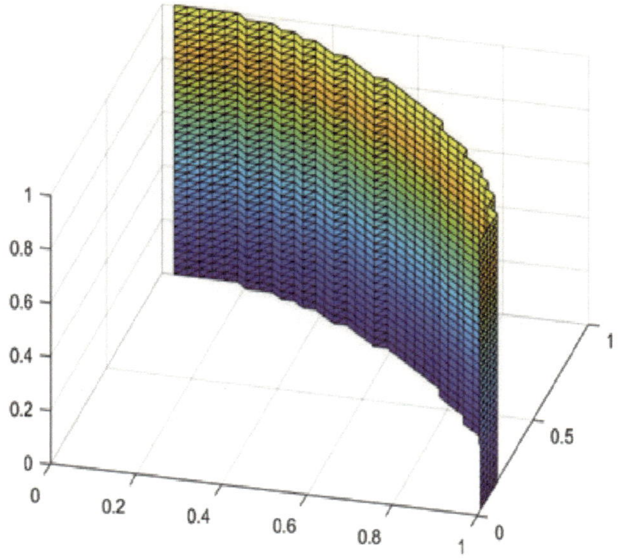

Figure 2.
Fuzzy set with n = 2.

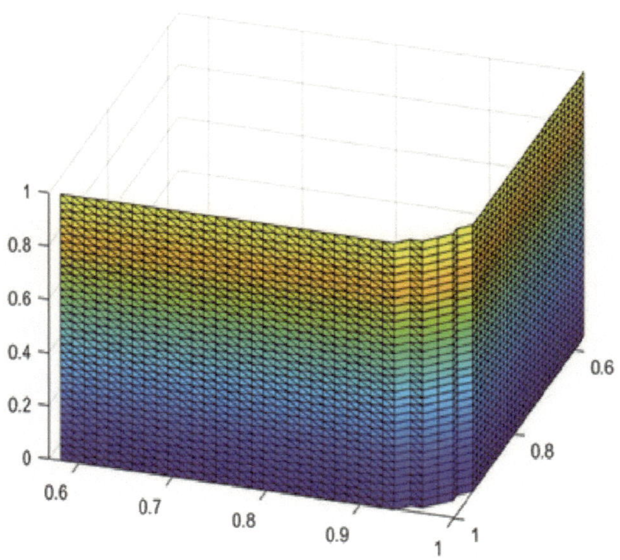

Figure 3.
Fuzzy set with n = 30.

6,00,000 cubic metre of air per day and can collect more than 75% of the particulate matter. It is a structure of concrete which has multiple layers of filters. The structure requires approximately 900 sq.m. in area for its installation. The device was designed to take in air from all angles and generate 1,30,000 cubic metres of clean pure air per hour. The 20 feet tall tower is fitted with exhaust fans to suck in polluted air and can remove upto 80% of the particulate matter ideally PM2.5 and PM10., which are the primary pollutants in Delhi's air.

The smog tower is expected to purify the air within a circumference area of almost 500 m to 750 m.

Chennai, the capital city of Southern state of Tamilnadu in India is plagued by air pollution. The sources of pollution in the city are due to transport, industries and open waste burning. The city also benefits from the land-sea breeze, limiting the contribution of sources outside the urban limit to contribute towards air pollution.

The state highway 49A popularly known as Rajiv Gandhi Salai or the IT corridor is a major road connecting Chennai with Mahabalipuram. It is a 45 km long road housing the prestigious TIDEL park, a home to a number of BPO and IT/ITES companies.

In the first 20 km stretch, 15 traffic signals are stationed, with two toll plazas. The traffic during peak hours cause a lot of pollution in spite of being close to the sea. The major junctions are the 15 signals that literally stall the vehicular movement on this road (**Figure 4**).

The first 20 km houses a small neighborhood called Perungudi. Being in the IT corridor, Perungudi is a preferred locality for booming business and software firms. It is also home to one of the two major landfills in Chennai. The dump yard is constantly in news for the burning of garbage spills despite it being banned. The area also has a sewage treatment plant. Thus, Perungudi faces the wrath of all kinds of pollution, mostly air pollution due to vehicles, dust from construction sites, stench from the sewage treatment plant along with burning of garbage. The area is also low on green cover and hence pollution has severe effect on the residents living there.

Thus, installation of smog free towers is very much required for a locality like Perungudi to fight air pollution.

Challenges and Uncertainty:

A locality like Perungudi is densely populated and lacks basic amenities even though it is part of the famous IT corridor. Ideally speaking, a smog tower should be installed in places where the vehicular movement is on a higher side. But, however such junctions with a space requirement of a minimum of 900 square metres is a big

Figure 4.
Map of Perungudi, Chennai, Tamilnadu, India – Google maps.

challenge. Further, the locality has already been allocated to builders who have taken over a majority of the area for construction purpose. Thus, a lot of uncertainty is involved in location selection considering the challenges faced in terms of space requirements. Hence, fuzzy system plays a crucial role in identifying the right place to install a smog tower considering all the challenges faced in a densely crowded locality. Since, possibility of an event occurring under uncertainty is being studied here, Zadeh's Z-numbers have been combined with WASPAS method to obtain optimal solution.

In order to select a proper location for installing the smog towers, we need to look into the following criteria and understand the feasibility of allocating a place for a tower that will be beneficial on a long run.

4.1 Criteria

Some of the main criteria to set up a smog tower in a locality are listed below;

C1: Minimum area of 900 sq.m.

C2: Continuous supply of electricity.

C3: Green cover in the locality to allow solar panels as an alternative.

C4: Pollution levels in that locality.

4.2 Method

Step 1: Perungudi locality can be broadly divided into four main zones which have maximum impact due to air pollution and noise pollution. The areas; Industrial estate, Srinivasa Nagar, Telephone Nagar and Venkateswara colony are densely populated and have industries contributing to air pollution.

Thus, Four alternatives were chosen for the location of the smog tower namely A1, A2, A3 and A4.

A1: Industrial Estate

A2: Srinivasa Nagar

A3: Telephone Nagar

A4: Venkateswara Colony

Step 2: The alternatives are mapped against the criteria using Zadeh's Z numbers as follows; the degrees of freedom of A1 with respect to criterion C1 is 0.2, 0.7 and 0.1 with probability of 0.8, 0.1 and 0.1 respectively. That is, finding an enclosed space of 900 sq.m. in a densely populated area is 0.2, with the strength of belief 0.8. Likewise, the degree of non-membership is 0.7, with a probability of 0.1 and the degree of hesitancy is 0.1, with a probability of 0.1.

The decision matrix is formed from the date collected from one of the residents and is tabulated as follows (**Table 1**).

Step 3: The maximum of all membership degrees of the alternatives, the minimum of the non-membership degrees of the alternatives and the average of the hesitancy degree of the alternatives are calculated and the decision matrix is normalized using (Eq. (2) and Eq. (3)). **Table 2** shows the normalized decision matrix.

Step 4: The weighted sum and weighted product are calculated.

The total weighted sum and product assessment is tabulated as below using.

(Eq. (1)) with $\lambda = 0.5$ and weights for C1 w1 = 0.3, C2:w2 = 0.1, C3: w3 = 0.3, C4: w4 = 0.3 (**Table 3**).

Step 5: The score function is calculated using the formula

$$s(x) = \frac{\mu_S(x) + 1 - \upsilon_S(x) - \Pi_S(x)}{3}.$$

Alternatives/criteria	C1	C2	C3	C4
A1	(0.2,0.8), (0.7,0.1), (0.1,0.1)	(0.6,0.2), (0.3,0.8), (0.1,0.5)	(0.8,0.8) (0.1,0.1) (0.1,0.1)	(0.8,0.8) (0.1,0.8) (0.1,0.8)
A2	(0.1,0.9) (0.7,0.8) (0.2,0.1)	(0.4,0.2) (0.5,0.8) (0.1,0.8)	(0.3,0.8) (0.6,0.2) (0.1,0.1)	(0.8,0.8) (0.1,0.8) (0.1,0.6)
A3	(0.6,0.7) (0.3,0.5) (0.1,0.1)	(0.7,0.6) (0.2,0.8) (0.1,0.1)	(0.4,0.4) (0.4,0.6) (0.2,0.1)	(0.8,0.9) (0.1,0.8) (0.1,0.8)
A4	(0.4,0.8) (0.6,0.8) (0,0.9)	(0.7,0.7) (0.2,0.8) (0.1,0.9)	(0.3,0.5) (0.3,0.6) (0.4,0.1)	(0.7,0.9) (0.1,0.9) (0.2,0.9)

Table 1.
Decision matrix for the alternatives.

Alternatives/criteria	C1	C2	C3	C4
A1	(0.33,0.8), (0.43,0.1), (1,0.1)	(0.86,0.2), (0.67,0.8), (1,0.5)	(1,0.8) (1,0.1) (0.5,0.1)	(1,0.8) (1,0.8) (0.8,0.8)
A2	(0.17,0.9) (0.43,0.8) (0.5,0.1)	(0.57,0.2) (0.4,0.8) (1,0.8)	(0.38,0.8) (0.17,0.2) (0.5,0.1)	(1,0.8) (1,0.8) (0.8,0.6)
A3	(1,0.7) (1,0.5) (1,0.1)	(1,0.6) (1,0.8) (1,0.1)	(0.5,0.4) (0.25,0.6) (1,0.1)	(1,0.9) (1,0.8) (0.8,0.8)
A4	(0.67,0.8) (0.5,0.8) (0,0.9)	(1,0.7) (1,0.8) (1,0.9)	(0.38,0.5) (0.33,0.6) (0.5,0.1)	(0.875,0.9) (1,0.9) (0.625,0.9)

Table 2.
Normalized decision matrix.

Alternatives/ criteria	Membership Z-number	Non-membership Z-number	Hesitancy Z-number
A1	(0.75,0.1024)	(0.769,0.0064)	(0.775,0.004)
A2	(0.47, 0.11)	(0.47, 0.1024)	(0.628, 0.0048)
A3	(0.83,0.23)	(0.717,0.192)	(0.9376,0.0008)
A4	(0.65,0.25)	(0.617,0.3456)	(0.2187,0.0729)

Table 3.
Matrix obtained using WASPAS in Z-number.

Hence the ranking of alternatives is obtained as shown below.

Table 4 clearly concludes that A4 > A2 > A1 > A3.

Further, as the values of $\lambda = 0.5$ were increased and decreased, the ranking of the alternatives remained unaltered.

Thus, the uncertainty involved in allocating suitable locations for installation of smog towers could be solved using Zadeh's Z-numbers and WASPAS and a feasible solution has been obtained.

Alternatives	Score function	Ranking
A1	0.067385	3
A2	0.123704	2
A3	0.058708	4
A4	0.272859	1

Table 4.
Score and ranking of the alternatives.

4.3 Challenges and limitations

This study focusses mainly on a specific locality which is densely populated in a small part of the city. On a large scale, a city like Chennai will require at least 20 such smog towers in each locality to control air pollution. Further, the money spent on these towers in installing and maintaining would cost a lot for the local government to manage. A smog tower of this capacity would require close to Rs.30,000 just for the maintenance. Hence, such smog towers cannot be the only solution to reduce air pollution. Further research is required to prove the effectiveness of these towers in reducing air pollution and providing clean air.

4.4 Awareness and suggestions

Air pollution is one of the key factors affecting the livelihood in a city like Chennai. Government should raise awareness about using green energy and install EV charging points at key locations for people to use electric vehicles. More research and funding needs to be given to develop low cost electric vehicles which are affordable for a common man.

5. Conclusion and future directions

In this chapter, a generalized fuzzy set was discussed. Zadeh's Z-numbers were combined with WASPAS method and the problem of location selection for smog towers in the locality of Perungudi was discussed. The most feasible solution was obtained which could provide some relief to the residents suffering because of air pollution. The concept can be extended to other decision-making techniques and better results can be obtained.

Author details

Janani Bharatraj
Chennai, India

*Address all correspondence to: jananichari@gmail.com

References

[1] Zadeh LA. Fuzzy Sets. Information and Control. 1965;**8**(3):338-353

[2] Atanassov KT. Intuitionistic Fuzzy Sets. Fuzzy sets and Systems. 1986; **20**(1):87-96

[3] Atanassov KT, Gargov G. Interval-valued Intuitionistic Fuzzy sets. Fuzzy Sets and Systems. 1989;**31**(3):343-349

[4] Cuong B C, Picture Fuzzy Sets. Journal of Computer Science and Cybernetics. 2015:30(4).

[5] Yager RR. Properties and applications of Pythagorean fuzzy sets, In: Angelov P, Sotirov S(eds) Imprecision and uncertainty in information representation and processing. Studies in Fuzziness and Soft Computing. 332. Springer.

[6] Torra V. Hesitant Fuzzy Sets. International Journal of Intelligent Systems. 2010;**25**(6):529-539

[7] Zhu B, Xu Z. Probability-hesitant fuzzy sets and the representation of preference relations. Technological and Economic Development of Economy. 2017;**24**(3)

[8] Gundogdu FK, Kahraman C. Spherical Fuzzy sets and Spherical Fuzzy TOPSIS method. Journal of Intelligent and Fuzzy Systems. 2019: 337-352

[9] Bellman RE, Zadeh LA. Decision making in fuzzy environment. Management Science. 1970;**17**(4): 141-164

[10] Zadeh LA. A note on Z-numbers. Information Sciences. 2011;**181**: 2923-2932t

[11] Zavadskas EK, Zenonas T, Antucheviciene J, Zakarevicius A. Optimization of weighted aggregated sum product assessment. Elektronika ir elektrotechnika. 2012;**122**(6):3-6

[12] Qiao D, Shen K, Wang J, Wang T. Multi-criteria PROMETHEE method based on possibility degree with Z-numbers under uncertain linguistic environment. Journal of Ambient Intelligence and humanized computing. 2020;**11**:2187-2220

[13] Aliev RA, Huseynov OH, Serdaroglu R. Ranking of Z-numbers and its application in decision making. Journal of Information Technology and Decision Making. 2016;**15**(6):1503-1519

[14] Peng H, Wang J, A multi-criteria group decision-making method based on the normal cloud model with Zadeh's Z-numbers. IEEE transactions on Fuzzy Systems.26(6):

[15] Wang F, Mao J, Approach to multicriteria group decision making with Z-numbers based on TOPSIS and Power aggregation operators. Mathematical problems in Engineering.2019:19pages.

[16] Chatterjee K, Kar S. A multi-criteria decision making for renewable energy selection using Z-numbers in uncertain environment. Technological and economic development of economy. 2018;**24**(2):739-764

[17] Chakraborty S, Zavadskas E, Antucheviciene J. Application of WASPAS method as a multi-criteria decision-making tool. In: Economic Computation and economic cybernetics studies and research. 2015

[18] Kahraman C, Otay I, Solar PV power plant location selection using Z-Fuzzy number based AHP. International Journal of the Analytical Hierarchy Process. 2018; 10(3).

[19] Mohammad Hossein Fazel Zarandi and Mohammad Mehdi Fazel Zarandi

(February 1st 2008). Fuzzy Multiple Agent Decision Support Systems for Supply Chain Management, Supply Chain, Vedran Kordic, IntechOpen, DOI: 10.5772/5343.

[20] Abdelwaheb Aydi (October 31st 2018). An Integrated Multicriteria and Fuzzy Logic Approach for Municipal Solid Waste Landfill Siting, Fuzzy Logic Based in Optimization Methods and Control Systems and Its Applications, Ali Sadollah, IntechOpen, DOI: 10.5772/intechopen.75161.

[21] Ali Sadollah (October 31st 2018). Introductory Chapter: Which Membership Function is Appropriate in Fuzzy System?, Fuzzy Logic Based in Optimization Methods and Control Systems and Its Applications, Ali Sadollah, IntechOpen, DOI: 10.5772/intechopen.79552.

[22] Voloşencu, C., Properties of Fuzzy Systems, WSEAS Transactions on Systems, Issue 2, Volume 8, February, 2009, ISSN: 1109–2777, pag. 210–228

Chapter 14

Fuzzy Modeling of Urban Water Supply Crisis

Welitom Ttatom Pereira da Silva
and Marco Antonio Almeida de Souza

Abstract

The chaotic growth of cities results in numerous problems related to public health and urban environment. One of these problems is the urban water supply system crisis. This research aims to develop a mathematical model for urban water supply crises (UWC) able to deal with the ambiguity of the real available data. The applied methodology comprises the following steps: (i) identifying the influencing factors in UWC; (ii) proposing a conceptual model for the description of UWC; (iii) collecting and simulating the necessary and available data; (iv) optimizing the conceptual model parameters; and (v) verifying the proposed model performance. The results indicate that there are many influencing factors in UWC. The model developed comprises two parts or two sub-models. The first sub-model explains water consumption, and the second sub-model explains water availability. In the first sub-model, the functions are related to the factors that influence water consumption. In the second sub-model, the functions are related to the factors that influence the availability of water. This research also aims to analyze the possibility of applying Fuzzy Logic to deal with the ambiguity of real data. It was concluded that, with the proposed model, the UWC was modeled appropriately. The model proposed can help to predict the impact of actions such as reducing losses, reducing pressure on the water supply network and intermittent supply on the intensity of water crisis cases in cities.

Keywords: water crisis, water scarcity, fuzzy logic, water consumption, water demand

1. Introduction

According to data from the World Meteorological Organization, global water consumption increased more than six-fold in less than a century, more than double the rate of the population growth, and continues to grow considering rising consumption in the agricultural, industrial and domestic sectors [1]. These data lead to the conclusion that in the coming years the global situation of water reserves will move towards a crisis, both in quantitative and qualitative aspects, if adequate water management actions are not taken. More recently, urban water supply crises (UWC) have been observed, a context characterized by water scarcity, as well as damage to the environment and population health, especially among poor populations. As human populations continue to grow, these problems are likely to become more frequent and serious. One example is the case of New York City's water supply, which is facing a crisis. The social and economic development of New York City from the 1970s led to a sudden crisis in the city's water supply system in the 1990s [2]. Other examples were also reported, such as in Palestine, where a UWC case was observed caused mainly by the inadequate access to freshwater resources and inappropriate management [3]. The city of Tijuana, in Mexico, has shown the highest rates of economic growth in the country, resulting in a rapid increase of water demand and consequently the emergence of a UWC [4]. From 1998 to 2000, the city of Campina Grande in Brazil faced a WSC caused by severe periods of drought and the complete absence of freshwater resource management [5]. This UWC caused serious water rationing in Campina Grande that lasted one year. This is not a unique case in Brazil as frequent water rationing has been observed in the cities of Recife and São Paulo [6]. In the Brazilian Federal District, rapid non-planned urbanization and land changes have had a considerable impact on water resources. From 2016 to 2018, the city of Brasilia also went through a UWC situation, and governance, regulation, as well as management support strategies in the urban and rural environment were implemented [7].

In another study, an investigation into Qatar's sustainability crisis, originating from high levels of water, electricity and food consumption was carried out [8]. The high levels of consumption were made possible by the significant wealth of hydrocarbons, redistributive water governance of a generous rentier state and structural dependence on imported food and subsidies on food production. In this state, the water crisis is silent because it does not cause interruptions in supply or public discontent. The possible solution comes from programs that integrate the water, energy and food sectors [8].

The imprecise and ambiguity are inherent to the water supply system, e.g., pressures, flow rate supply and consumption, age and characterization of pipe resistance. Some researches that addressed this imprecise using Fuzzy Logic are briefly described below. A study to accommodate aleatory uncertainty was performed using stochastic analysis to represent the input uncertainties and to estimate resulting uncertainty in nodal pressures and pipe flows) [9]. Results of Fuzzy analyses for two realistically sized water distribution networks show that the proposed method performs with an acceptable level of accuracy and greatly reduces computational time [9].

The need to improve predictive models of hydraulic transients in water systems (water supply networks) was the subject of another study [10]. For this purpose, triangular Fuzzy numbers are used to represent the input uncertainties. Then, to obtain the extreme pressure heads in each location of the network and at each level of uncertainty, four independent optimization problems are solved. The results is found computationally fast and promising for real applications [10].

The Fuzzy Logic is a perfect tool among white-box methods (models based on laws of physics, chemistry, others, that govern the dynamic behavior of the system) for risk analysis due to the capabilities in dealing with limited data, subjective and temporal variables, and modeling expert opinions, among others [11]. For this reason, Fuzzy Logic was used as comprehensible framework for assessing water supply risk based on existing and under-construction projects in Mashhad city, Iran. The results showed that the framework based on the Fuzzy and possibility theory is befitting to information gathered from experts. Such a framework can provide a simple method to apply the proposed methodology to other water management projects, where, despite the high level of investment, there is no clear idea of the risks and their consequences [11].

Mathematical modeling is a well-known tool for water management. However, when considering UWC, there are some limitations of the conventional mathematical modeling that are related to vague and ambiguous data (e.g. real water-loss, real water availability, setting water tariffs, among others). In theory, the UWC problem can be formulated as a mathematical problem. In practice, rules are considered to be a more practical method. However, the combination of mathematical programming and Fuzzy rule has rarely been discussed in the literature [12]. The aim of this chapter is to describe the develop a mathematical model for UWC, dealing with this ambiguity and real data uncertainty.

2. Theoretical background

UWC modeling was essentially based on the definitions of water crisis and Fuzzy Logic. The topics are presented briefly below.

2.1 Urban water supply crisis (UWC)

A UWC case was presented in Iran, and information about its installation process, climatic characteristics of the region, range of per capita water consumption, among others was provided [13]. As the authors mention, up to 1990, water supply was not a critical problem and there was an acceptable relationship between water demand and distribution. Over the last decade, however, the problem has become critical and the reasons identified include rapid population growth. In addition, there is a reduction in the number of water supply systems (due to the loss of financial resources) and the widespread occurrence of droughts. Considering these observations, one can arrive at the concept of UWC, basically translated as a mismatch between water supply and consumption rates, according to Eq. (1).

$$UWC = C - A \tag{1}$$

In which: C is water consumption; A is the water availability.

According to [14], when clarifying the terms of water consumption (C) and water availability (A) and the definition of constraints, some issues then need answers. Is there a more suitable form of a mathematical model to represent water consumption (C)? Is there a mathematical model that is the water availability (A)? What and how many factors are there (socioeconomic, environmental, cultural, urban, and management conditions) that should be considered in these models? What restrictions should be imposed on the optimization model? The literature review suggests that Fuzzy Logic can help by incorporating imprecision and ambiguity in the translation of influential factors and field characteristics.

2.2 Fuzzy logic

The advent of Fuzzy Logic originates from the need of a method that can systematically express inaccurate, vague, ill-defined quantities [15]. For example, instead of using a complex mathematical model, industrial controllers based on Fuzzy Logic can be implemented using knowledge from human operators, or heuristic knowledge. This makes the control action using Fuzzy Logic as good as using the raw knowledge (generally better) and always consistent. In decision analysis, Fuzzy Logic can be used in tasks in which individual variables are not defined in exact terms. Some examples of this have been provided in the literature [16, 17]. In the environmental area, the population's preference for water conservation action, the assessment of the performance of environmental education programs and policies in preventing forest fires are examples of individual variables not defined in exact terms. The properties of Fuzzy Systems are demonstrated in several publications, as, for example, in [18]. An introduction to the theory of Fuzzy sets is presented below.

In general, a Fuzzy Set can be summarized as follows [15, 18, 19]:

- Definition 1: a subset A of a set X is said to be Fuzzy Set if $\mu_A : X \to [0,1]$, where μ_A denote the degree of belongingness of A in X.
- Definition 2: a fuzzy set A of set X is said to be normal if $\mu_A(x) = 1, \forall x \in X$.
- Definition 3: the height of A is defined and denoted as $h(A) = \sup \mu_A, \forall x \in X$.
- Definition 4: the α-cut and strong α-cut is defined and denoted respectively as $\alpha_A = \{x/\mu_A(x) \geq \alpha\}, \mu_A^+ = \{x/\mu_A(x) > \alpha\}$.
- Definition 5: let $\tilde{a}, \tilde{b}$ be two fuzzy numbers, their sum is defined and denoted as $\mu_{\tilde{a}+\tilde{b}}(z) = \sup_{z=u+v} \min \{\mu_{\tilde{a}}(u), \mu_{\tilde{b}}(v)\}$, where $0 \leq \lambda \in R$.
- Definition 6: if a Fuzzy number $\tilde{a}$ is fuzzy set A on R, it must possess at last following three properties: (i) $\mu_A(x) = 1$; (ii) $\{x \in R/\mu_{\tilde{a}}(x) > \alpha\}$ is a closed interval for every α∈ [0,1]; (iii) $\{x \in R/\mu_{\tilde{a}}(x) > 0\}$ is a bounded and it is denoted by $[a_\lambda^L, a_\lambda^R]$.
- Theorem 1: a fuzzy set A on R is convex if and only if $\mu_A(\lambda x_1 + (1-\lambda)x_2) \geq min\,[\mu_A(x_1), \mu_A(x_2)]$, for all $x_1, x_2 \in X$ and for all $\lambda \in [0,1]$ where min denotes the minimum operator.
- Theorem 2: let $\tilde{a}$ be a fuzzy set on R, the $\tilde{a} \in f(R)$ if and only $\mu_{\tilde{a}}$ satisfies (Eq. 2):

$$\mu_{\tilde{a}}(x) = \begin{cases} 1, & for\, x \in [m,n] \\ L(x), & for\, x < m \\ R(x), & for > n \end{cases} \tag{2}$$

Where: $L(x)$ is the right continuous monotone increasing function, $0 \leq L(x) \leq 1$ and $\lim_{x \to \infty} L(x) = 0$; $R(x)$ is a left continuous monotone decreasing function, $0 \leq R(x) \leq 1, \lim_{x \to \infty} R(x) = 0$.

The Fuzzy Linear Programming Problem (FLPP) with decision variables and coefficient matrix of constraints are in Fuzzy nature (Eqs. 3–6).

$$\tilde{z} = max \sum_{j=1}^{n} \tilde{c}_j x_j \tag{3}$$

subject to

$$\sum_{j=1}^{n} \tilde{A}_{ij} x_j \leq \tilde{B}_i \tag{4}$$

$$1 \leq i \leq m \tag{5}$$

$$x_i > 0 \tag{6}$$

The triangular Fuzzy numbers A which can be represented by three crisp numbers s, l, r (Eqs. 7–10).

$$f_i(x_j) = max \sum_{j=1}^{n} \tilde{c}_j x_j \tag{7}$$

subject to

$$\sum_{x \geq 0} (s_{ij}, l_{ij}, r_{ij}) x_{ij} \leq (t_i, u_i, v_i) \tag{8}$$

$$1 \leq i \leq m \tag{9}$$

$$1 \leq j \leq n \tag{10}$$

Where: $\tilde{c} = (c_s, c_l, c_r)$, $\tilde{A} = (s_{ij}, l_{ij}, r_{ij})$ and $\tilde{B} = (t_i, u_i, v_i)$ are Fuzzy numbers.

- Theorem 3: For any two triangular Fuzzy numbers $\tilde{A} = (s_1, l_1, r_1)$ and $\tilde{B} = (s_1, l_2, r_2)$, $\tilde{A} \leq \tilde{B}$ if and only if $s_1 \leq s_2, s_s - l_1$ and $s_1 + r_1 \leq s_2 + r_2$. Above problem can be rewritten (Eqs. 11–15).

$$f_i(x_j) = max \sum_{j=1}^{n} \tilde{c}_j x_j \tag{11}$$

subject to

$$\sum_{j=1}^{n} s_{ij} x_{ij} \leq t_i \tag{12}$$

$$\sum_{j=1}^{n} (s_{ij} - l_{ij}) x_j \leq t_i - u_i \tag{13}$$

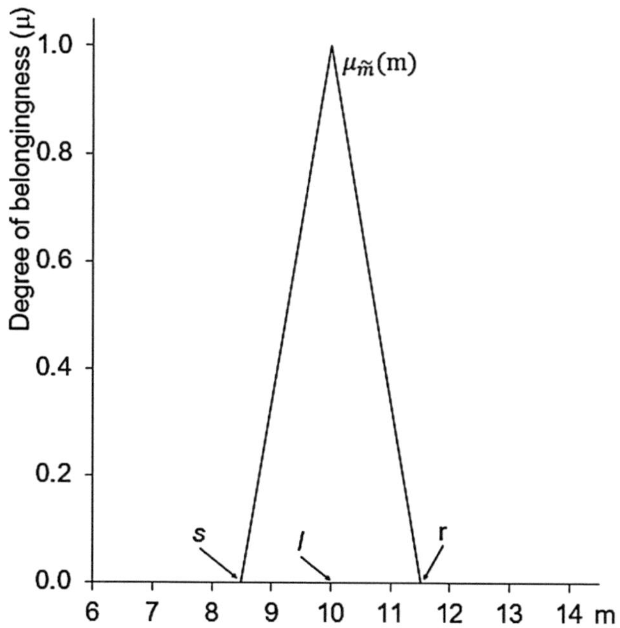

Figure 1.
Fuzzy number $\tilde{m} = (m_s, m_l, m_r)$, triangular membership function.

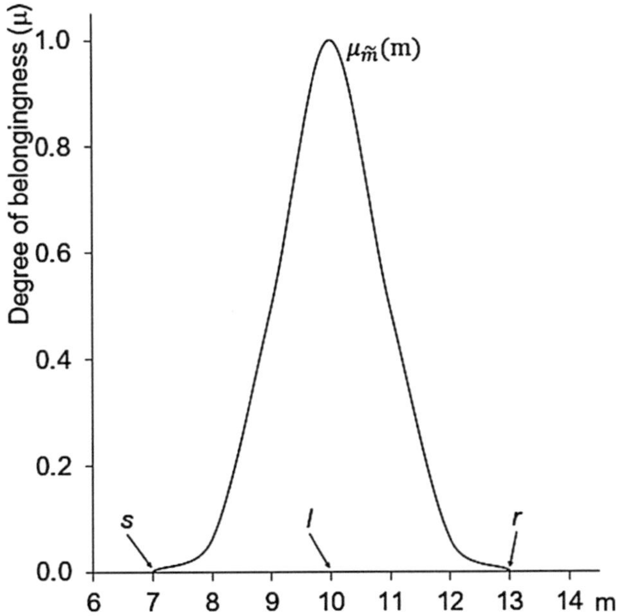

Figure 2.
Fuzzy number $\tilde{m} = (m_s, m_l, m_r)$, bell-shaped membership function.

$$\sum_{j=1}^{n}(s_{ij} - l_{ij})x_j \leq t_i - v_i \tag{14}$$

$$x_i \geq 0 \tag{15}$$

As examples of membership functions for a Fuzzy number $\tilde{m}$, such as approximately $m = 10$, a triangular membership function (Eq. 16) and a bell-shaped membership function (Eq. 17) is widely used.

$$\mu_{\tilde{m}} = \left(0, 1 - \frac{|x - m|}{a}\right), a > 0 \tag{16}$$

$$\mu_{\tilde{m}} = e^{-b(x-m)^2}, b \geq 1 \tag{17}$$

Such membership functions are illustrated in **Figures 1** and **2**. More information about Fuzzy Mathematical Programming is available in the literature [15–23].

3. Methodology

The methodology of this work comprised the following steps: (1) identifying the influencing factors in UWC; (2) proposing a conceptual model for UWC; (3) data collection and data simulation; (4) optimizing the proposed conceptual model parameters (calibration); and (5) assessing conceptual model performance (verification).

For step (1), a literature review was conducted related to WSC management.

For step (2), the definition of UWC found in the literature was taken and the Fuzzy Non-Linear Programming (FNLP) was used [12, 15–24].

For the third step (step 3), a case study was simulated in Brazil, more precisely in the Federal District, taking into account predictions made by some researchers.

The necessary data were collected from the Brazilian Federal District's water supplier and sanitation company (CAESB), from the National Institute of Meteorology in Brazil (INMET) and from the Brazilian Federal District's Government (GDF). The simulation was conducted assuming the hypothesis of a significant increase in the water consumption in coming years, as well as managerial and political stagnation of the water supplier and sanitation company. The time series analysis refers to three years (2007, 2008 and 2009). The period for model calibration was considered for two years and the verification period was considered for one year. Data were normalized prior to optimization of conceptual model parameters. This was done to restrict their range within the interval of −1.0 to +1.0 to eliminate the difference among scales measuring the influencing factors, according to Eq. (18).

$$x_{norm} = 1 - \frac{2 \times (x_{max} - x_0)}{x_{max} - x_{min}} \tag{18}$$

Where: x_{norm} is the normalized value; x_0 is the original value; x_{max} is the maximum value; x_{min} is the minimum value.

For step (4), the minimization of the sum of squared errors and the Differential Evolution & Particle Swarm Optimization algorithm (DEPS) was adopted, using the spreadsheet from Open Office (Calc-Solver). As a justification for using the DEPS algorithm, it presents good performance to obtain global optimums, based on the synergy between Particle Swarm Optimization (PSO) and Differential Evolution (DE) [25].

In step (5), some tests used to model performance assessment were proposed, among them, the correlation coefficient (r), the determination coefficient (R^2), the average relative error percentage (AREP) and graphic observed values *versus* estimated values.

4. Results and discussions

A bibliographic review was conducted by [26] and resulted in a list of influential factors presented below:

- Population growth rate.
- Human population density.
- Socio-economic level.
- Education level.
- Industry level.
- Ambient temperature.
- Relative humidity.
- Rainfall.
- Seasonality.
- Size and topographic characteristics of the city.

- Percentage of water metering.
- Water tariffs.
- Type of water tariff policies.
- Existence of wastewater collection systems.
- Human Development Index.
- Pressure in the water distribution network.
- Existence of conservation habits.
- Number/type of hydro-sanitary equipment per household.
- Constructed area per household.
- Number of rooms.
- Abundance or scarcity of water sources.
- Water-loss.
- Social representation and identification of each family.
- Existence and type of municipal water resources policy.
- Acceptance of the population to water conservation and rational use actions.
- Typology of land use.
- Type of consumers.
- Type of municipality.
- Predominant function of urban environment.
- Existence of policies to promote water conservation.
- Intermittence in the water supply system.
- Energy consumption.
- Existence of regulatory policy on water consumption.
- Existence of environmental education program.
- Dissemination of the belief: water is an inexhaustible resource and low-priced.

The UWC has appeared as an inadequate ratio between water consumption and water supply [12]. Water consumption and water availability are vague and ambiguous terms, because they are dependent on haziness measures, qualitative factors, scarcity data and low-quality data [27].

Design and analysis of water distribution networks (WDNs) are laden with uncertainty. There is natural randomness, such as variations in reservoir elevation heads, and there is epistemic randomness, i.e., incomplete knowledge, imprecise data, and linguistic ambiguity. Both are associated with the characterization of pipe resistance, nodal demands, and hydraulic responses [9]. The analysis of water distribution networks has to take into account the variability of users' water demand and the variability of network boundary conditions, e.g. the presence of local private tanks and intermittent distribution [28].

Some of these factors were considered in this paper, the faulty water metering ($\tilde{C}$ and $\tilde{A}$), the imprecise value of the average pressure ($\tilde{p}$) in the water distribution network and the imprecise value of water-loss ($\tilde{l}$) in the water supply system. A fuzzy mathematical model of the UWC is written as Eq. (19).

$$UWC = \tilde{C} - \tilde{A} \qquad (19)$$

Where: $\tilde{C}$ is a fuzzy mathematical model that represents the average value of the water consumption; $\tilde{A}$ is a fuzzy mathematical model that represent the average value of the water availability.

For a mathematical representation of the $\tilde{C}$, a fuzzy mathematical model was proposed considering some influencing factors in water consumption: the ambient temperature, the relative humidity, rainfall, collected revenues, unemployment indicator, and average pressure in the water distribution network. The $\tilde{C}$ model was based on some assumptions, as follows: the existence of a non-linear relationship among the influencing factors, the existence of a haziness of $\pm 10\%$, that is $\tilde{C} = (C_s, C_l, C_r)$, in observed values of water consumption (error of household's water meter), the increase in the water consumption with the increase in the ambient temperature, the existence of a relationship between relative humidity and the rainfall, the reduction of water consumption with the increase relative humidity, the increase in the water consumption with the increase in the revenues collected, the reduction in water consumption with the increase of the unemployment indicator, the increase in water consumption with the increase of the average pressure (p) in the water distribution network and the existence of a haziness of $\pm 5\%$ in observed values of the average pressure in the water distribution network, that is $\tilde{p} = (p_s, p_l, p_r)$.

Likewise, a Fuzzy mathematical behavior of the $\tilde{A}$ was proposed, considering the total water-loss in water supply system and the intermittence in the water supply system. The $\tilde{A}$ model was based on the following assumptions as influencing factors: the existence of a non-linear relationship among the influencing factors, the existence of a haziness of $\pm 10\%$, that is $\tilde{A} = (A_s, A_l, A_r)$, in observed values of the water supply (error of waterworks' water meter), the reduction of the water supply with the increase in the intermittence in the water supply system, the reduction in the water supply with the increase in the total water-loss (l) in the water supply system and the existence of a haziness of $\pm 10\%$ in observed values of the total water-loss, that is $\tilde{l} = (l_s, l_l, l_r)$.

The membership functions for the fuzzy numbers $\tilde{C}$ and $\tilde{A}$ are approximately equal to the observed values of water consumption and water supply, following a triangular membership function as shown in **Figures 3** and **4**.

The membership functions for the fuzzy numbers ($\tilde{p}$) and ($\tilde{l}$) were considered approximately equal to the observed values of the average pressure in the water distribution network and total water-loss in the water supply system, following a pattern of a triangular membership function as shown in **Figures 5** and **6**.

All assumptions, from $\tilde{C}$ and $\tilde{A}$, made are according to previous research [4, 27–31]. Eqs. (20) to (27) compose the proposed FNLP model.

$$\tilde{z}_C = min \sum_{i=1}^{n} \left(\tilde{C}_{E,i} - \tilde{C}_{O,i}\right)^2 \tag{20}$$

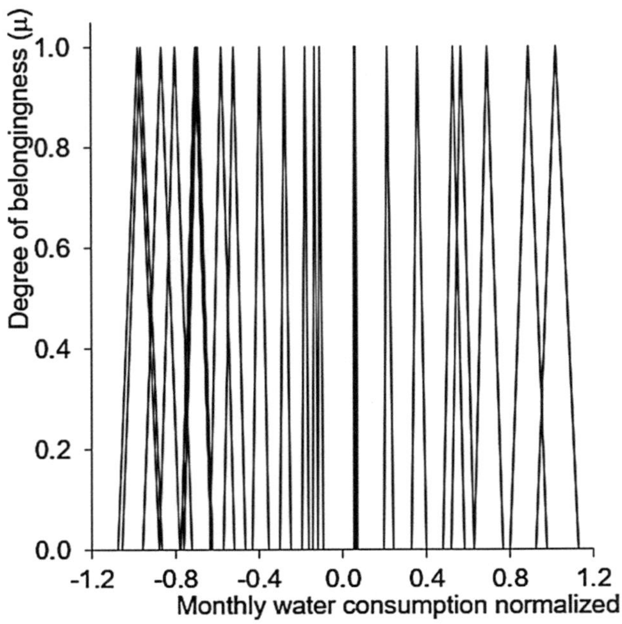

Figure 3.
Fuzzy number for water consumption, $\tilde{C}$.

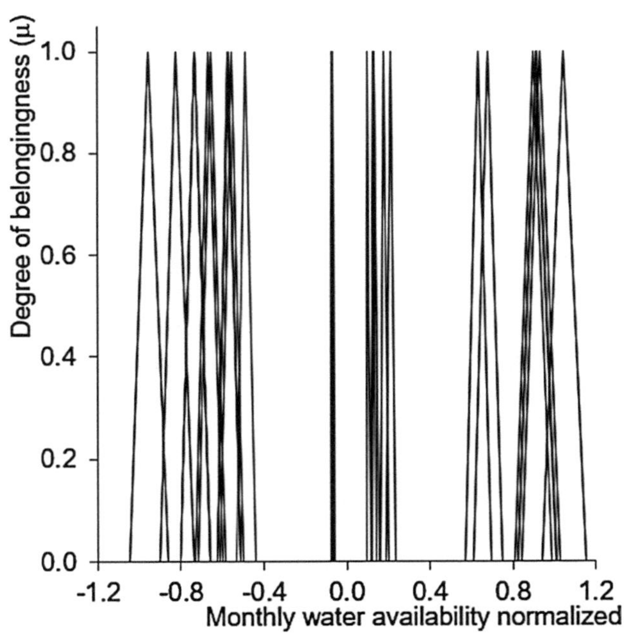

Figure 4.
Fuzzy number for water availability, $\tilde{A}$.

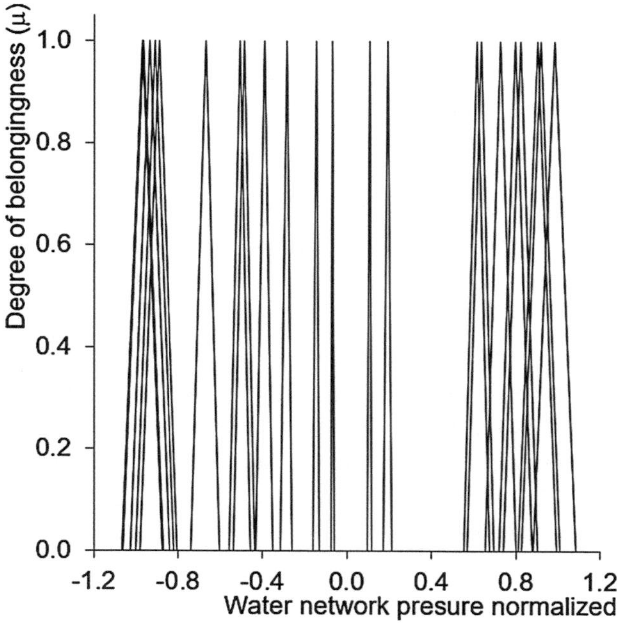

Figure 5.
Fuzzy number for pressure in the water distribution network, $\tilde{p}$.

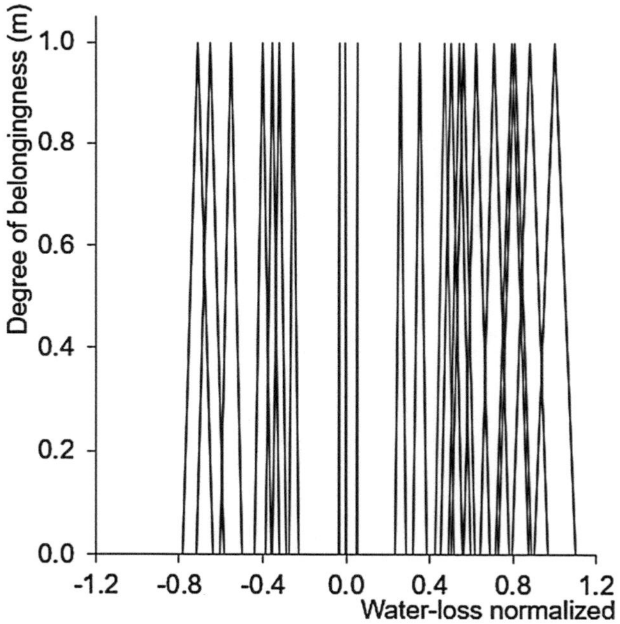

Figure 6.
Fuzzy number for total water-loss in water supply system, $\tilde{l}$.

subject to

$$\tilde{C}_{E,i} = \beta_0 + (\beta_1)^{x1_i} - (\beta_2)^{x2_i+x3_i} + (\beta_3)^{x4_i} - (\beta_4)^{x5_i} + (\beta_5)^{\tilde{p}_i} \tag{21}$$

$$\tilde{p} = (p_s, p, p_r) \tag{22}$$

$$\tilde{C}_{O,i} = (C_{O,i,s}, C_{O,i}, C_{O,i,r}) \tag{23}$$

$$\tilde{z}_A = \acute{m}in \sum_{i=1}^{n} \left(\tilde{A}_{E,i} - \tilde{A}_{O,i}\right)^2 \tag{24}$$

subject to

$$\tilde{A}_{E,i} = \beta_6 - \beta_7^{\tilde{l}} - \beta_8^{x6} \tag{25}$$

$$\tilde{l} = (l_s, l, l_r) \tag{26}$$

$$\tilde{A}_{O,i} = (A_{O,i,s}, A_{O,i}, A_{O,i,r}) \tag{27}$$

Where: $\tilde{z}_C$ is the objective function representing the calibration of $\tilde{C}$; $\tilde{C}_{E,i}$ is a fuzzy number representing the estimated water consumption; $\tilde{C}_{O,i}$ is a Fuzzy number representing the observed water consumption; $C_{O,i,s}$ and $C_{O,i,r}$ are the lower bound and upper bound of Fuzzy triangular numbers in the set $\tilde{C}_{O,i}$, as is shown in **Figure 1**; $\tilde{z}_A$ is the objective function representing the calibration of $\tilde{A}$; $\tilde{A}_{E,i}$ is a Fuzzy number representing the estimated water availability; $\tilde{A}_{O,i}$ is a Fuzzy number representing the observed water availability; $\tilde{A}_{O,i,s}$ and $\tilde{A}_{O,i,r}$ are the lower bound and upper bound of Fuzzy triangular numbers in the set $\tilde{A}_{O,i}$, as is shown in **Figure 1**; $\beta 0$, $\beta 1$, … , $\beta 8$ are parameters, $x1$ is the ambient temperature; $x2$ is the relative humidity; $x3$ is the rainfall; $x4$ is the amount of collected revenues; $x5$ is the unemployment indicator; $\tilde{p}$ is the average pressure in the water distribution network; $\tilde{l}$ is the total water-loss in the water supply system; $x6$ is the intermittence in the water supply system.

Eq. (28) and (29) show the results of the parameter optimization in the proposed conceptual model when applying the previously described data normalized according to Eq. 4.

$$\tilde{C}_P = -0.9859 + 1.2350^{x1} - 1.4188^{x2+x3} + 0.8302^{x4} - 1.4983^{x5} + 1.4620^{\tilde{p}} \tag{28}$$

$$\tilde{A}_P = 2.1559 - 1.7744^{\tilde{l}} - 1.4819^{x6} \tag{29}$$

Collected and simulated data	Average ± sd*	Symbol	Unit	Data source
Water consumption	14,381.2 ± 773.8	$\tilde{C}$	$\times 10^3$ m^3.month^{-1}	Simulated
Water availability	15,234.0 ± 325.3	$\tilde{A}$	$\times 10^3$ m^3. month $^{-1}$	Caesb** [32]
Ambient temperature	21.1 ± 1.15	$x1$	°C	INMET** [33]
Relative humidity	67.83 ± 10.10	$x2$	%	INMET** [33]
Rainfall	113.21 ± 82.26	$x3$	mm.month^{-1}	INMET** [33]
Collected revenues	681.70 ± 41.05	$x4$	million R$.month^{-1}	GDF** [34]
Water distribution network pressure	54.63 ± 6.62	$\tilde{p}$	mH_2O	Simulated
Unemployment indicator	16.85 ± 1.13	$x5$	%	GDF [34]
Total water-loss in water supply system	18.99 ± 4.41	$\tilde{l}$	%	Simulated
Intermittence in water supply system	35.32 ± 6.95	$x6$	hours.month^{-1}	Simulated

**sd is the standard error*
***availability of data on the Internet, partially simulated (fault fill)*

Table 1.
Collected and simulated data.

The collected and simulated data and its respective descriptions are shown in **Table 1**. **Table 2** and **Figures 7–12** show the results of the model performance.

The calibration and verification results indicated that the proposed conceptual model has shown good agreement for the collected and simulated data, when considering some previous research. For instance, a study on water demand developed in the cities of Oklahoma and Tulsa Oklahoma State, USA, resulted in a statistical model to explain water demand with R^2 range within the interval of 0.140 a 0.920 [35].

In a household water demand study in the northwest of Spain, the price, billing, climatic, and sociodemographic variables were used as explanatory variables and the results showed a R^2 range within the interval of 0.198 to 0.891 [36].

Another study aiming to predict future water consumption from Istanbul, Turkey, was developed using the Takagi Sugeno Fuzzy method for modeling monthly water consumption, and the overall prediction presented an AREP of less than 10% [37]. In this study, AREP ranged from −40% to 5% indicating an opportunity to improve the model developed. In another regional water study, in the case of Tijuana, in Northwest Mexico, the purpose was to analyze monthly water

	Calibration			**Verification**		
Models	**r**	**R^2**	**AREP**	**r**	**R^2**	**AREP**
Water consumption,	0.8777	0.7704	−7.95	0.5994	0.3593	−40.15
Water availability, $\tilde{A}$	0.9755	0.9516	5.32	0.9412	0.8858	−16.95

r: correlation coefficient
R^2: determination coefficient
AREP: average relative error percentage

Table 2.
Results of calibration and verification of models.

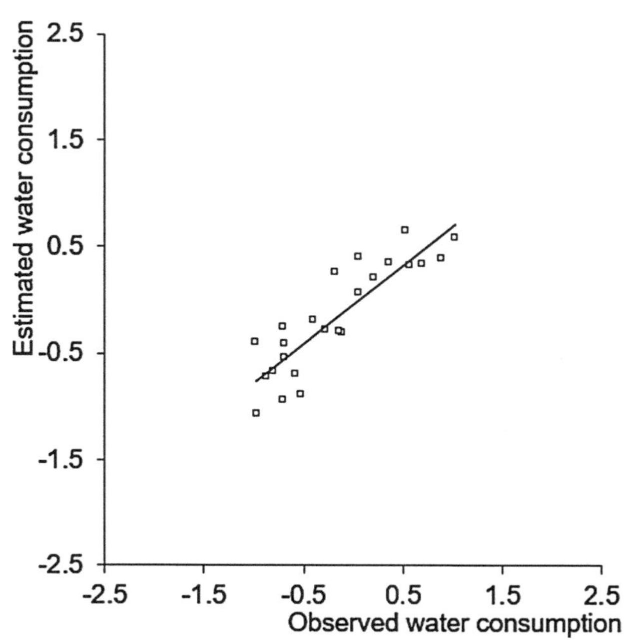

Figure 7.
Calibration $\tilde{C}$ model.

consumption dynamics, and the empirical estimation results were considered fairly satisfactory with the R^2 range from 0.4582 to 0.5932 [4].

The analysis in **Figure 12**, agreeing with the AREP of −40%, reinforces the difficulty of the model developed in the $\tilde{C}_P$ adjustment. The reasons for this difficulty are varied, including problems with data quality (simulation, filling in gaps), the need to incorporate influential variables, the need to improve Fuzzy modeling.

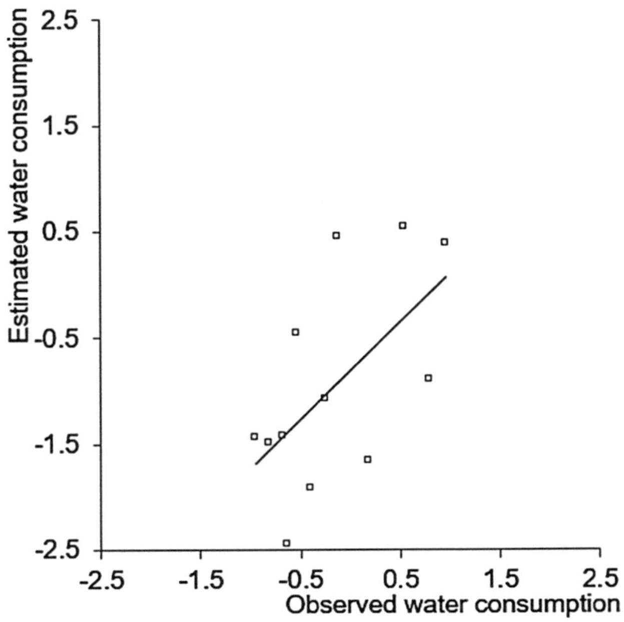

Figure 8.
Verification $\tilde{C}$ model.

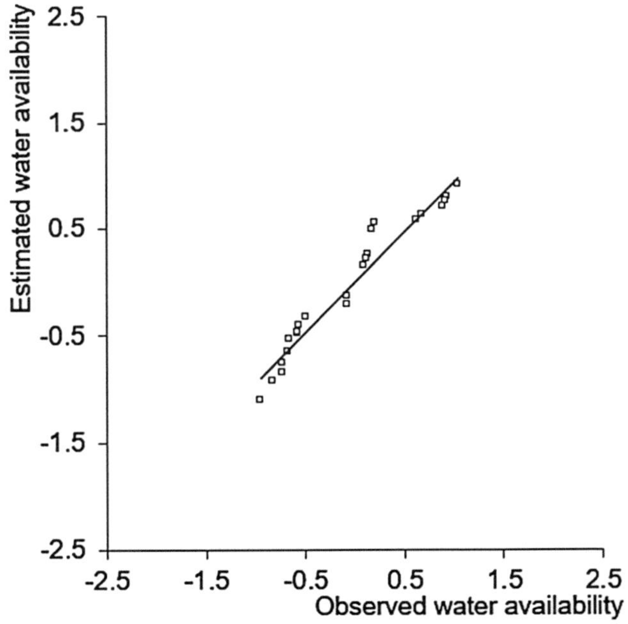

Figure 9.
Calibration $\tilde{A}$ model.

Therefore, we encourage continuous improvement of water consumption forecasting models focusing on optimizing the pertinence functions for the case (format, error rate rates, others).

In general, when considering the quality of the models developed ($\tilde{C}_P$ e $\tilde{A}_P$), according to **Table 2** and **Figures 7–12**, the viability of the FNLP can be verify. The Fuzzy Logic allowed inaccuracies inherent to the water system to be captured and

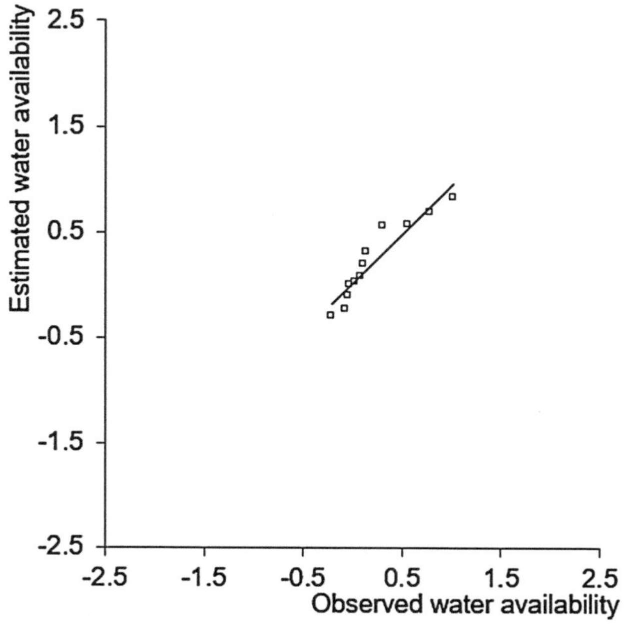

Figure 10.
Verification $\tilde{A}$ model.

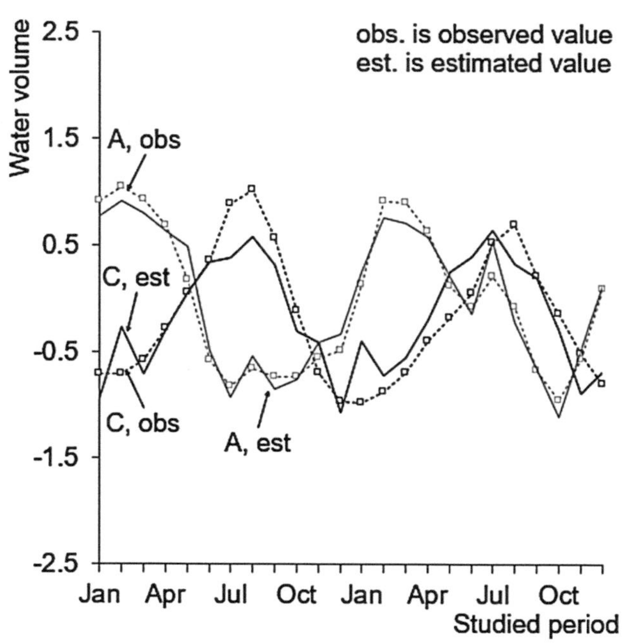

Figure 11.
Calibration model.

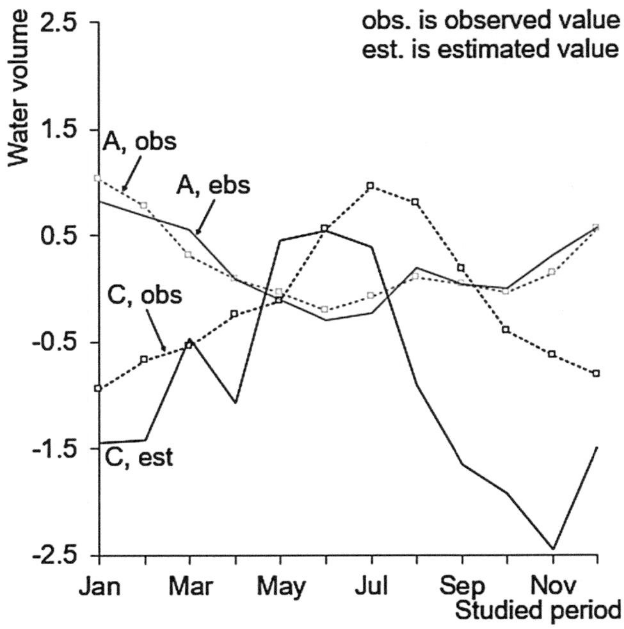

Figure 12.
Verification model.

incorporated into conventional mathematical programming techniques. The satisfactory results agree with previous studies [9, 27, 28].

5. Conclusions and recommendations

5.1 Conclusions

It can be concluded that the UWC was adequately modeled, the ambiguity and the lack of precision of the real data availability was acceptably managed, and the fuzzy approach showed to be adequate for the problem studied. The conceptual model developed in this research can contribute to the water conservation in an urban environment, which is an important tool for water resource planning. More specifically, the model helps to predict the impact of actions such as reducing losses, reducing pressure on the water supply network and intermittent supply on the intensity of water crisis cases in cities.

The methodology used for developing the model can be replicated in other cases. Influential variables can also be adjusted according to the desired responses and available resources. For example, variables such as type of tariff policy, implementation of environmental education programs and incentives to reduce water consumption are possible to be considered in this model. Thus, Fuzzy modeling is a very promising tool and should be encouraged so as to deepen and expand similar models.

5.2 Recommendations

The ambiguity and haziness index of the real data should be considered in further studies. In the future, more influencing factors in water crises should be included in the model being developed, in order to improve prediction results.

Finally, studies focusing on selecting the best forms of representation of Fuzzy variables (Fuzzy membership functions and optimized parameters) should be considered.

Acknowledgements

The authors would like to express their gratitude for the financial support from the Brazilian agencies CNPq (project No. 556084/2009-8), CAPES and DPP-UnB.

Conflict of interest

The authors declare no conflict of interest.

Author details

Welitom Ttatom Pereira da Silva[1] and Marco Antonio Almeida de Souza[2]*

1 Federal University of Mato Grosso (Post-graduate Program on Water Resources), Cuiabá, Brazil

2 University of Brasilia (Post-Graduate Program on Environmental Technology and Water Resources), Brasília, Brazil

*Address all correspondence to: marcantoniosouza@gmail.com

References

[1] Freitas M.A.V.A. O estado das águas no Brasil: perspectivas de gestão e informação de recursos hídricos. Brasília: ANEEL/MME/MMA-SRH/OMM/PNUD; 1999. 334 p.

[2] Gandy M. The making of a regulatory crisis: restructuring New York city's water supply. Blackwell Publishing on Behalf of the Royal Geographic Society. 1997; 22(3): 338–358.

[3] Zahra B.A.A.A. Water crisis in Palestina. Desalination. 2001; 136: 93–99. DOI: 10.1016/S0011-9164(01)00169-2

[4] Fullerton Jr. T.M., Tinajero R., Cota J. E.M. An empirical analysis of Tijuana consumption. Atlantic Economic. 2007; 35: 357–369. DOI: 10.1007/s11293-007-9074-x

[5] Rêgo J.C., Ribeiro M.M.R., Albuquerque J.P.T., Galvão, C.O. Participação da Sociedade na Crise 1998–2000 no Abastecimento d'água em Campina Grande – PB, Brasil. In: Proceedings of the IV Diálogo Interamericano de Gerenciamento de Águas; 2–6 September 2001; Foz do Iguaçu. Porto Alegre: ABRH/IWRA; 2001. p. 1–10

[6] Tucci C.E.M., Hespanhol I., Cordeiro Netto O.M. Cenários da gestão da água no Brasil: uma contribuição para a visão mundial da água. Revista ABRH. 2000; 5 (3): 31–43. DOI: 10.21168/rbrh.v5n3.p31-43

[7] Lima J.E.F.W., Freitas G.K., Pinto M. A.T., Salles P.S.B.A. Gestão da crise hídrica 2016–2018: Experiências do Distrito Federal. Brasília: Adasa / Caesb / Seagri / Emater; 2018. 328 p.

[8] Hussein H., Lambert L.A. A Rentier state under blockade: Qatar's water-energy-food predicament from energy abundance and food insecurity to a silent water crisis. Water. 2020; 12(4): 1051. DOI: 10.3390/w12041051

[9] Moosavian N., Lence J. Approximation of Fuzzy Membership Functions in Water Distribution Network Analysis. Journal of Hydraulic Engineering. 2018; 144(7): 04018039-1–04018039-8. DOI: 10.1061/(ASCE)HY.1943-7900.0001483

[10] Sabzkouhi A. M., Haghighi A. Uncertainty analysis of transient flow in water distribution networks. Water Resources Management. 2018; 32: 3853–3870. DOI: 10.1007/s11269-018-2023-4

[11] Ghandehari A., Davary K., Khorasani H. O., Vatanparast M., Pourmohamad Y. Assessment of urban water supply options by using Fuzzy possibilistic theory. Environmental Processes. 2020; 7: 949–972. DOI: 10.1007/s40710-020-00441-8

[12] Tsai H-R., Chen T. A Fuzzy nonlinear programming approach for optimizing the performance of a four-objective fluctuation smoothing rule in a wafer fabrication factory. Journal of Applied Mathematics. 2013. DOI: 10.1155/2013/720607

[13] Motiee H., Manouchehri G.H., Tabatabai M.R.M. Water crisis in Iran: codification and strategies in urban water. In: Proceedings of the Frontiers in urban water management: deadlock or hope? 18–20 June 2001; Marseille. Paris: UNESCO; 2001. p. 55–62

[14] Silva W.T.P., Souza M.A.A. A decision support model to aid the management of crises in urban water supply systems (the UWCMODEL). Urban Water Journal. 2017; 14(6): 612–620. DOI: 10.1080/1573062X.2016.1223861

[15] Zadeh L.A. Fuzzy sets. Information and control. 1965; 8: 338–353. DOI: 10.1016/S0019-9958(65)90241-X

[16] Volosencu C, editor. Fuzzy Logic. London: IntechOpen; 2020. 213 p. DOI: 10.5772/intechopen.77460

[17] Liu K.R.F. Evaluating Environmental Sustainability: An Integration of Multiple-Criteria Decision-Making and Fuzzy Logic. Environmental Management. 2007; 39: 721–736. DOI: 10.1007/s00267-005-0395-8

[18] Volosencu C. Properties of Fuzzy Systems. WSEAS Transactions on Systems. 2009; 2(8): 210–228.

[19] Thakre P.A., Shelar D.S., Thakre S. P. Solving fuzzy linear programming problem as multi objective linear programming problem. In: Proceedings of the World Congress on Engineering Vol II; 1–3 July 2009; London. Hong Kong: IAENG; 2009. p. 1–4

[20] Slowinski R., Teghem J. Stochastic versus fuzzy approaches to multiobjective mathematical programming under uncertainty. Dordrecht: Kluwer Academic Publishers; 1990. 426 p.

[21] Sakawa M. Fuzzy sets and interactive multiobjective optimization. New York: Plenum Press; 1993. 308 p.

[22] Peraei E.Y., Maleki H.R., Mashinchi, M. A method for solving a fuzzy linear programming. Korean Journal of Computational and Applied Mathematics. 2001; 8(2): 347–356.

[23] Cao B-Y. Fuzzy geometric programming. 2nd ed. Dordrecht: Kluwer Academic Publishers; 2002. 267 p.

[24] Lu T., Liu S-T. Fuzzy nonlinear programming approach to the evaluation of manufacturing processes. Engineering Applications of Artificial Intelligence. 2018; 72: 183–189. DOI: 10.1016/j.engappai.2018.04.003

[25] Das S., Abraham A., Konar, A. Particle swarm optimization and differential evolution algorithms: technical analysis, applications and hybridization perspectives. Studies in Computational Intelligence. 2008; 116: 1–38.

[26] Silva W.T.P., Souza M.A.A. Mathematical modelling of urban water supply system crisis. ENGEVISTA. 2011; 13(1): 46–56. DOI: 10.22409/engevista.v13i1.331

[27] Sim P., McDonald A., Parsons J., Rees P. The options for UK domestic reduction: a review [Internet]. 2005. Available from: http://eprints.whiterose.ac.uk/5002/1/domestic_water_reduction.pdf [Accessed: 2009-8-15]

[28] Freni G., Marchis D. M., Napoli E. Implementation of pressure reduction valves in a dynamic water distribution numerical model to control the inequality in water supply. Journal of Hydroinformatics. 2014; 16(1): 207–217. DOI: 10.2166/hydro.2013.032

[29] Lertpalangsunti N., Chan C.W., Mason R., Tontiwachwuthikul P. A toolset for construction of hybrid intelligent forecasting systems: application for water demand prediction. Artificial Intelligence in Engineering. 1999; 13(1): 21–42. DOI: 10.1016/S0954-1810(98)00008-9

[30] Froukh M.L. Decision-support system for domestic water demand forecasting and management. Water Resources Management. 2001; 15: 363–382. DOI: 10.1023/A:1015527117823

[31] Fernandes Neto M.L., Naghettini M., Von Sperling M. Avaliação da relevância dos parâmetros intervenientes no consumo per capita de água para os municípios de Minas Gerais. Revista Engenharia Sanitária e Ambiental. 2004; 9(2): 100–107.

[32] Companhia de Saneamento Ambiental do Distrito Federal (Caesb).

Sisagua: Sinopse do Sistema de Abastecimento de Água. Brasília: Caesb; 2008. 143 p.

[33] Ramos A.M., Santos L.A.R., Fortes L.T.G. Normais Climatológicas do Brasil 1961–1990: INMET; 2009. 4656p.

[34] Governo do Distrito Federal (GDF). Secretaria de Estado de Fazenda do Distrito Federal: Relatório de Atividades, 2007: GDF; 2007. 33 p.

[35] Cochram R., Cotton A. Municipal water demand study Oklahoma city and Tulsa Oklahoma, Oklahoma water resource board [Internet]. 1984. Available from: http://www.owrb.ok.gov/studies/reports/reports_pdf/pub_123_municipal_water.pdf [Accessed: 2010-2-18]

[36] Martinez-Espiñeira R. Residential water demand in the Northwest of Spain. Environmental and Resource Economics. 2002; 21: 161–187. DOI: 10.1023/A:1014547616408

[37] Altunkaynak A., Özger M., Çakmakci M. Water consumption of Istanbul city by using logic fuzzy. Water Resources Management. 2005; 19: 641–654. DOI: 10.1007/s11269-005-7371-1

Section 8

Applications in Medicine

Chapter 15

Fuzzy Multi-Attribute Decision Making (FMADM) Application on Decision Support Systems (SPK) to Diagnose a Type of Disease

Sugiyarto Surono and Mustika Sari

Abstract

Fuzzy logic is widely applied to daily life with various methods. One method is fuzzy multi-attribute decision making (FMADM). FMADM is able to select the best alternative from a number of alternatives. In FMADM there is a supporting method so that the results obtained are accurate and optimal, namely the classic MADM method. One method in classic MADM is the Simple Additive Weighting (SAW) method. The SAW method is precisely used to minimize diagnostic errors, but if a decision support system is made, the SAW method still requires a further development method, one of which is the FMADM method with its development. The purposes of this study are to describe the steps of SAW method and the development of FDM in theory, implement SAW method and the development of FDM to diagnose a type of disease and implement it in a decision support system using GUI matlab. The completion step of those two methods is through two stages, the first one will go through FMADM stage with SAW, which is weighted sum, then the output will be used as input to the FDM method based on total integral values. The result of this study is proven by patient experienced initial symptoms of high fever at a temperature of 39.5° C - 40° C, very much spots appear in rumple leed test (> 50 petheciae), bleeding gums, rarely got nausea and headache, as well as diarrhea. Accuracy for the decision support system using MAPE was obtained 93% so that the decision support system with FMADM method to diagnose the disease was feasible to use.

Keywords: diagnosing a type of disease, FDM, FMADM, SAW

1. Introduction

Decision making is a problem solving process that produces a goal of factors such as subjectivity and linguistics which tend to be presented in real life to a lower or greater level [1]. Difficulties are often encountered when a problem involves several alternatives and the factors that influence it (criteria), to overcome this problem, it is able to use the Multi-Attribute Decison Making (MADM) method. The results of these methods still contain uncertainty so that in this case fuzzy logic plays an important role in overcoming problems that contain uncertainty. Fuzzy logic is the basis of a system that can implement a problem and solve sharp

IntechOpen

problems [2]. However, Fuzzy MADM is only able to solve the problem of uncertainty in the data presented and numbers of diverse attributes is usually conflicting, thus to make a decision there needs to be a classic MADM method, so that decisions are more precise and more accurate [3], besides this method can also be used to provide input to the doctor so that there is no mistake in diagnosing dengue disease. One of the classic MADM methods that can be used is Simple Additive Weighting.

Simple Additive Weighting is often referred as a method with weighted sum. The basic concept of SAW method is to find a weighted sum of performance branches on each alternative of all attributes [4]. One of the problems that can be solved by this method is the misdiagnosis of DHF. DHF is a type of infectious disease caused by the degue virus which is transmitted through the bite of the *Aedes aegypti* and *Aedes albopictus* mosquitoes. DHF is often misdiagnosed with Typoid Fever, Morbili, ARI, Ensafalitis and Acute Pharyngitis. These errors occur because the initial symptoms that arise from the five diseases are almost the same as DHF [5]. However, in this case the application of SAW method is less effective if a Decision Support System is made so that a development method is needed. The development method that can be used is the FMADM method with its development or often called Fuzzy Decision Making (FDM). This method is development method of the classic MADM method. The results of SAW method will be used as a level of importance or input on the FDM method. The combination of these two methods will produce more optimal output.

2. Methodology and realization

2.1 Designing FMADM with SAW and FDM

The data used are primary and secondary data, primary data obtained from the results of doctor interviews and secondary data is data on patients with DHF, secondary data will be used to validate the system. Completion of cases of dengue diagnosis will be through SAW method then the results of SAW method are used in the FDM method.

The first method will use one crisp value with 1 degree membership and use preference weight multiplication while the second method uses 3 crisp values namely right boundary, left boundary and crisp value with 1 membership degree which will later go through the aggregation process and total integral value.

2.2 The FMADM method with SAW to diagnose a type of disease

Completion using the FMADM method with SAW:

2.2.1 Determine alternative sets and criteria

Alternative (A_i) is a_1 = Morbili, a_2 = DBD, a_3 = ARI, a_4 = Typoid fever, a_5 = Acute pharyngitis, a_6 = Ensafalitis. C_i criteria are c_1 = Fever, c_2 = Spots, c_3 = Bleeding gum, c_4 = Nausea, c_5 = Headache, c_6 = Defecation Disorders

2.2.2 Determine the criteria weight

The weight of the criteria is obtained from triangular fuzzy numbers which are then converted into the form of crisp.

2.2.2.1 Fever

The author defines the universal value for the criteria for fever is [0,1] and divides it into 5 categories of fuzzy triangle sets, which are normal (N), low fever (DR), moderate fever (DS), high fever (DT), very high fever (DST).

By using the concept of the Likert scale and the defuzzy method, Large of Maximum, **Table 1** is obtained as the weight of the criteria for fever.

2.2.2.2 Spots (Petheciae)

The author defines the universal value for the criteria of spots is [0,1] and divides them into 5 categories of fuzzy triangle sets which are none (TA), few (SDK), somewhat a lot (ABYK) many (BYK), very much (SBYK). By using the concept of the Likert scale and the defuzzy method, Large of Maximum, the **Table 2** is obtained as the weight of the criteria for spots:

2.2.2.3 Bleeding gum

We are defines the universal value for bleeding gum criteria is [0,1] and divides it into 2 fuzzy triangle set categories namely never (TP), ever (P). By using the concept of the Likert scale and the defuzzy method, Large Of Maximum, the **Table 3** is obtained as the weight of the bleeding gum criteria.

2.2.2.4 Nausea

The author defines the universe value for the nausea criteria is [0.1] and divides it into 4 fuzzy triangle set categories namely never (TP), ever (P), rare (J) and often (S). By using the concept of the Likert scale and the defuzzy method, Large of Maximum, **Table 4** is obtained as the weight of the criteria for nausea.

Fever	Fuzzy Set	Crisp Value (weight)
36°C-37,5°C	Normal (N)	0
37,5–38°C	Low Fever (DR)	0.25
38°C-39,5 °C	Moderate Fever (DS)	0.5
39,5–40°C	High Fever (DT)	0.75
>40°C	Very High Fever (DST)	1

Table 1.
Weight of fever.

Spots	Fuzzy Set	Crisp Value (weight)
0–10 spots	None (TA)	0
10–20 spots	few (SDK)	0.25
20–30 spots	Somewhat a lot (ABYK)	0.5
30–50 spots	Many (BYK)	0.75
>50 spots	Very Much (SBYK)	1

Table 2.
Weigth of spots.

Bleeding Gums	Fuzzy Sets	Crisp Value (weight)
0 No	Never (TP)	0
Once or More	Ever (P)	1

Table 3.
Weight of bleeding gum.

Nausea	Fuzzy Sets	Crisp Value (weight)
0	Never (TP)	**0**
1 time a day	Ever (P)	**0.25**
2–3 times a day	Rare (J)	**0.5**
>3 times a day	Often (S)	**0.75**

Table 4.
Weigth of nausea.

2.2.2.5 Headache

The author defines the universal value for the headache criteria is [0,1] and divides it into 4 fuzzy triangle set categories namely never (TP), ever (P), rarely (J) and often (S). By using the concept of the Likert scale and the defuzzy method, namely Large Of Maximum, **Table 5** is obtained as the weight of the headache criteria.

2.2.2.6 Defecation disorder

The author defines the universal value for the criteria for defecation disorder is [0,1] and divides it into 3 categories of fuzzy triangles, namely normal (N), difficult to do defecation (SB) and diarrhea (D). By using the concept of the Likert scale and the defuzzy method, Large of Maximum, **Table 6** is obtained as the weight of the criteria for BAB defects.

2.2.3 Determine the suitability rating of each alternative on each criterion

Interview results from an expert (doctor) on **Table 7**.
From the **Table 8**, the match rating value is obtained as follows:

Headache	Fuzzy Set	Crisp Value (weight)
0	Never (TA)	**0**
1 time a day	Ever (P)	**0.25**
3–4 times a day	Rare (J)	**0.5**
4–5 times a day	Often (S)	**0.75**

Table 5.
Weight of headache.

Defecation Disorder	Fuzzy Set	Crisp Value (weight)
1–2 times a day	Normal (N)	0.5
1–2 days unable to do defecation	Hard to do Defecation (SB)	0.75
>3 times a day	Diarrhea (D)	1

Table 6.
Defecation disorder weight.

	c_1	c_2	c_3	c_4	c_5	c_6
a_1	DT	SDK	TP	J	P	SB
a_2	DT	SBYK	P	J	J	D
a_3	DT	TA	TP	T	J	N
a_4	DST	SDK	TP	P	S	D
a_5	DS	TA	TP	S	P	N
a_6	DST	TA	TP	J	S	SB

Table 7.
Linguistics data.

	c_1	c_2	c_3	c_4	c_5	c_6
a_1	0.75	0.25	0	0.5	0.25	1
a_2	0.75	1	1	0.5	0.5	0.75
a_3	0.75	0.25	0	0	0.5	0.5
a_4	1	0	0	0.75	0.75	0.75
a_5	0.5	0	0	0.5	0.25	0.5
a_6	1	0	0	0.5	0.75	1

Table 8.
Match rating value.

The compatibility rating in this method is also called the decision matrix which will be normalized.

2.2.4 The determination of the preference weight

The determination of the preference weight is stated in **Table 9** as follows:

2.2.5 Normalization of the matrix

$$R = \begin{bmatrix} 0.75 & 0.25 & 0 & 0.5 & 0.25 & 1 \\ 0.75 & 1 & 1 & 0.5 & 0.5 & 0.75 \\ 0.75 & 0.25 & 0 & 0 & 0.5 & 0.5 \\ 1 & 0 & 0 & 0.75 & 0.75 & 0.75 \\ 0.5 & 0 & 0 & 0.5 & 0.25 & 0.5 \\ 1 & 0 & 0 & 0.5 & 0.75 & 1 \end{bmatrix} \tag{1}$$

c_1	c_2	c_3	c_4	c_5	c_6	**Total**
30%	30%	10%	10%	10%	10%	100%
0.3	0.3	0.1	0.1	0.1	0.1	1

Table 9.
Preference weight (W).

To find a matrix you can use the following formula:

$$r_{ij} = \begin{cases} \dfrac{x_{ij}}{max_i\, x_{ij}}, j \text{ is benefit attribute} \\ \dfrac{min_i\, x_{ij}}{x_{ij}}, j \text{ is cost attribute} \end{cases} \tag{2}$$

2.2.6 Finding preference values obtained from multiplication of weights W with normalized matrix R

$$V_j = \sum_{j=1}^{n} w_j r_{ij} \tag{3}$$

The results of the calculation are shown in **Table 10** as follows.

The highest value achieved by the second alternative (V_2) is DBD so someone will be stated to suffer from DHF if they experience symptoms of high fever, spots (petheciae) very much, have experienced bleeding gums if they have entered a severe stage, rarely nausea, rarely headaches and have diarrhea, but to be sure to be able to use laboratory tests again.

In this case, SAW method is not appropriate if it is used to make a decision support system thus the author tries to use a method developed by Joo (2004) [6], namely the FMADM method with development or FDM.

2.3 The FMADM method with SAW to diagnose a type of disease

2.3.1 Representation of the problem

Consists of 3 stages, namely:

a. Objective Identification

The purpose of this decision is to determine or diagnose an illness that is suffered based on the initial symptoms experienced.

b. Identification of Criteria and Alternatives.
The criteria used are still 6 types of diseases and 6 criteria (symptoms).

V_1 **(Morbili)**	V_2 **(DBD)**	V_3 **(ARI)**	V_4 **(Typhoid Fever)**	V_5 **(Acute pharyngitis)**	V_6 **(Encephalitis)**
0.5	0.83	0.42	0.58	0.30	0.57
Rank 4	Rank 1	Rank 6	Rank 2	Rank 5	Rank 3

Table 10.
Preference value.

c. The hierarchical structure that determines the disease is shown in the **Figure 1**.

2.3.2 Evaluation of Fuzzy Sets

Consists of 4 stages, namely:

a. Selecting the set of ratings for the criteria weights. There are two things that must be done, namely determining the degree of importance and determining the degree of compatibility. T (importance) W = {c_1 = {N, DR, DS, DT, DST}, c_2 = {TA, DK, ABYK, BYK, SBYK}, c_3 = {TP, P}, c_4 =, c_5 = {TP, P, J, S}, c_6 = {NR, D, SB}}. T (match) S = {Very Low (SR), Low (R), Enough (C), High (T), Very High (ST)}.

   The parameters of each level of interest are as follows:

$$
\begin{aligned}
N &= (0, 0, 0.25), & TA &= (0, 0, 0.25),\\
DR &= (0, 0.25, 0.5), & SDK &= (0, 0.25, 0.5),\\
DS &= (0.25, 0.5, 0.75), & ABYK &= (0.25, 0.5, 0.75),\\
DT &= (0.5, 0.75, 1), & BYK &= (0.5, 0.75, 1)\\
DST &= (0.75, 1, 1), & SBYK &= (0.75, 1, 1)\\
TP &= (0, 0, 1), & NR &= (0.25, 0.5, 0.75),\\
P &= (0, 1, 1), & D &= (0.5, 0.75, 1)\\
J &= (0.25, 0.5, 0.75), & SB &= (0.75, 1, 1)\\
S &= (0.5, 0.75, 1), & &
\end{aligned}
$$

The degree of compatibility of each decision criteria as follows:

$$
\begin{aligned}
\text{Very Low (SR)} &= (0, 0, 0.25),\\
\text{Low (R)} &= (0, 0.25, 0.5),
\end{aligned}
$$

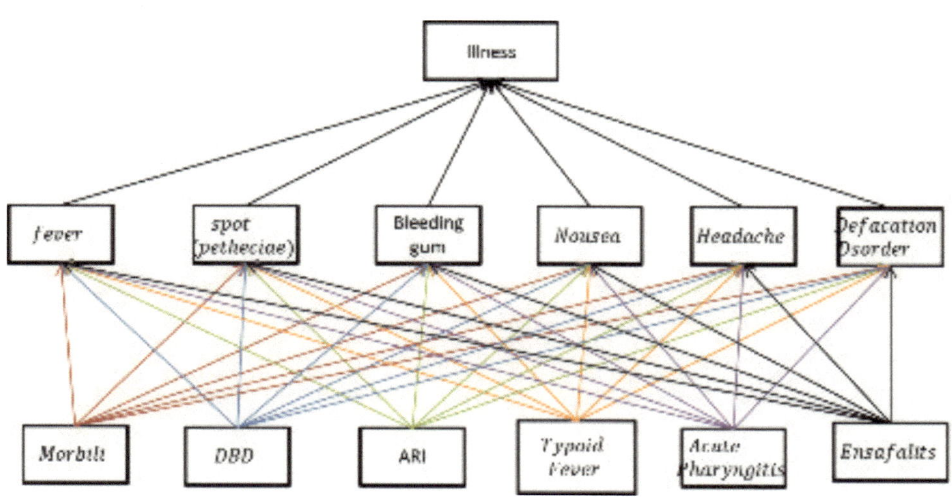

Figure 1.
Hierarchy Structure.

$$\text{Enough (C)} = (0.25, 0.5, 0.75),$$
$$\text{Height (T)} = (0.5, 0.75, 1)$$
$$\text{Very High (ST)} = (0.75, 1, 1)$$

Based on this, the degree of compatibility of each alternative is obtained to the decision criteria in **Table 11** and the branch of interest for the decision criteria in **Table 12.**

b. Aggregate the weight of criteria and the degree of compatibility of each alternative with its criteria, using the following equation:

$$Y_i = \left(\frac{1}{k}\right) \sum_{t=1}^{k} (o_{it} a_i) \tag{4}$$

$$Q_i = \left(\frac{1}{k}\right) \sum_{t=1}^{k} (p_{it} b_i) \tag{5}$$

$$Z_i = \left(\frac{1}{k}\right) \sum_{t=1}^{k} (q_{it} c_i) \tag{6}$$

The result is compatibility index obtained from the aggregation of the weight of the criteria and the degree of compatibility of each alternative with its criteria that's shown in **Table 13.**

2.3.3 Selecting optimal alternatives

Prioritizing decision alternatives based on aggregation results by substituting the fuzzy match index value into the following equation:

$$I_T^{\alpha}(F) = \left(\frac{1}{2}\right)(\alpha c + b + (1-\alpha)a) \tag{7}$$

	c_1	c_2	c_3	c_4	c_5	c_6
a_1	T	R	SR	C	R	ST
a_2	T	ST	R	C	C	T
a_3	T	ST	SR	SR	C	C
a_4	ST	R	SR	T	T	T
a_5	C	SR	SR	C	C	C
a_6	ST	SR	SR	C	T	ST

Table 11.
The degree of compatibility of each alternative to the decision criteria.

Fever	Spot	Bleeding gum	Nausea	Headache	Defecation Disorder
High	Very Much	Ever	Rare	Rare	Diarrhea
(0.5,0.75, 1)	(0.75,1, 1)	(0, 1, 1)	(0.25, 0.5, 0.75)	(0.25, 0.5, 0.75)	(0.5, 0.75, 1)

Table 12.
Branch of interest for decision criteria.

Alternative	Compatibility Rate						Fuzzy Compatibility Index		
	c_1	c_2	c_3	c_4	c_5	c_6	Y_i	Q_i	Z_i
a_1	T	R	SR	C	R	ST	0.1146	0.3229	0.6146
a_2	T	ST	R	C	C	T	0.1979	0.4792	0.7708
a_3	T	ST	SR	SR	C	C	0.1667	0.3646	0.6250
a_4	ST	R	SR	T	T	T	0.1458	0.3854	0.7083
a_5	C	SR	SR	C	C	C	0.0625	0.2083	0.5208
a_6	ST	SR	SR	C	T	ST	0.1563	0.3542	0.6354

Table 13.
Compatibility index.

By taking optimism degree (α), namely: α = 0 (not optimistic), α = 0.5 (optimistic) and α = 1 (very optimistic). The following results are obtained on **Table 14**.

Based on the results above, it can be seen that regardless of the degree of optimism, the alternative a2 is that DHF has the greatest value compared to other alternatives.

2.4 System Implementation

2.4.1 Algorithm of decision support system

The following figure (**Figure 2**) is a flowchart that shows how decision support system works.

2.4.2 Implementation in MATLAB

Based on the *matlab* program algorithm, we must first do the FMADM process with SAW by making a coding in the editor according to the FMADM algorithm with SAW, then the results of the method will be used as input for the next method using the Graphical User Interface (GUI) that will be shown in the **Figure 3**.

Figure 4 is the appearance of the two *matlab* programs with a GUI that contains: self-identity, symptoms experienced, save, clean, close, diagnosis, output, for self-identity and symptoms must be filled. The second display looks like the following picture:

Alternative	Integral Total Value		
	α = 0	α = 0.5	α = 1
a_1	0.22	0.34	0.47
a_2	0.34	0.48	0.63
a_3	0.27	0.38	0.49
a_4	0.27	0.47	0.55
a_5	0.14	0.25	0.36
a_6	0.26	0.37	0.49

Table 14.
Integral total value.

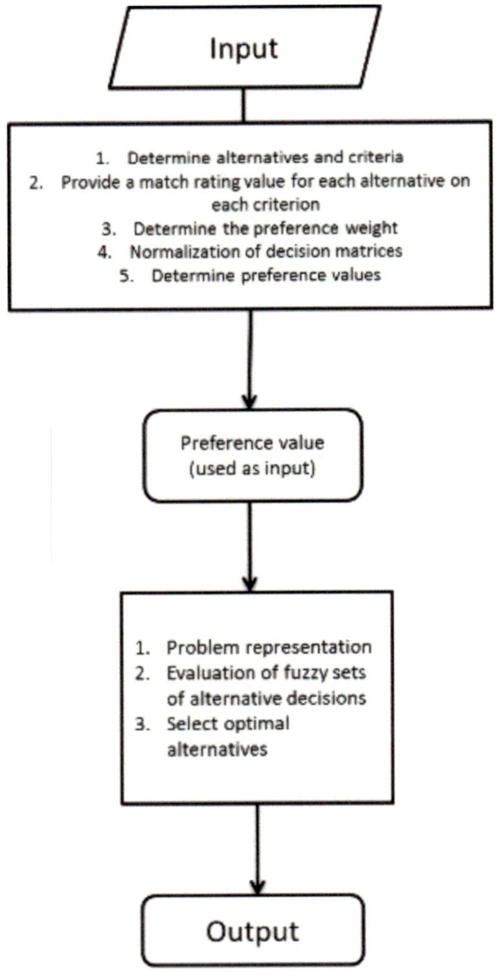

Figure 2.
Decision Support System Algorithm.

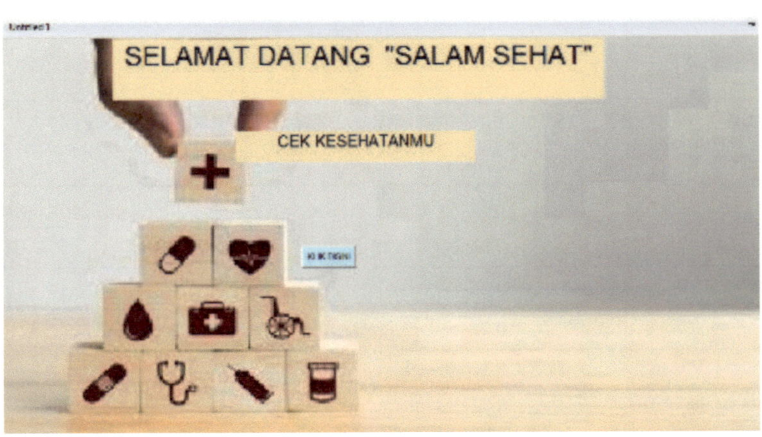

Figure 3.
View of GUI (Opening).

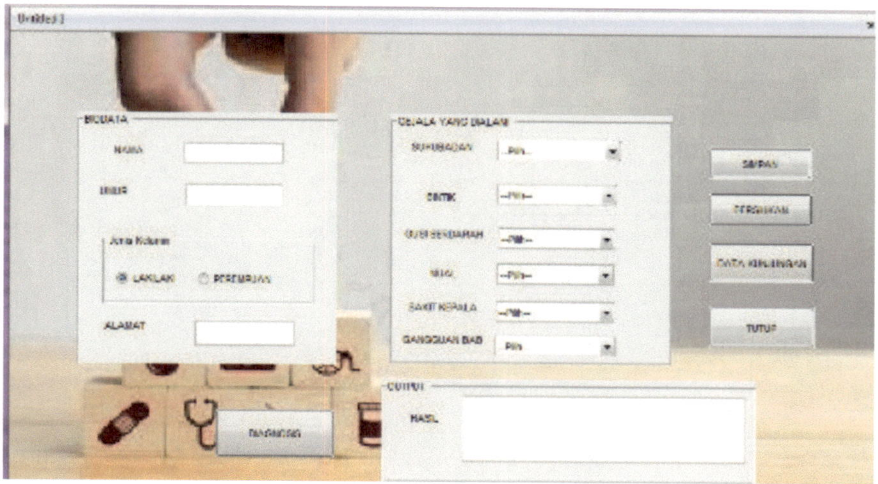

Figure 4.
GUI Display (Form Filling).

Following are the steps to diagnose a type of disease: Fill in the biodata form and symptoms, then click the diagnosis button then click the save button. The results of the diagnosis are obtained as follows like what's shown in **Figure 5**.

The storage results are displayed in a form of what's shown in **Figure 6**.

2.4.3 System accuracy testing

The accuracy of the FMADM decision support system with MAPE obtained the following equation

$$\text{The Accuracy} = \frac{\sum \text{Dataujibenar}}{\sum \text{Totaluji}}\ x\ 100\% \tag{8}$$

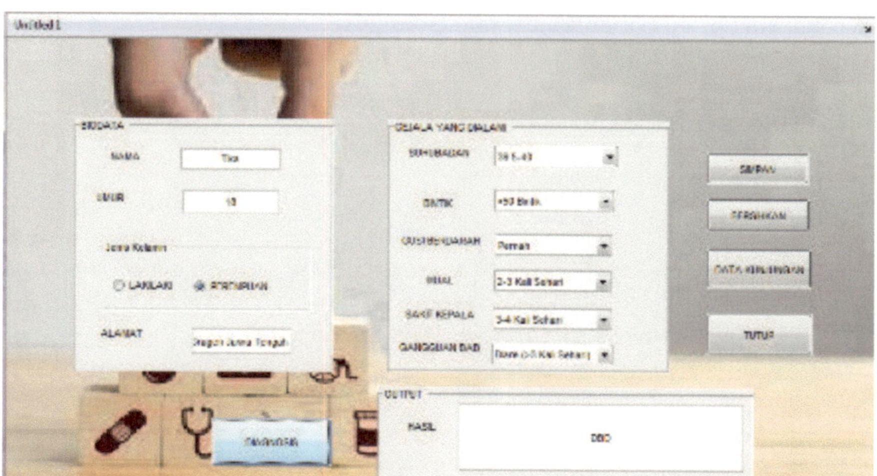

Figure 5.
GUI Display (Diagnose Result).

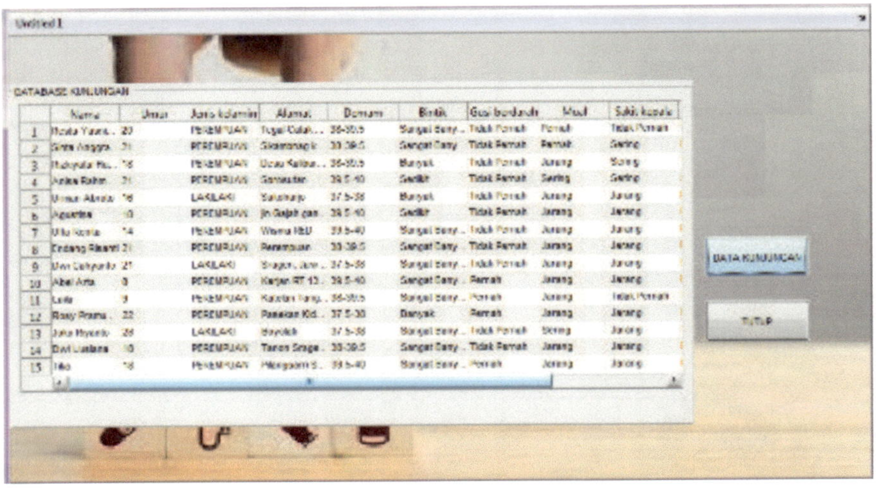

Figure 6.
GUI Display (Data Base).

Obtained from 30 data is as follows:

$$\text{The accuracy} = \frac{30-2}{40}\ x\ 100\% = 93\% \tag{9}$$

3. Conclusion

Based on the method in the first stage, the FMADM method with SAW rank 1 was obtained in the second alternative (V_2) so that someone can be confirmed to suffer from dengue if they experience the initial symptoms of high fever at 39.5° C - 40° C many spots appear during the lumple leed test (> 50 petheciae), bleeding gums, rarely experiencing nausea and headaches, then experiencing diarrhea. In the second method, the results of the first method will be the input for the second method, then the total integral value will be obtained with the degree of optimism α = 1, from the second method or FMADM with Development (FDM). Then the results of the accuracy of the decision support system with MAPE obtained 93% of 100% consisting of 40 patients suffering from DHF.

Author details

Sugiyarto Surono* and Mustika Sari
Ahmad Dahlan University Yogyakarta, Indonesia

*Address all correspondence to: sugiyarto@math.uad.ac.id

References

[1] Kusumadewi Sri. 2003, Artificial Intelligence (Teknik dan Aplikasinya). Yogyakarta : Graha Ilmu.

[2] J. Ross,Timothy. 1997. Fuzzy Logic With Engneering Applications. Singapore ; Mc. Grow. Hill, Inc.

[3] Chen, Shu- Jen & Chin- Lai Hwang. 1992.Fuzzy Multi Attribute Decision Making : Methods and Applications, Berlin : Spinger- Varlag

[4] Kusumadewi, Sri, Sri Hartati, Agus Harjoko & Retantyo Wardoyo. 2006. Fuzzy Multi-Attribute Decision Making (Fuzzy MADM). Yogyakarta : Graha Ilmu

[5] Masjoer, Arif. 2000. Kapita Selekta Kedokteran. Jakarta:Media Aesculapus.

[6] Joo. Hyun Monn & Kang, Chang Sung. 2009. Application of Fuzzy Decision Making Method to the Evaluation of spent fuel stronge Options: Korea